$200.00

R.B H

S0-ARA-942

Human Embryonic Stem Cells

Human Embryonic Stem Cells

The Practical Handbook

Editors

**Stephen Sullivan, Chad A. Cowan
and Kevin Eggan**

Harvard University, Cambridge, MA, USA.

John Wiley & Sons, Ltd

Copyright © 2007   John Wiley & Sons Ltd, The Atrium, Southern Gate, Chichester,
West Sussex PO19 8SQ, England

Telephone (+44) 1243 779777

Email (for orders and customer service enquiries): cs-books@wiley.co.uk
Visit our Home Page on www.wileyeurope.com or www.wiley.com

All Rights Reserved. No part of this publication may be reproduced, stored in a retrieval system or
transmitted in any form or by any means, electronic, mechanical, photocopying, recording, scanning or
otherwise, except under the terms of the Copyright, Designs and Patents Act 1988 or under the terms of a
licence issued by the Copyright Licensing Agency Ltd, 90 Tottenham Court Road, London W1T 4LP,
UK, without the permission in writing of the Publisher. Requests to the Publisher should be addressed to
the Permissions Department, John Wiley & Sons Ltd, The Atrium, Southern Gate, Chichester,
West Sussex PO19 8SQ, England, or emailed to permreq@wiley.co.uk, or faxed to (+44) 1243 770620.

Designations used by companies to distinguish their products are often claimed as trademarks. All brand
names and product names used in this book are trade names, service marks, trademarks or registered
trademarks of their respective owners. The Publisher is not associated with any product or vendor
mentioned in this book.

This publication is designed to provide accurate and authoritative information in regard to the subject
matter covered. It is sold on the understanding that the Publisher is not engaged in rendering professional
services. If professional advice or other expert assistance is required, the services of a competent
professional should be sought.

Other Wiley Editorial Offices

John Wiley & Sons Inc., 111 River Street, Hoboken, NJ 07030, USA

Jossey-Bass, 989 Market Street, San Francisco, CA 94103-1741, USA

Wiley-VCH Verlag GmbH, Boschstr. 12, D-69469 Weinheim, Germany

John Wiley & Sons Australia Ltd, 42 McDougall Street, Milton, Queensland 4064, Australia

John Wiley & Sons (Asia) Pte Ltd, 2 Clementi Loop #02-01, Jin Xing Distripark, Singapore 129809

John Wiley & Sons Canada Ltd, 6045 Freemont Blvd, Mississauga, Ontario, L5R 4J3, Canada

Wiley also publishes its books in a variety of electronic formats. Some content that appears
in print may not be available in electronic books.

Anniversary Logo Design: Richard J. Pacifico

Library of Congress Cataloging-in-Publication Data

Human embryonic stem cells : the pratical handbook / edited by Stephen
Sullivan, Kevin Eggan, and Chad Cowan.
       p. ; cm.
  Includes bibliographical references.
  ISBN 978-0-470-03356-2
  1. Embryonic stem cells — Research. 2. Human embryo — Research. I.
Sullivan, Stephen, Dr. II. Eggan, Kevin. III. Cowan, Chad.
  [DNLM: 1. Embryonic Stem Cells — physiology. 2. Cell Culture
Techniques — methods. 3. Cell Differentiation — physiology. 4. Cell
Line — physiology. QU 325 H9185 2007]
  QH588.S83H863 2007
  616'.02774 — dc22

                                                   2007011301

British Library Cataloguing in Publication Data

A catalogue record for this book is available from the British Library

ISBN 978-0-470-03356-2

Typeset in 10/12pt Times by Laserwords Private Limited, Chennai, India
Printed and bound in Great Britain by Antony Rowe Ltd, Chippenham, Wiltshire
This book is printed on acid-free paper responsibly manufactured from sustainable forestry
in which at least two trees are planted for each one used for paper production.

This book is dedicated to the memory of Professor John Clark OBE FRSE (1951–2004), former director of the Roslin Institute (Edinburgh), a passionate scientist and a pioneer of human embryonic stem cell research.

Contents

Foreword

A good "how-to" book is difficult to write and at the same time it must be easy to read. For a field/subject about which we actually know relatively little, producing a how-to book is indeed especially difficult.

Human embryonic stem (ES) cells have been around for less than ten years and, despite their popularity, the number of scientists possessing extensive working experience with them is still limited. Moreover, as in any emerging field, standard procedures are non-existent or in the process of being developed. The human ES cell field certainly suffers from "there are many ways to skin a cat" syndrome and progress in this field is unlikely while such a situation prevails. So what is somebody who has just decided to start working with human ES cells to do? Well, buying this book would be a good first step.

The editors assembled a group of people with extensive (or at least as extensive as possible) experience in working with various aspects of human ES cell isolation, culture, characterization and differentiation in preparation for clinical applications. They also kept the authors firmly away from extensive theorizing and steered them toward describing practical and workable procedures. This does not mean that one could not quibble about details and I, and probably all other researchers, do some things differently than described in these protocols. It does not matter; as far as I can tell every protocol should work as it stands and a novice in the human ES cell field would be well-advised to stick, at least at the beginning, to every letter. I am sure that in five to ten years many of these protocols, especially those in the second half of the book, will have substantially changed. By then the first clinical trials of human ES cell-based interventions will be under way or will even have been completed and we will have a much better idea about what is necessary and what is possible. Until that time this book is going to be an indispensable bench tool and everybody involved in its making should be commended and thanked for their effort.

Davor SOLTER
Department of Developmental Biology
Max-Planck Institute of Immunology
Stübeweg 51
79108 Freiburg, Germany

Preface

The optimal conditions for isolating, growing and manipulating human embryonic stem (ES) cells are still being identified. In the human ES cell research field presently, a major obstacle to progress is the lack of standardized protocols across the community. It can often be difficult to repeat work done in other laboratories, especially where work is published in journals with space constraints that do not facilitate a full disclosure of the experimental method.

So forms the inspiration for creating this book. We invited the best scientists in the field to submit detailed experimental protocols for their particular area of expertise. Unlike other protocol books, we also asked them to detail what lines they used these protocols with as current data suggest different lines have different proclivities and levels of robustness in culture.

We edited so as to produce a book we ourselves would use in the laboratory. There was a need for a simple and concise book of protocols to isolate, grow, characterize, and manipulate human ES cells. We set out to produce a protocol book that, because of its simplicity of format and reliability of protocols described therein, would be attractive to the researcher, be they a novice or a veteran.

Unique to this book are the troubleshooting sections for human ES cell protocols. We felt that experimental protocols, as normally published, fail to give advice on what to do when experiments fail to work, what variables are likely to be the cause, and what to change when repeating or rescuing an experiment.

The book itself follows a natural chronological order, **Section 1** deals with obtaining and culturing human ES cells, **Section 2** describes how human ES cells can be characterized *in vitro* and *in vivo*, and the final section (**Section 3**) discusses how to manipulate these cells (both in terms of directed differentiation and genetic manipulation). This includes differentiating cells into trophectoderm and the three embryonic germ layers. Also included are chapters dealing with genetic manipulation of human ES cells.

We thank Adam Wilkins for initial encouragement in outlining and proposing the book. We are also indebted to Anne McLaren, Azim Surani, Gabriela Durcova-Hills, Steve Pells, Ian Freshney and Tom Maniatis for advice on the book's contents and format.

Joan Marsh, Andrea Baier, Kate Pamphilon and Fiona Woods at Wiley Publishers are acknowledged for all their work and for their sound judgement, wise council, and enthusiasm.

We especially thank all authors contributing to this book. Their willingness to share what they have learned through their own efforts will no doubt assist the human ES cell community.

We would like to thank readers for using our book; we hope it will assist them. We invite them to visit www.HumanESCellBook.com, the website associated with this book, as it includes a series of discussion boards where users can ask questions and share their experiences of human ES cell culture and manipulation with other users.

Finally, we urge readers to carefully consider the ethical, legal and social implications of experiments using human ES cells before they carry them out, as in many cases such work will have no precedent. While such deliberation falls outside the remit of this simple handbook, we feel it essential to highlight this issue, and to recommend the following to assist the reader in this regard:

'Guidelines for Human Embryonic Stem Cell Research' by National Research Council (U.S.), Board on Life Sciences, National Research Council (U.S.) ISBN: 0-3090-9653-7.

Stephen Sullivan, Chad A. Cowan and Kevin Eggan
Harvard University, 2007

List of Contributors

Please note that many contributors have multiple affiliations; these are listed in full at the start of their respective chapters.

Hidenori Akutsu
National Research Institute for Child Health and Development
Department of Reproductive Biology
2-10-1 Okura
Setagaya
Tokyo 157–8535
Japan

Peter W. Andrews
Centre for Stem Cell Biology
University of Sheffield
Western Bank
Sheffield S10 2TN
UK

Nissim Benvenisty
Department of Genetics
The Hebrew University
Jerusalem 91904
Israel

Mickie Bhatia
McMaster Stem Cell and Cancer Research
Institute
Faculty of Health Sciences
McMaster University
Hamilton, Ontario L8N 3Z5
Canada

Laurie A. Boyer
Whitehead Institute for Biomedical
Research

Nine Cambridge Center
Cambridge, MA 02142
USA

Ali H. Brivanlou
Laboratory of Molecular Vertebrate
Embryology
Rockefeller University
New York, NY 10021
USA

Chantal Cerdan
McMaster Stem Cell and Cancer Research
Institute
Faculty of Health Sciences
McMaster University
Hamilton, Ontario L8N 3Z5
Canada

Alice E. Chen
Department of Molecular and Cellular
Biology
Harvard University
7 Divinity Ave
Cambridge, MA 02138
USA

Hiram Chipperfield
ES Cell International Pte Ltd
11 Biopolis Way
05-06 Helios
Singapore 138667

Julie Hsu Clark
Department of Chemistry and
Cell Biology
The Scripps Research Institute
La Jolla, CA 92037
USA

Ondine Cleaver
Department of Molecular Biology
University of Texas Southwestern Medical
Center
Dallas, TX 75390-9148
USA

Alan Colman
ES Cell International Pte Ltd
11 Biopolis Way
05–06 Helios
Singapore 138667

Chad A. Cowan
Harvard Stem Cell Institute
and Massachusetts General Hospital
Center for Regenerative Medicine
Cardiovascular Research Center
Richard B. Simches Research Center
Charles River Plaza
Boston, MA 02114
USA

Jeremy M. Crook
ES Cell International Pte Ltd
11 Biopolis Way
05–06 Helios
Singapore 138667

George Q. Daley
Division of Hematology/Oncology
Children's Hospital Boston
Karp Family Research Laboratories
300 Longwood Avenue
Boston, MA 02115
USA

Emily A. Davis
Howard Hughes Medical Institute
Department of Cellular and Molecular

Medicine
University of California San Diego
La Jolla, CA 92093
USA

Chris Denning
Wolfson Centre for Stem Cells,
Tissue Engineering and Modeling
University of Nottingham
Queens Medical Centre
Nottingham NG7 2UH
UK

Sheng Ding
Department of Chemistry and Cell Biology
The Scripps Research Institute
La Jolla, CA 92037
USA

Jonathan S. Draper
Centre for Stem Cell Biology
University of Sheffield
Western Bank
Sheffield S10 2TN
UK

Kevin Eggan
Stowers Medical Institute and Harvard
Stem Cell Institute
Department of Cellular and Molecular
Biology
Harvard University
7 Divinity Ave
Cambridge, MA 02138
USA

Dieter Egli
Stowers Medical Institute and Harvard
Stem Cell Institute
Department of Cellular and
Molecular Biology
Harvard University
7 Divinity Ave
Cambridge, MA 02138
USA

Lino S. Ferreira
Department of Chemical Engineering
Massachusetts Institute of Technology
Cambridge, MA 02139
USA

Lawrence S.B. Goldstein
Howard Hughes Medical Institute
Department of Cellular and
Molecular Medicine
University of California San Diego
La Jolla, CA 92093
USA

Rachel Horne
ES Cell International Pte Ltd
11 Biopolis Way
05-06 Helios
Singapore 138667

Bao-Yang Hu
Departments of Anatomy and Neurology
University of Wisconsin
Madison, WI 53705
USA

Ole Isacson
Center for Neuroregeneration Research
Harvard Medical School
McLean Hospital
Belmont, MA 02478
USA

Rudolf Jaenisch
Whitehead Institute for Biomedical
Research
Nine Cambridge Center
Cambridge, MA 02142
USA

Jeffrey M. Karp
Department of Chemical Engineering
Massachusetts Institute of Technology
Cambridge, MA 02139
USA

Dan S. Kaufman
Stem Cell Institute and Department of
Medicine
University of Minnesota
Minneapolis, MN 55455
USA

Ali Khademhosseini
Division of Health Sciences
and Technology
Massachusetts Institute of Technology
Cambridge, MA 02139
USA

Robert Langer
Department of Chemical Engineering
Massachusetts Institute of Technology
Cambridge, MA 02139
USA

Neta Lavon
Cedars-Sinai International Stem Cell
Institute
Cedars-Sinai Medical Center
8700 Beverly Blvd.
Los Angeles, CA 90048
USA

M. William Lensch
Division of Hematology/Oncology
Children's Hospital Boston
Karp Family Research Laboratories
Boston, MA, USA

Shulamit Levenberg
Faculty of Biomedical Engineering
Technion-Israel Institute of Technology
Technion City
Haifa 32000
Israel

Alborz Mahdavi
Department of Chemical Engineering
Massachusetts Institute of Technology
Cambridge, MA 02139
USA

Shannon McKinney-Freeman
Division of Hematology/Oncology
Children's Hospital Boston
Karp Family Research Laboratories
Boston, MA
USA

Douglas A. Melton
Howard Hughes Medical Institute and
Department of Molecular and Cellular
Biology
Harvard University
7 Divinity Ave
Cambridge, MA 02138
USA

Stephen L. Minger
Stem Cell Biology Laboratory
Wolfson Centre for Age-Related Diseases
Kings College London
London SE1 1UL
UK

Maisam Mitalipova
Whitehead Institute for Biomedical
Research
Nine Cambridge Center
Cambridge, MA 02142
USA

Christine Mummery
Hubrecht Laboratory
Interuniversity Cardiology Institute of the
Netherlands and the Heart Lung Centre
University of Utrecht Medical School
Utrecht 3584 CT
The Netherlands

Scott A. Noggle
Laboratory of Molecular Vertebrate
Embryology
Rockefeller University
New York, NY 10021
USA

Kenji Osafune
Department of Molecular and Cellular
Biology
Harvard University
7 Divinity Ave
Cambridge, MA 02138
USA

Robert Passier
Hubrecht Laboratory
Uppsalalaan 8
3584CT Utrecht
The Netherlands

Minal J. Patel
Stem Cell Biology Laboratory
Wolfson Centre for Age-Related
Diseases
Kings College London
London SE1 1UL
UK

R. Douglas Powers
Boston IVF
130 Second Avenue
Waltham, MA 02451
USA

Jan Pruszak
Center for Neuroregeneration Research
Harvard Medical School
McLean Hospital
Belmont, MA 02478
USA

A. Henry Sathananthan
Monash Immunology and Stem Cell
Laboratories
Monash University
Melbourne Victoria 3800
Australia

Thorsten M. Schlaeger
Division of Hematology/Oncology
Children's Hospital Boston

Karp Family Research Laboratories
Boston, MA
USA

Cheryle A. Séguin
Department of Developmental Biology
Hospital for Sick Children
Toronto, Ontario MG5 1×8
Canada

David A. Shaywitz
Neuroendocrinology Division
Bullfinch 4
Massachusetts General Hospital
Boston, MA 02114
USA

Carrie Soukup
Department of Molecular Biology
University of Texas Southwestern Medical
Center
Dallas, TX 75390-9148
USA

Francesca M. Spagnoli
Laboratory of Molecular Vertebrate
Embryology
Rockefeller University
New York, NY 10021
USA

Glyn Stacey
UK Stem Cell Bank
National Institute for Biological Standards
and Control
South Mimms
Hertfordshire EN6 3QG
UK

Emma L. Stephenson
Stem Cell and Regenerative Medicine Bio-
processing Unit
Advanced Centre for Biochemical
Engineering
University College London

London WC1E 7JE
UK

Stephen Sullivan
Stowers Medical Institute and Harvard
Stem Cell Institute
Department of Cellular and Molecular
Biology
Harvard University
7 Divinity Ave
Cambridge, MA 02138
USA

Alan Trounson
Monash Immunology and Stem Cell
Laboratories
Monash University
Melbourne Victoria 3800
Australia

Asmin Tulpule
Division of Hematology/Oncology
Children's Hospital Boston
Karp Family Research Laboratories
Boston, MA, USA

Andrew J. Washkowitz
Columbia University
Sherman Fairchild Center, MC 2427
New York, NY 10027
USA

Jeannine Witmyer
Boston IVF
130 Second Avenue
Waltham, MA 02451
USA

Petter S. Woll
Stem Cell Institute and Department of
Medicine
University of Minnesota
Minneapolis, MN 55455
USA

Holm Zaehres
Division of Hematology/Oncology
Children's Hospital Boston
Karp Family Research Laboratories
Boston, MA, USA

Su-Chun Zhang
Departments of Anatomy
and Neurology
University of Wisconsin

Madison, WI 53705
USA

Thomas P. Zwaka
Center for Cell and Gene Therapy and the
Department of Molecular and Cellular
Biology
Baylor College of Medicine
Houston, TX 77030
USA

Section 1
Obtaining and Culturing Human Embryonic Stem Cells

1

Organization and good aseptic technique in the human embryonic stem cell laboratory

MINAL J. PATEL[1,2], EMMA L. STEPHENSON[2] AND STEPHEN L. MINGER[1]

[1]*Stem Cell Biology Laboratory, Wolfson Centre for Age-Related Diseases, Kings College London, London SE1 1UL, UK*
[2]*Stem Cell and Regenerative Medicine Bioprocessing Unit, Advanced Centre for Biochemical Engineering, University College London, London WC1E 7JE, UK*

Introduction

The derivation and long-term propagation of human embryonic stem (ES) cells requires specialized facilities and equipment. For instance, many groups passage human ES cells by manual dissociation ("cutting") which requires an expensive stereomicroscope and either a laminar flow cabinet or Class II cabinet with a built-in microscope. This chapter will provide an overview of how to establish and equip a human ES cell laboratory as well as provide practical suggestions for aseptic technique and screening for microbial contamination.

Laboratory layout

The physical positioning of a dedicated tissue culture facility for human ES cell work is important. A single self-contained room is preferable, with minimal traffic flow. Additionally, a windowless room may be preferred because this avoids potential fluctuations in temperature, incursion of micro organisms due to insufficient sealing, and also prevents ultraviolet (UV) denaturation of tissue culture media.

Equipment

The following is a partial list of what is required to equip a laboratory for both derivation and long-term propagation of human ES cells:

- Class II biosafety cabinets with UV lights (located preferably adjacent to the cell
  operating surface). The tissue culture cabinets should have connecting ducts to the
  exterior of the building to facilitate decontamination and to improve air circulation.
  If air conditioning systems are located within the tissue culture room itself, it is
  best to increase as much as is possible the distance between the air circulating vents
  and the tissue culture cabinets, thus reducing the risk of particle contamination. See
  Figure 1 for typical Biosafety cabinet layout.

- P20, P200, P1000 pipettes, and a pipetteman that can accommodate 1, 2, 5, 10,
  and 25–50 mL serological pipettes.

- A vacuum system (either in-built or using floor-mounted external system) for aspirating medium, filtering solutions. Small floor mounted aspirators are very useful
  for speed of routine tasks although when working with these external systems

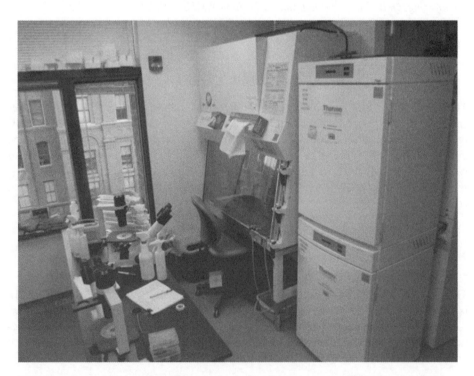

Figure 1 Typical biosafety cabinet set up. The cabinet is organized so that the user
work space is uncluttered while items such as pipettes, tissue paper for cleaning, waste
containers, etc. are all close to hand.

additional precautions are required which include the use of a collection vessel containing bleach or equivalent solution for disinfection, ensuring the collection tube is secured against falling to the ground and cleaned thoroughly with ethanol before use, and the use of a filter between the operator and the vacuum line to avoid liquid or vapor contamination.

- Laminar flow cabinet and high powered zoom stereo microscope with a temperature-controlled heated stage — this is useful for manipulating embryos, isolation of the inner cell mass using either immunosurgery or physical means, and for manual passaging of human ES cell colonies ("cutting"). See **Figure 2** for typical embryo manipulation set up.

- Tissue culture incubators, preferably with 4–6 individual compartments per incubator. Such division limits air flow, and temperature and humidity fluctuations which may adversely affect cells particularly in busy laboratories with many operators where the incubator door may be opened several times an hour.

- Phase contrast microscope with 4×, 10×, 20× and 40× objective lenses for examining cells. This should be minimally equipped with a small digital camera for photomicroscopic documentation of cells, but a dedicated integrated system including

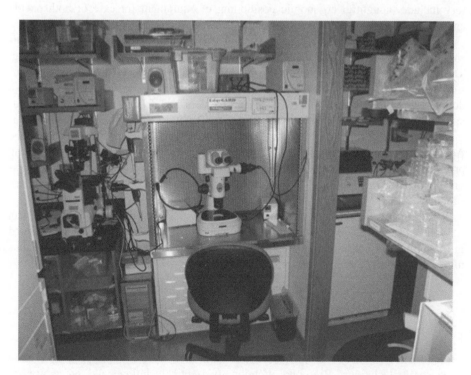

Figure 2 Typical embryo manipulation set up. A dissection microscope with heated stage is positioned in a laminar air hood to minimize the risk of microbial contamination while working.

a digital camera, computer and imagining software is preferred. Software that can also take measurements of colony size and shape is often very useful in monitoring growth characteristics, or the development of embryos.

- Fridges and freezers for storage of growth factors, medium, serum etc. Glass paneled refrigerators should be avoided as many components of human ES cell culture media are light sensitive. There should be easy access to −80°C freezer for long-term storage of labile materials.

- Tabletop centrifuge for pelleting cells.

- Glass Pasteur pulling machine or naked flame (e.g. Bunsen burner) to prepare glass needles for manual passaging of human ES cell colonies. Note special stem cell knives and autoclavable metal cutting instruments are also available but these are generally very expensive.

- Liquid nitrogen container i.e. Dewar flask.

- Emergency power system (e.g. back up generator) for tissue culture incubators, freezers and other important equipment.

General principles to consider when physically setting up a human ES cell laboratory include thoughtful ergonomic positioning of equipment for ease of both people and air flow, with care to avoid potential vibration of incubators and tissue culture cabinets from adjoining equipment. There should be an area adjacent to, but outside of, the human ES cell laboratory itself for changing of shoes and hand washing. There must be easy access to storage at different temperatures for sterile and non-sterile reagents, consumables, glassware and other essential reagents. If possible, a double-sided storage area where goods are replenished from one side (dirty area) and are used from the opposite side (clean area) within the human ES cell laboratory is preferred, although not every laboratory will be able to accommodate this. Additionally, easy access to liquid nitrogen stores is essential, especially if using primary feeder cells for maintaining human ES cell lines. An alarm system warning of low liquid nitrogen levels is recommended to prevent loss of valuable frozen samples.

Work in the human ES cell culture facility should ideally be dedicated only to human ES cells and feeder cells. Minimizing the number of people and procedures performed in the human ES cell laboratory will ensure more sterile and clean environment, thus minimizing the risk of cell culture contamination.

Aseptic technique and good practice

Growth of high quality human ES cell lines and their differentiated progeny require good sterile technique. Rigorous attention to aseptic technique and procedures to minimize possible sources of contamination is imperative. There are a number of potential sources of microbial contamination that can adversely affect human ES cells, including those that originate from tissue culture operators, work surfaces, air supply, and reagents (e.g. tissue culture plastics, media or individual medium additives).

Maintaining the highest standard of aseptic technique will significantly reduce most microbial infections, but unfortunately this is a problem even with the best technique. Low-level contamination can remain undetected for long periods, particularly if cells are grown in the presence of antibiotics. Abnormal cell growth patterns are usually a good indicator of contamination, as well as changes in medium turbidity and color as bacterial or fungal infection affects pH of media and can be easily observed if the medium contains a pH indicator such as phenol red.

The nutrient rich media required for routine human ES cell culture encourages the rapid growth of most microbial contaminants. The most common problems in any cell culture laboratory are bacterial or fungal contaminants which are normally visible, particularly if antibiotic-free medium is used. However, the infection by mycoplasma is harder to detect under the microscope and if untreated can retard cell growth, cause morphological changes, alter amino acid and nucleic acid metabolism and induce chromosomal aberrations within the cells. Hence an established and thorough microbial screening routine, employing appropriate assays, should be instituted by all stem cell laboratories.

Reducing the risk of contamination

The following procedures reduce the risk of bacterial, fungal and mycoplasma contamination:

Cleaning

- The work surfaces of Class II biosafety cabinets and the laminar flow hoods should be wiped down with 70% ethanol prior to and at the end of cell culture. All items which enter sterile work areas which are not sterile (e.g. tissue culture medium bottles) should be sprayed with 70% ethanol and wiped dry with clean tissue paper before placing them in the work area.

- A regular cleaning and disinfection schedule for the culture facility should be implemented and rigorously adhered to, including daily washing of Class II biosafety cabinets, bench tops and equipment with 70% ethanol and biocidal spray as well as weekly mopping of floor surfaces with a bleach solution. Additionally, each Class II biosafety cabinets should be thoroughly cleaned and carefully inspected at least weekly, in particular checking under the work surface for any tissue culture medium spillages which may have gone undetected.

- Clean tissue culture incubators every 3 months, running the decontamination/sterilization cycle if the incubators have this feature. Make sure to dry the rubber seal on the inside of the incubator door, where over time fungus can get a hold.

- Remember to clean the floor of the facility when it will not be used for some time (e.g. last thing on a Friday afternoon). Sweeping or mopping the floor will disturb settled microbes and so it is best to let everything resettle after cleaning before using the facility again.

Other measures

- Apply antimicrobial agents (e.g. Clear Bath Algicide, VWR Catalog # 13272-031) to the water trays of incubators and to water baths. This reduces bacterial and fungal contamination but should not to be used as a substitute for regular cleaning and disinfection procedures.

- Service contracts should be in place for all equipment to ensure that the equipment is working optimally — for example the filters in the Class II biosafety cabinets should be changed at 6 month intervals by properly trained personnel.

- Remove all consumables and reagents from their exterior packaging prior to bringing these items into the cell culture facility. Cardboard packaging is frequently contaminated with fungal spores.

- All non-sterile reagents which are used in human ES cell culture (e.g. pipette tips, glass pipettes, and deionized water for humidifying the incubators) should be autoclaved at 180°C or gamma irradiated to destroy any potential microbial contamination before bringing these items into the human ES cell culture facility.

- Use adhesive floor mats (e.g. adhesive floor mat, SIGMA Catalog # M6625) at the point of entry into the facility. This will reduce particulate matter from shoes or cart wheels carried into the facility. Position the mats such that people cannot easily step over them, and change the disposable sheets when they become saturated with dirt.

Cell culture practice

Microbial contamination of cell cultures frequently occurs in laboratories where one or more users have poor aseptic technique and/or where there is not enough screening for culture contamination. It is important to have a code of practice that all staff using the facility rigorously adhere to in order to minimize introduction and spread of contaminants. The facility will only be as clean as its dirtiest user.

- All new staff should be trained by experienced cell culture personnel so that there is consistent practice among all users of the facility. To save time, there is a CD-ROM available '*Freshney's Culture of Animal Cells: A Multimedia Guide*' (http://www.wiley.com/legacy/products/subject/life/freshney/order.htm) that can be used as both a primer of good aseptic technique for new staff (before they are trained specifically) in-house and also as a revision of technique for trained users having protracted problems with contamination.

- Prior to entering the culture facility, each tissue culture operator should don shoe covers or else remove street shoes and put on cell culture clogs. They should then remove jewellery, thoroughly scrub their hands and forearms with a antiseptic hand gel or soap.

- Dedicated cell culture lab coats should be put on and buttoned to the top. Immediately upon entering the laboratory, sterile latex gloves should be put on, pulled up totally over the cuffs of the lab coat and gloved hands should be thoroughly rinsed with 70% ethanol spray to completely disinfect hands. Only after these procedures have been completed should the operator begin to work in the Class II tissue culture cabinet or open the tissue culture incubator. If any procedures are to be performed in the laminar flow hood, the additional step of placing a sterile surgical mask over the face should be employed.

- The routine use of antibiotics in human ES cell culture is to be avoided, both because this has long-term adverse effects on human ES cell viability and growth (Cohen et al., 2006) and because antibiotic-free cultures allow the early detection of contamination rather than masking the problem for prolonged periods. Such long term, low level contamination allows the selection of a resistant form of bacteria, which is extremely hard (and costly) to get rid of.

- Observe your cultures regularly for signs of microbial contamination. These include cell death, slowing of cell proliferation, and changes in cell morphology. Early detection of contamination is essential to avoid widespread problems.

- When beginning work in the Class II biosafety cabinet, only essential items related to the task at hand should be placed onto the work surface. This reduces the potential for touching non-sterile items when manipulating cell culture dishes. It also prevents interruption of laminar air flow.

- Avoid working at the interface of filtered and external air flow (i.e. the periphery of the cabinet). Sit close to the cabinet and perform all manipulations well within the sterile work area.

- When aspirating medium from cells, change the pipette tip between wells to reduce the risk of cross-contamination.

- Do not leave tissue culture plates, flasks or media bottles open any longer then necessary. Organize the work area so that you do not have to reach over open culture vessels or media bottles. This allows microbes to fall onto the cultures or into the stock medium.

- Work within your line of vision to ensure sterile tips are not contaminated by touching them against potentially non-sterile surfaces, for example the work surface, the outside or top of the medium bottle or the outside of tissue culture vessels. If there is any doubt that this has occurred immediately dispose of the plastics and replace with sterile ones.

- If any medium is spilt, clean the relevant area immediately with 70% ethanol and allow to air dry.

- Do not shake or invert bottles or tubes containing tissue culture medium or other tissue culture components as liquid can be retained in the thread of the lid and be a source of contamination.

- When changing the medium, only completely sterile tips should ever be introduced into sterile media components. Under no circumstances should "double-dipping" occur. This is where a tip that has previously been used to feed a plate of cells is reinserted into a bottle of tissue culture medium to feed another plate of cells. Double-dipping will almost certainly result in contamination across a number of culture vessels. Mouth pipetting should also be avoided, even if there is a filter between the operator and the tissue culture vessel, not only because it represents extremely poor aseptic technique but also a health and safety concern for the tissue culture operator.

- Personnel should make up, label, and use only their own tissue culture medium and other components (e.g. proteases). This will prevent cross contamination between users.

- Tissue culture medium for human ES cells and fibroblast feeder layers should only be used for a maximum of 2 weeks, at which time any remaining medium should be discarded and replaced with freshly made-up medium. β FGF and L-glutamate are particularly labile.

- If any tissue culture reagent is inadvertently sucked into the cotton wool plug of the serological pipette or into the filter of the pipetteman, immediately discard the serological pipette and/or change the filter immediately before proceeding with cell culture.

- Do not allow the waste collection tube to fall on the ground and routinely spray it with 70% ethanol before use.

- Once work in the biosafety cabinet is completed, the tissue culture work surface should be cleaned with 70% ethanol and when the cabinet is not in use the UV lights should be switched on. UV lights in each cabinet should be left on overnight and whenever the cabinet is not in use.

- Finally, a systematic and reliable procedure for testing of mycoplasma should be instituted when the human ES cell laboratory is first set up. Minimally all cell lines (each human ES cell line, each set of feeder cells) should be routinely tested, preferably at least every 3 months, to ensure no widespread mycoplasma contamination can occur, which can significantly affect research productivity for several months. We recommend the Mycoalert mycoplasma detection kit from CAMBREX (Catalog # LT07-118).

Reference

Cohen S, A Samadikuchaksaraei, JM Polak and AE Bishop. (2006). Antibiotics reduce the growth rate and differentiation of embryonic stem cell cultures. *Tissue Eng* **12**(7): 2025–2030.

2

Sourcing established human embryonic stem cell lines

GLYN STACEY

UK Stem Cell Bank, National Institute for Biological Standards and Control, Blanche Lane, South Mimms, Herts. EN6 3QG, UK

Introduction

Selecting your cell line

It is clear that the potential of human embryonic stem (ES) cells to differentiate into cells of almost any tissue of the human body provides exciting new opportunities for *in vitro* research and the development of therapies for the future. However, it is vital that in these pursuits researchers attend carefully to certain issues in sourcing cell lines to ensure that their work provides data that is reproducible and informative.

It is good practice in any research involving cell lines to start with a number of candidate cells that will help to ensure more accurate interpretation of the data and avoid wasting time and resources on a single cell line that subsequently is revealed to be functionally abnormal or otherwise inappropriate for the purpose it is being used in experiments. Internationally there is now thought to be a resource of human ES cell lines in excess of 300, which should service the needs for many areas of research. However, not all of these lines are published, methods and media used for ES cell line derivation and culture vary and the degree and nature to which individual cell lines have been characterized is not uniform. The selection of cell lines on which to work will be affected by these variables.

The history of *in vitro* cell culture has shown any cell line can, if unchecked, fall foul to microbial contamination, cross-contamination with other cell lines, or culture-related abnormalities (e.g. cell transformation in culture, loss of cell function during prolonged periods in culture, or changes associated with drastically and abruptly changing culture conditions) (Stacey, in press). When selecting which human stem cell lines to work with, there are certain technical criteria that should be considered (e.g. their origin, their culture history, their purity, and the degree to which they have been functionally characterized, and their functional stability). The specific sources of cell lines will determine how much information a researcher will receive pertaining to such criteria, and so the different sources of human ES cells are now distinguished from each other.

Stem cell banks and other sources of human ES cells

What are "cell banks"?

A cell bank can be described as a depositary of material (usually frozen in a viable state) which provides secure stocks of cells. Banks of cells provided for clinical applications or manufacturing purposes will have very different requirements and resource issues compared to those for research purposes as illustrated in **Table 1**. Cell banks will also vary depending on the number of laboratories to be supplied and whether the bank is intended to be a local, national or international resource.

Naturally, the early provision of human ES cells for research has been led by the originators of the lines. NIH funding promoted the early exchanges of the cell lines from the suppliers of cells posted on the NIH registry (http://stemcells.nih.gov/research/registry/). The UK Stem Cell Bank was established in 2003 at the National Institute for Biological Standards and Control by the Medical Research Council and the Biotechnology and Biological Sciences Research Council to provide a public service collection for sourcing human ES cells (www.ukstemcellbank.org.uk, Healy *et al.*, 2005). Currently, over 40 human ES cell lines have been approved for deposit in the UK Stem Cell Bank and four were available for distribution at the time of publication. More recently, WiCell, a private American company, were awarded a grant from the NIH to establish a US stem cell bank (www.wicell.org) which currently (at the time of submitting this publication) lists ten human ES cell lines available from a catalogue of 13.

Sources of stem cell lines

There are several sources of human ES cell lines (**Table 2**). The providers of cell lines, or repositories working on their behalf, should be able to deliver documentation for source of cells, traceability to original ethically sourced seed stocks and evidence of careful cell banking and quality control procedures. Cells grown in culture are at a high risk of cross-contamination. Evidence for this is discussed later in this chapter and has been most powerfully identified by MacLeod *et al.* (1999) who showed that 18% of cell lines deposited at the DSMZ (Deutsche Sammlung von Mikroorganismen und Zellkulturen) cell culture collection were cross-contaminated or switched with other cells. Notably, half of these were from originators. Accordingly, it is recommended to obtain lower passage sources of cells from the originator or centers committed to provision of quality controlled cells.

A number of methods of making human ES cells more widely available have been explored including:

- Commercial provision e.g. ESI, WiCell.

- Private or NIH funding has ensured early distribution of approved lines by providing resources to the originating centers e.g. Harvard Stem Cell Institute.

- UK research council funding for a central bank has been established (www.ukstemcellbank.org.uk) and now the US has also funded a central source via WiCell.

Table 1 Stem cell banks for research, production and clinical use

Type of organization	Primary activities
Local laboratory stocks for research	Satisfying demand for undifferentiated cells for experimental use
Centralized Institute stocks for research	Provision of cultures or frozen cells to a number of laboratories Basic quality control
National supply of human ES cells for a broad range of researchers	Establish large stocks of reproducible, well characterized and quality controlled cells Provide safe storage as a back up for originators of lines Distributed cells must meet national best practice safety guidelines (e.g. screened for serious common blood born pathogens) Operate under a published Code of Practice Ensure recipients and their use of lines meet demands of national regulation Engage in other national resource and training programmes
International Stem Cell Bank	As for national resource but with additional safeguards: 1. ensure compliance of deposited lines and distribution with all national guidance, regulation and laws of depositors and recipients (ideally thorough a high level oversight body) 2. a rigorous quality assurance system that will assist in the delivery of reproducible and reliable cells and service. 3. coordination with other national banks to promote best practice
Tissue bank supplying stem cells for transplantation (e.g. bone marrow, cord blood, mobilized peripheral blood stem cells)	Donor screening including blood tests for viral markers Fully informed consent on donated material Rigorous quality assurance under national guidelines Validation of bank processing procedures Compliance with regulation including appropriate national or international inspections
Master and Working cell banks for Pharmaceutical production of medicinal products from cell lines	Delivery of cells for a specific end product Donor selection and informed consent if appropriate (new human cell lines) Preparation under a product license in accredited certified good medical practice (cGMP) facilities Exhaustive accredited safety testing and quality control

Table 2   Information pertaining to human ES cell lines cited in this book

Abbreviations: UKSCB = UK stem cell bank, HSCB = Harvard stem cell bank, NSCB = National stem cell bank, NSCB = Singapore stem cell bank, ESI = ES Cell international, CIHR = Canadian Institute of Health Research.
mF = mouse feeder layer
hF = human feeder layer

Cell line	Link	Provider	Available in which cell banks	Passages available at time of going to press	Originally reported genotype	Disaggregation	Reported to grow on	Cost
HUES1	www.mcb.harvard.edu/melton/hues	Harvard	UKSCB, HSCB	19	46XX	Enzymatic	MEF	Shipping only
HUES3	www.mcb.harvard.edu/melton/hues	Harvard	UKSCB, HSCB	26	46XY	Enzymatic	MEF	Shipping only
HUES6	www.mcb.harvard.edu/melton/hues	Harvard	UKSCB, HSCB	17	46XX	Enzymatic	MEF	Shipping only
HUES8	www.mcb.harvard.edu/melton/hues	Harvard	UKSCB, HSCB	21	46XY	Enzymatic	MEF or matrigel	Shipping only
HUES9	www.mcb.harvard.edu/melton/hues	Harvard	UKSCB, HSCB	18	46XX	Enzymatic	MEF or matrigel	Shipping only
H1	www.wicell.org	WiCell	UKSCB, NSCB	57	46XY	Enzymatic	MEF	Shipping only
H1.1 subclone	www.wicell.org	WiCell	NSCB	68	46XY	Enzymatic	MEF	Shipping only
H7	www.wicell.org	WiCell	NSCB	49	46XX	Enzymatic	MEF	Shipping only
H9	www.wicell.org	WiCell	NSCB	31	46XX	Enzymatic	MEF	Shipping only
H9.1 subclone	www.wicell.org	WiCell	NSCB	48	46XX	Enzymatic	MEF	Shipping only
H9.2 subclone	www.wicell.org	WiCell	NSCB	51	46XX	Enzymatic	MEF	Shipping only
H14	www.wicell.org	WiCell	NSCB	69	46XY	Enzymatic	MEF	Shipping only
hES-1	http://www.escell-international.com/	ESI	NSCB	78	46XX	Mechanical	MEF	Shipping only
hES-2	http://www.wicell.org	ESI	NSCB	87	46XX	Mechanical	MEF	US$500
hES-3	http://www.wicell.org	ESI	NSCB	83	46XX	Mechanical	MEF	US$500
hES-4	http://www.wicell.org	ESI	NSCB	52	46XY	Mechanical	MEF	US$500
hES-5	http://www.escell-international.com/	ESI	NSCB	78	46XY	Mechanical	MEF	US$500
hES-6	http://www.escell-international.com/	ESI	NSCB	To be determined	46XX	Mechanical	MEF	US$500
NL-hES1	www.niob.knaw.nl	Hubrecht Lab/ESI	none	To be determined	46XX	Mechanical	MEF	US$500+shipping
NL-hES2	www.niob.knaw.nl	Hubrecht Lab/ESI	none	To be determined	46XX	Mechanical	MEF	US$500+shipping

Table 2 (continued)

Cell line	Link	Provider	Available in which cell banks	Passages available at time of going to press	Originally reported genotype	Disaggregation	Reported to grow on	Cost
NL-hES3	www.niob.knaw.nl	Hubrecht Lab/ESI	none	To be determined	46XY	Mechanical	hF	US$500+shipping
NL-hES4	www.niob.knaw.nl	Hubrecht Lab/ESI	none	To be determined	46XX	Mechanical	hF	US$500+shipping
ESI-014	http://www.escell-international.com/	ESI	SSCB	To be determined	46XX	Enzymatic	hF	Shipping only
ESI-017	http://www.escell-international.com/	ESI	SSCB	To be determined	46XX	Enzymatic	hF	Shipping only
ESI-035	http://www.escell-international.com/	ESI	SSCB	To be determined	46XX	Enzymatic	hF	Shipping only
ESI-049	http://www.escell-international.com/	ESI	SSCB	To be determined	46XY	Enzymatic	hF	Shipping only
ESI-051	http://www.escell-international.com/	ESI	SSCB	To be determined	46XX	Enzymatic	hF	Shipping only
ESI-053	http://www.escell-international.com/	ESI	SSCB	To be determined	46XX	Enzymatic	hF	Shipping only
CA1	www.mshri.on.ca/nagy	CIHR	None, request from Nagy lab directly	>45	46XY	Mechanical	mF	Shipping only
CA2	www.mshri.on.ca/nagy	CIHR	None, request from Nagy lab directly	>52	46XX	Mechanical	mF	Shipping only
HSF1	http://escells.ucsf.edu/cells/nih_home.asp	UCSF	NSCB	>60	46XY	Enzymatic	mF	$5000+shipping
HSF6	http://escells.ucsf.edu/cells/nih_home.asp	UCSF	NSCB	>60	46XX	Enzymatic	mF	$5000+shipping
BG01	www.wicell.org	Bresagen Inc.	NSCB	>70	46XY	Mechanical	mF	Shipping only
BG02	www.wicell.org	Bresagen Inc.	NSCB	>40	46XY	Mechanical	mF	Shipping only
BG03	www.wicell.org	Bresagen Inc.	NSCB	>35	46XX	Mechanical	mF	Shipping only
RUES1	Xenopus.rockefeller.edu	Rockefeller	None, request from Brivanlou lab directly	17	46XY	Enzymatic	MEF or matrigel	Shipping only

At time of going to press the UK Stem Cell Bank (www.ukstemcellbank.org.uk) had permission to distribute the following lines: MEL 1, MEL 2 (Melbourne University, Australia); NCL 1–5 (University of Newcastle, UK); RH 1–4 (Roslin Institute, UK); EDI-1 (Edinburgh University, UK); WT-3, WT4, CF1 (King College London, UK); Shef 1–6 (University of Sheffield, UK); and HUES 1–17 (Harvard University, USA).

- Other research centers are also planning to make cells available under varying conditions.

Table 2 provides sourcing information for all human ES cell lines cited in this book in addition to all the lines currently distributed by the UK Stem Cell Bank.

Whatever source of cells is used, critical issues will be the quality of cells provided, the nature of agreements under which they are supplied and any consequent constraints on the use of the cells. Naturally, the scientists deriving cell lines will wish to facilitate exchanges of cells in their early collaborations. However, in the longer term significant benefits can be had by both originators and collaborators by providing cell lines from cell banks. These benefits include:

- Safe depositories of cells for depositors.

- Depositors are released from the burden of culture, preservation and dispatch.

- Common system of materials transfer agreements.

- Consistent documented regimes for quality control of stem cell lines.

- Focus on careful maintenance and ongoing characterization of seed stocks.

Such resource centers should promote non-restrictive access to stem cell lines and dissemination of current best practices (Stacey and Hunt, 2006; Healy *et al.*, 2005).

Criteria for selection of cell lines for research purposes

The selection of cell lines should be based on the fundamental elements of purity, authenticity and stability: these are now discussed at length. However, there are additional factors addressed elsewhere including safety issues for laboratory workers (See **Chapter 1**), culture history of cells (**Chapters 3–6**), cell line genotypic and phenotypic characteristics (See **Chapters 7 and 8**), as well as the ethical sourcing of human ES cells.

Cell line purity

Chapter 1 has previously outlined how the risk of microbial contamination can be reduced in a human ES cell laboratory. Cells being received from outside sources (however reputable) should be tested for microbial contamination before routinely used in a cell culture facility. This is particularly important in the case of human ES cells as they are conventionally grown in rich medium not supplemented with antibiotics. The different forms of microbial contamination are now highlighted.

Bacteria and fungi

Contamination by bacteria or fungi is generally readily detected by turbidity of the culture medium and causes catastrophic loss of affected cultures. Transient use of antibiotics may be helpful to avoid this in circumstances where contamination is inevitable, as in primary mouse embryonic feeder cultures. However, long-term use of antibiotics is not recommended as this may affect the cells and selects for "superbug" organisms that are resistant to the antibiotic, leaving no fall back treatment for protection of critical cultures. In certain circumstances, antibiotics may inhibit contaminants without eliminating them thus allowing for their re-emergence in the future. Bacterial and fungal contamination can arise from a variety of sources in the laboratory environment and their exclusion is most effectively achieved using good aseptic technique. However, the most experienced of workers may suffer from occasional contamination and frozen stocks of cells should be tested for contamination by inoculation of culture supernatant into bacteriological broth media followed by incubation at both standard cell culture temperature (typically 35–37°C) and room temperature to reveal growth of contaminants with different optimal growth temperatures. Refer to **Chapter 2** for more on testing for and avoiding contamination of human ES cell cultures.

Recommended methods include those published in national pharmacopoeia (European Phamacopoeia, 2006a; US Food and Drugs Administration, 2005a). However, such methods will not detect certain organisms that are known to arise in cell culture but will not grow in standard sterility testing conditions. By far the most common of these are mycoplasma. Specific tests for the presence of these organisms should be applied to stem cell lines.

Mycoplasma

Mycoplasma are microorganisms, generally smaller than bacteria, which often grow in close association with the cell membrane. They may not affect growth of a cell line. Thus, persistent infections can change cells genetically and phenotypically (Del Guidice and Gardella, 1984; McGarrity et al., 1993; Rottem and Naot, 1998). This combined with the ability of these organisms to survive treatment with some antibiotics and not to cause culture medium turbidity, enables them to survive and spread undetected between cultures. They enter cell culture via contaminated materials of animal origin (e.g. serum, trypsin, primary feeder cell cultures) or even possibly from human operators. Today, prohibition of mouth pipetting in laboratories and improvements in quality control of commercial cell culture reagents have largely eliminated these original sources of mycoplasma contamination. There is at least one notable exception: primary animal cells (e.g. mouse embryonic feeder cells). However, historical contamination, along with a failure to identify and destroy contaminated cultures, have led to an unacceptably high incidence, often quoted at 15–30%, in cell culture laboratories (Uphoff and Drexler, 2002).

Clearly, it is important to identify and isolate cells contaminated with mycoplasma. It is important to always destroy affected cultures unless they are irreplaceable. Curing cells of mycoplasma infection can be attempted with certain antibiotics but is not recommended due to the poor success rate in eradication and the genotoxicity of many anti-mycoplasma agents. Any laboratory that will supply stem cell lines to other laboratories should perform mycoplasma testing on the stocks they release. Discussion on detection methods and recommended regimes for testing can be found in Stacey (in press). Common methods used include enzymatic assays, PCR, DNA staining of test cells and culture in selective media. The last two techniques have been established in pharmacopoeia and provide useful reference methods (European Pharmacopoeia, 2006b; US Food and Drugs Administration, 2005b). A number of kits for detection of Mycoplasma are available commercially including a rapid detection system called 'Mycoalert' from Cambrex (www.cambrex.com). Note that all cultures should be grown in the absence of antibiotics for a minimum of 2 days prior to the test to maximize sensitivity of this assay.

Viruses

Viruses are dependent on host cells for their replication and in doing so take over and alter the biology of the cell. Some viruses can produce persistent infection of cell lines and it is likely that they will alter the functions and responses of the cells. Thus, any kind of viral infection, even by non-pathogenic virus, would be undesirable in a stem cell culture where the researcher is expecting to make interpretations from the performance of the culture that relate to the activity of stem cells *in vivo*. In short once a cell line has become infected with a virus it may be assumed to have transformed characteristics.

Mouse cells in culture may also harbor viruses and in rare circumstances these may be pathogenic in humans (European Medicines Evaluation Agency, 1997; Stacey *et al.*, 1998). Thus, cultures of primary embryonic mouse fibroblast feeder cells (Stacey *et al.*, 2006) and other preparations, including antibodies and cell culture supernatants (Nicklas *et al.*, 1993), could be affected by viral contamination that may affect both the performance and safety of stem cell lines. Sources of animals used for primary cell preparation should therefore meet current best practice for animal husbandry (e.g. infection control, colony testing) to minimize such risks.

Clearly, cells can be the source of viral contamination. Accordingly, it is appropriate to test each stem cell line for viral contamination. The virus tests to be included for stem cell lines have been discussed by Cobo *et al.* (2005) and should certainly include blood borne viruses capable of causing serious human infection (outlined in **Table 3**). In general, whatever testing has been performed, it is wise to treat all cell cultures as potentially infectious and use appropriate containment procedures (Coecke *et al.*, 2005).

Thus, there are a numerous factors to be considered in the selection of cell lines based on the potential presence of viral and other contaminants. An approach to the selection of cell lines based on biosafety issues and other effects of viral contamination is summarized in **Figure 1**.

Cell line authenticity

Cross-contamination between human ES cell lines is undesirable as it can lead to false data and misleading conclusions. The early years of the isolation and culture of human cell lines were fraught with problems of cross-contamination of cell lines.

Table 3 Viruses that may be considered when screening donors of blood, tissues and cells[a]

Product type	Typical viral contaminants to be considered in testing regimes
Blood products	Blood borne pathogens causing serious human disease including HIV 1&2, hepatitis B, hepatitis C[b]
Blood products contaminated with some cellular material	The above plus 'cell associated' viruses such as hCMV, HTLV I&II, hepatitis A & E
Organs and tissues	Donor screening considerations would include those viruses above and additional testing may be applied for certain indications (e.g. toxoplasma for bone marrow transplant)
Cellular products from cell culture	All of the above plus a range of potential contaminants, depending on risk assessments, including a range of factors including susceptibility of the culture system to infection with different viruses including human herpes viruses (herpes simplex virus, Epstein–Barr virus, HV-6, HHV-7, HHV-8), human polyoma viruses ('JC' and 'BK' virus), parvovirus B19 and Transfusion Transmitted Virus (TTV)

[a]Examples of the testing required by different regulatory authorities are given in Cobo *et al.* (2005).
[b]Non-viral screening would also include *Treponema paillidum* (causative organism of syphilis).

Originally, this problem was associated with accidental cross-contamination by the HeLa cell line (Nelson-Rees *et al.*, 1981) but it is now clear that this is a more general issue involving a number of rapidly growing cell types (Masters *et al.*, 2001, MacLeod *et al.*, 1999). It has been shown that up to 30% of cell lines donated to certain public service collections are contaminated in this way. Furthermore, a significant proportion of the cross-contaminated lines were offered in good faith by their originators (Drexler *et al.*, 2003).

The selection process for stem cell lines should include the requirement that they are authentic. The first step is to ensure that there is a documented trail from the available culture to the original source of the cells. This tracking process should enable proof that the line has been obtained with appropriate ethical approval (see below), and also will verify the source laboratory (i.e. provenance of the line) and details of derivation.

It is also important that the line has been characterized for identity at an early stage so that later stocks can be checked against the unique features of the original isolate. This can be achieved by a number of techniques (**Table 4**) but the most routinely used today is molecular genotyping. Numerous polymerase chain reaction (PCR)-based techniques are available for DNA typing based on similar hypervariable satellite DNA sequences and single nucleotide polymorphisms (SNPs).

Numerous companies now offer DNA identification services at low cost (see table in Stacey *et al.*, 2007). The commercially available DNA typing kits often use common

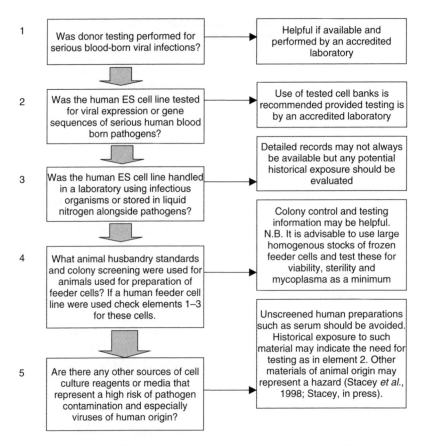

Figure 1   Some key elements for selection of stem cell lines based on risk of contamination with infectious agents.

alleles for cell identification purposes and it should therefore be possible to exchange data on DNA typing between laboratories to verify the authenticity of cells in each laboratory.

Cell line stability

Cell lines in general are prone to genetic and/or phenotypic change when passaged for extended periods (Yu *et al.*, 1997; Briske-Anderson *et al.*, 1997). Genetic abnormalities often occur with extended culture of human ES cell lines (Draper *et al.*, 2002; Hanson and Caisander, 2005). As a general principle one should perform experiments with cells of as low passage as possible. However, cells available at very low passage levels such as "P3" may not become properly established as cell lines. When sourcing new cultures it is important that the cells supplied are capable of long-term passage (e.g. master and working stocks have been established) and have good records available for

Table 4   Comparison of identity testing techniques for cell lines

Technique	Evaluation for cell lines
Isoenzyme analysis (general reference Doyle and Stacey, 2000)	Rapid confirmation of species of origin although profiles may be affected by loss of certain chromosomes
Karyology (general reference Doyle and Stacey, 2000)	Confirms species of origin and may give information on culture stability and genetic aberrations. Requires consider able experience to assess and interpret Giemsa banded preparations
Cytochrome oxidase sequencing (Folmer *et al.*, 1994)	A relatively rapid means of confirming species of origin with valuable validation data through the US National Center for Biotechnology Information (http://www.ncbi.nih.gov)
HLA molecular typing (Christensen *et al.*, 1993)	Can give high level discrimination between different human cell lines and detection of mixed cell lines
Multilocus DNA fingerprinting (Jeffreys *et al.*, 1985)	Provides individual-specific identification of human cell lines. Also provides valuable discrimination amongst cell lines from other species. Requires relatively long technical processing and inter-gel comparisons require very careful standardization and experience
Multiplex PCR DNA profiling (e.g. Masters *et al.*, 2001)	Rapid and highly specific identity testing but commercially available kits are only useful for human cells

their passage history. Cells available only at very high passage numbers (e.g. >P60) may well have undergone permanent genetic and epigenetic changes.

Conclusions and general protocol for selection of human stem cell lines

When selecting human embryonic stem cells it is clear that researchers must take care to address a range of issues (**Figure 2**). Key technical issues include the authenticity of the cells, potential contamination and demonstration of pluripotent characteristics. Good quality guidance and reliable central sources of these cells will be vital to promote the appropriate governance and technical standards for stem cell research. However, it is important for stem cell researchers to bear in mind that the science and ethics are still highly dynamic and periodic review of technology and ethical regulation will be vital to sustain the quality of embryonic stem cell research and its continued public support.

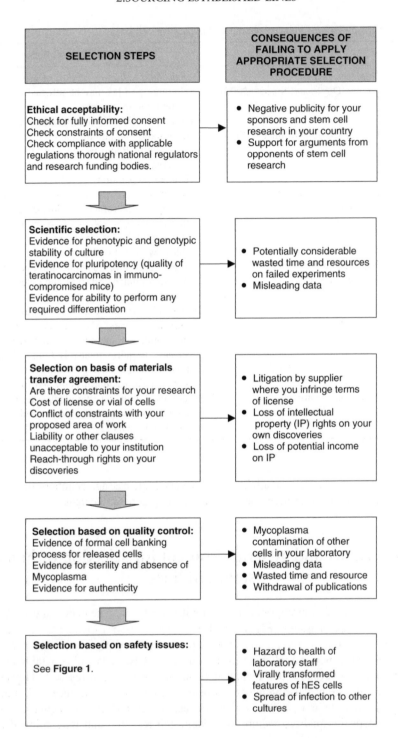

Figure 2   Summary scheme for the selection of human stem cell lines.

References

Briske-Anderson MJ, JW Finley and SM Newman. (1997). The influence of culture passage number on morphology and physiological development of Caco-2 cells. *Proc Soc Exp Biol Med* **4**: 248–257.

Christensen B, C Hansen, M Debiek-Rychter, J Kieler, S Ottensen and J Schmidt. (1993). Identity of tumourigenic uroepithelial cell lines and 'spontaneously' transformed sublines. *Br J Can* **68**: 879–884.

Cobo F, GN Stacey, C Hunt, C Cabrera, A Nieto, R Montes, JL Cortes, P Catalina, A Barnie and A Concha. (2005). Microbiological control in stem cell banks: approaches to standardisation. *Appl Microbiol Biotechnol* **68**: 456–66.

Coecke S, M Balls, G Bowe, J Davis, G Gstraunthaler, T Hartung, R Hay, O-W Merten, A Price, L Shechtman, GN Stacey and W Stokes. (2005). Guidance on good cell culture practice. A report of the second ECVAM Task Force on Good Cell Culture Practice, *ATLA* **33**: 1–27.

Del Guidice RA and RS Gardella. (1984). Mycoplasma infection of cell culture: effects, incidence and detection, In: *In vitro Monograph 5, Uses and standardisation of vertebrate cell cultures*. Tissue Culture Association, Gaithersberg MD, USA, 1984, pp 104–115.

Doyle A and GN Stacey. (2000). Cell and cell line characterisation. In: *Encyclopedia of Cell Technology*. R Spier, ed. John Wiley & Sons, New York, pp 285–293.

Draper JS, C Pigott, JA Thomson and PA Andrews. (2002). Surface antigens of human embryonic stem cells: changes upon differentiation in culture. *J Anat* **200**: 249–258.

Drexler HG, WG Dirks, Y Matsuo and RA MacLeod. (2003). False leukemia-lymphoma cell lines: an update on over 500 cell lines. *Leukemia* **17**: 416–426

European Medicines Evaluation Agency (1997) Note for guidance on quality of biotechnological products: viral safety evaluation of biotechnology products derived from cell lines of human or animal origin (CPMP/ICH/295/95). European Medicines Evaluation Agency, 7 Westferry Circus, Canary Wharf, London, E14 4HB UK. (http://www.eudra.org/emea.html).

European Pharmacopeia (2006b). European Pharmacopeia section 2.6.7 (Mycoplasma) (5th Edition). Maisonneuve SA, Sainte Ruffine.

Folmer O, M Black, W Hoeh, R Lutz and R Vrijenhoek. (1994). DNA primers for amplification of mitochondrial cytochrome c oxidase subunit I from diverse metazoan invertebrates. *Mol Mar Biol Biotechnol* **3**: 294–299.

Hanson C and C Caisander. (2005). Human embryonic stem cells and chromosome stability. *APMIS* **113**: 751–755.

Healy L, C Hunt, L Young and G Stacey. (2005). The UK Stem Cell Bank: Its role as a public research resource centre providing access to well-characterised seed stocks of human stem cell lines. *Adv Drug Deliv Rev* **57**: 1981–1988.

Jeffreys AJ, V Wilson and S-L Thein. (1985). Individual specific fingerprints of human DNA. *Nature* **316**: 76–79.

MacLeod RAF, WG Dirks, Y Matsuo, M Kaufman, H Milch and HG Drexler. (1999). Widespread intra-species cross-contamination of human tumour cell lines arising at source. *Int J Cancer* **83**: 555–563.

McGarrity GI, H Kotani and GH Butler. (1993). Mycoplasma and tissue culture cells. In: *Mycoplasmas: Molecular Biology and Pathogenesis*. J Maniloff, ed. ASM Press, pp 445–454.

Nelson-Rees WA, DW Daniels and RR Flandermeyer. (1981). Cross-contamination of cells in culture. *Science* **212**: 446–452.

Nicklas W, V Kraft and B Meyer. (1993). Contamination of transplantable tumour cell lines and monoclonal antibodies with rodent viruses. *Lab Anim Sci* **43**: 296–300.

Rottem S and Y Naot. (1998). Subversion and exploitation of host cells by mycoplasma. *Trends Microbiol* **6**: 436–440.

Stacey GN. (in press). Risk assessment of cell culture procedures. In: *Medicines from Animal Cells*. GN Stacey and J Davis, eds. John Wiley and Sons, Chichester, UK.

Stacey GN, ED Byrne and JR Hawkins. (2007). DNA Fingerprinting and characterisation of Animal Cell Lines, Chapter 4, In: *Animal Cell Biotechnology Methods and Protocols*, Second Edition, ED Ralf Portner, Humana Press, Totowa, NJ, pp 123–145.

Stacey GN, DA Tyrrel and A Doyle. (1998). Source materials. In: *Safety in Cell and Tissues Culture*, GN Stacey, PJ Hambleton and A Doyle, eds. Kluwer Academic Publishers, Dordrecht, Netherlands, pp 1–25.

Stacey GN, F Cobo, A Nieto, P Talavera, L Healy, A Concha. (2006). The development of 'feeder' cells for the preparation of clinical grade hES cell lines: challenges and solutions. *J Biotechnol* **125**: 583–8.

Stacey GN and C Hunt. (2006). The UK Stem Cell Bank: a UK government funded International resource centre for stem cell research. *Regen Med* **1**: 139–142.

Uphoff CC and HG Drexler. (2002). Comparative PCR analysis for detection of mycoplasma infections in continuous cell lines. *In Vitro Cell Dev Biol Anim* **38**: 79–85.

US Food and Drugs Administration. (2005a). Title 21, Code of Federal Regulations, Volume 7, revised April 2005, CFR610.12 (Sterility). FDA, Department of Health and Human Services.

US Food and Drugs Administration. (2005b). Title 21, Code of Federal Regulations, Volume 7, revised April 2005, CFR610.30 (Test for Mycoplasma). FDA, Department of Health and Human Services.

Yu H, TJ Cook and PS Sinko. (1997). Evidence of diminished functional expression of intestinal transporters in Caco-2 monolayers at high passage. *Pharm Res* **14**: 757–762.

3

Culture of human embryos for stem cell derivation

R. DOUGLAS POWERS AND JEANNINE WITMYER[1,2,3]

[1]*Boston IVF, 130 Second Avenue, Waltham, MA 02451, USA*
[2]*Harvard Medical School, 25 Shattuck Street, Boston, MA 02115, USA*
[3]*Harvard Stem Cell Institute, Harvard University, Cambridge, MA 02138, USA*

Introduction

The methods and results described in this chapter come from our experience thawing and culturing human embryos that were donated to the stem cell research program at Harvard University. We obtained consent for donation of the embryos from patients who had finished with their fertility treatment and wished to donate their remaining cryopreserved embryos for research. The consent form followed the guidelines suggested by President Clinton's Bioethics Commission and was approved by the Harvard University Institutional Review Board.

Overview of protocol

This chapter describes protocols for thawing frozen human embryos, culturing such embryos (either from cleavage or blastocyst stages) and for grading human embryos.

Materials, reagents and equipment

- Prior to cryopreservation the embryos were cultured in SageBiopharma culture media (SageBiopharma, Cooper Surgical/Sage Biopharma, Trumball, CT, USA). We used Forma incubators (Forma, Thermo Electron, Waltham, MA, USA) with 5% CO_2/air atmosphere at a pH of 7.3 ± 0.1. The oocytes were inseminated in Sage Fertilization Medium (**Table 1**) supplemented with 6% Plasmanate (Talecris Biotherapeutics, ResearchTriangle Park, NC USA). Important constituents of the medium are: alanyl-glutamine — a stable form of glutamine that does not break

Table 1   Components of fertilization medium

Sodium chloride
Potassium chloride
Magnesium sulfate
Potassium phosphate
Calcium lactate
Sodium bicarbonate
Glucose
Sodium pyruvate
Alanyl-glutamine
Taurine
L-Asparagine
L-Aspartic acid
Glycine
L-Proline
L-Serine
Sodium citrate
EDTA
Gentamicin
Phenol red

down into ammonium, also providing alanine which is necessary for embryo development, taurine — acts as an anti-oxidant, eliminating the external effects of cell metabolism, glucose — needed for sperm capacitation, citrate — an optimal energy source, non-essential amino acids — to aid in embryo development.

- All insemination and embryo culture was performed in microdrops of culture medium under oil (Sage) in plastic tissue 60 mm culture dishes (Falcon Catalog #3004).

- After fertilization the embryos were cultured in Sage Cleavage Medium supplemented with 6% Plasmanate (Plas/C), which has the same ingredients as the fertilization medium except for the elimination of phosphate and the alteration of the concentrations of the amino acids, glucose and citrate.

- After thawing, embryos were cultured in Global Life Medium (IVF Online) supplemented with 15% Plasmanate (Plas/G) from the eight-cell stage to blastocyst.

Protocols

We have found that embryos frozen at the expanded blastocyst stage show the highest efficiency for human ES stem cell derivation.

When using cleavage stage embryos, Global medium (one step) showed a higher efficiency of stem cell derivation compared to the Sage/Biopharma two-step culture system. The one step system allows the embryo to determine the timing of the transition from lactose to glucose metabolism rather than the arbitrary timing with the two-step system.

Thawing and culturing embryos

Pre-thaw preparation

1. The afternoon prior to the thaw set up twelve 10 µl drops of Plas/G medium in a 60 mm tissue culture dish (Falcon Catalog #3004) and overlay with 11 mL of mineral oil. Repeat this for a total of five dishes. Label the bottom of the dishes using a diamond tip pen. Do not use markers as labeling can become smudged and unreadable. Incubate the dishes overnight at 37°C and 5% CO_2 in air.

2. Aliquot 3 mL of Plas/G medium into ten 30 mm cell culture dishes (Falcon Catalog #3801). Incubate the dishes overnight at 37°C and 5% CO_2 in air.

Thaw procedure for embryos

As adapted from Cohen et al. (1988); Freidler et al. (1988); Veeck (1998)

1. Pre-warm the water bath to 30°C.

2. Label six 30 mm cell culture dishes (Falcon Catalog #3801) for each of the solutions listed in **Table 2**.

3. Remove 1.5 M PROH/0.2 M sucrose and 0.0 M PROH/0.2 M sucrose stock solutions from the refrigerator and use them to prepare the working solutions as shown in **Table 2**.

4. Allow the solutions to equilibrate to room temperature (20–22°C) for a minimum of 30 min.

5. Fill a Dewar flask with liquid nitrogen to within 1 inch of the top. Transfer the cane containing the vial(s) to be thawed from the storage tank to the Dewar flask. Verify the information labeled on the vial.

Table 2   Working solutions for cleavage stage thaws

	1.5 M PROH/suc	0.0 M PROH/suc
1.5 M PROH/suc	3 mL	—
1.0 M PROH/suc	2 mL	1 mL
0.75 M PROH/suc	1.5 mL	1.5 mL
0.5 M PROH/suc	1.0 mL	2.0 mL
0.25 M PROH/suc	0.5 mL	2.5 mL
0.0 M PROH/suc	—	3.0 mL

6. Remove the appropriate vial from the cane. Place the vial on the bench top for
    1 min then place in the 30°C water bath. Swirl the vial gently until the ice crystals
    have disappeared. Do not submerge the vial in the event that the seal has been
    broken. (If the fluid volume in the vial is 1 mL the vial will generally thaw in
    2 min, if the volume is 500 μl the vial will generally thaw in 90 s.) Take care
    not to allow the vial to continue warming once the ice crystals have disappeared.

7. Using a sterile glass pipette transfer the fluid from the vial into an empty small
    culture dish. Locate the embryos under the microscope. If embryos are missing
    use some of the 1.5 M PROH/0.2 M sucrose solution to rinse the vial and locate
    the remaining embryos. Transfer the embryos into the first solution shown in
    Table 3.

8. Follow the rehydration steps shown in **Table 3** for cleavage stage embryos. Fol-
    low the rehydration steps in **Table 4** for blastocyst stage embryos.

9. Remove a Plas/G rinse and culture dish from the incubator. Rinse the embryos.
    Place them individually into separate microdroplets.

10. Assess the embryos using an inverted microscope (200× magnification). Record
    the cell #, % lysis and embryo grade on a Thaw Summary Record.

11. Return the dish to the incubator.

Table 3 Steps for rehydrating cleavage stage embryos

Solutions		Time
1.0 M	PROH/suc	3 min
0.75 M	PROH/suc	3 min
0.5 M	PROH/suc	3 min
0.25 M	PROH/suc	3 min
0.0 M	PROH/suc	3 min

Table 4 Steps for rehydrating blastocysts

Solutions	Time
10%glycerol/0.4 M sucrose	30–40 s
5%glycerol/0.4 M sucrose	3 min
0.4 M sucrose	3 min
0.2 M sucrose	3 min
0.1 M sucrose	3 min
20%Plas/mHTF	rinse
1:1 (20%Plas/mHTF and 20%Plas/B)	rinse

Post-thaw culture for cleavage stage embryos

As adapted from Veeck and Zaninovic (2003).

- Assess the embryos on the afternoon of Day 4 (24–27 hr post-thaw). Record the results on the Thaw Summary Sheet.

- Assess the embryos on the morning of Day 5 (44–48 hr post-thaw). Record the results on the Thaw Summary Sheet.

- Assess the embryos at 12 pm on Day 6 (68–72 hr post-thaw). Record the results on the Thaw Summary Sheet.

- Those blastocysts showing a well defined inner cell mass should be selected for immunosurgery.

Post-thaw culture of thawed blastocysts

1. Monitor the blastocysts for re-expansion over a period of 4–6 hr. Use an inverted microscope (200× magnification) to score the re-expansion, evaluate the trophectoderm and the inner cell mass.

2. Record the observations.

3. Those blastocysts showing a well defined inner cell mass were selected for immunosurgery.

Embryo grading

As described in Van Royen et al. (1999); Witmyer and Powers (in preparation).

1. Day 2 or Day 3 stage embryos should be scored for cell number, percent of fragmentation and presence of multinucleated blastomeres (**Figure 1**).

2. Day 4 embryos should have completed the compaction process by mid to late afternoon (**Figure 2**).

3. Day 5 blastocysts should exhibit low to moderate levels of cavitation (**Figures 3a, 3b**).

4. Day 6 blastocysts should exhibit a fully expanded blastocoel with a well defined inner cell mass (**Figures 4a, 4b, 4c**).

Later steps of embryo culture and human ES cell derivation are covered in **Chapter 4**.

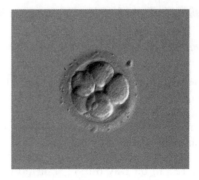

Figure 1   A typical eight-cell human
embryo on the third day of culture.

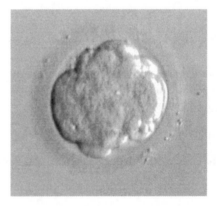

Figure 2   A morula on day 4 of culture.

Troubleshooting

I want to minimize the risk of microbial contamination, what can I do?

Mainly have a well laid out facility, practice good aseptic technique and screen all
feeders and media for contamination beforehand. Read and follow all the directions
in **Chapters 1 and 2**.

I have a low survival and development rate after embryo resuscitation. Why?

There are several possible causes for poor survival and development of
frozen/thawed human embryos. It is important to realize that the post-thaw
development will depend greatly on the quality of the sperm, oocyte, culture
conditions prior to cryopreservation and the cryopreservation procedure itself. A
review of the documentation provided with the frozen embryo should be performed

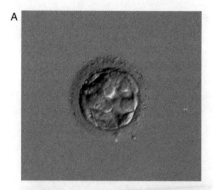

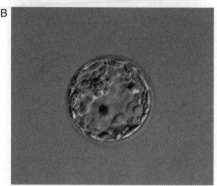

Figure 3   (a) A day 5 human embryo at
the blastocyst stage with a minimal blastocoel cavity. (b) A blastocyst with a larger
blastocoel.

prior to the thaw to determine if the embryo had been developing at a normal rate
until it was frozen. For example, an embryo frozen on day 3 of culture should
have been an eight-cell embryo. It is also important to determine the quality of the
frozen embryo. Quality is typically judged by fragmentation and multinucleation
of the cells. The best quality embryos are mononucleated and have less than
10% fragmentation. The documentation with the frozen embryo should give this
information and can be used to estimate post-thaw success rates. Poor development
prior to freezing will inevitably lead to poor success rates in terms of blastocyst
formation. Culture conditions after thawing cannot rescue poor embryos.

How many embryos should survive thawing?

It is important to review the record of development in order to distinguish poor
pre-freeze embryo development from a poor freeze or thaw process. Embryos that
have developed well should survive thawing at a rate of 85%. Survival is judged by
comparing the number of intact post-thaw cells and their quality with the pre-thaw
embryo. Freeze/thaw damage will be apparent within 10 min of thawing. While
it is too late to do anything about the freezing process at that point, immediate

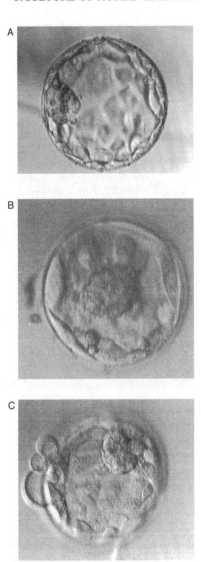

Figure 4 Three day 6 blastocysts
with fully expanded cavities. The inner
cell mass is clearly visible.

post-thaw damage might be due to poor thawing technique, equipment or media. It
is recommended to prepare new solutions, verify the temperature in the water bath
and insure that the protocol is followed in detail.

What factors are generally important for in vitro embryo development?

A successful thaw should result in more than 50% of day 3 embryos surviving
to blastocyst. If there is poor blastocyst development and quality of the initial

Table 5 Comparison of two-step and one-step culture media % based on # survived

	Cleavage stage two-step media	Cleavage stage one-step medium	Blastocyst
# thawed	293	105	58
# survived	201 (68.6%)	76 (72.4%)	45 (77.5%)
% blastocyst	46.3 (93/201)	42.1 (32/76)	NA
% Immunosurgery	32.8 (66/201)	26.3 (20/76)	80.0 (36/45)
% Outgrowth	12.4 (25/201)	17.1 (13/76)	44.4 (20/45)
% Stem cell line	2.5 (5/201)	14.5 (11/76)	26.7 (12/45)
% Day 3 arrest	38.8 (78/201)	30.2 (23/76)	NA
% Day 4 arrest	14.4 (29/201)	27.6 (21/76)	NA

frozen embryos is good, then one should verify that the culture medium and any additives are correct, sterile and are used within their expiration date. As shown in **Table 5**, we found that different media can significantly alter the rate of blastocyst development. If blastocyst development is consistently poor, one should check the type of media used for pre-freeze culture to be certain that the post-thaw medium is compatible. Since most human IVF media is purchased rather than made in each laboratory, it can be very useful to contact the supplier of the media for specific recommendations.

General laboratory conditions are very important for successful development of human embryos. Air quality is very important and should be checked for high levels of volatile organic compounds (VOCs) if development is poor. Unfortunately there are no reference standards for these compounds and no list of specific embryotoxic molecules beyond those usually considered toxic to humans. Most testing laboratories can provide general guidelines.

The conditions inside the incubators are very important. The pH must be maintained in a narrow range of 7.2–7.4 by the appropriate balance of CO_2 in the atmosphere (usually 5–6%) and bicarbonate in the culture media. Ideally, the pH of the media should be measured inside the incubator. If this is not possible, then the CO_2 level should be measured daily. The incubator should be opened as infrequently as practical and the atmosphere should be maintained at 100% humidity. A reasonable test for media and incubator quality is to culture mouse one-cell embryos to the blastocyst stage. A success rate of 85% indicates a satisfactory environment.

References

Cohen J, GW DeVane, CW Elsner, CB Fehilly, HI Kort, JB Massey and TG Turner. (1988). Cryopreservation of zygotes and early cleaved human embryos. *Fertil Steril* **49**: 282–289.

Freidler S, LC Guidice and EJ Lanb. (1988). Cryopreservation of embryos and ova. *Fertil Steril* **49**: 743–764.

Van Royen E, K Mangelschots, D De Neubourg, M Valkenburg, M Van de Meerssche, G Ryckaert, W Eestermans and J Gerris. (1999). Characterization of a top quality embryo, a step towards single-embryo transfer. *Hum Reprod* **14**: 2345–2349.

Veeck LL. (1998). *Embryology Laboratory Manual*. Cornell University, New York Hospital.

Veeck LL and N Zaninovic. (2003). *An Atlas of Human Blastocycsts*. Parthenon Publishing Group, New York, USA.

4

Derivation of human embryonic stem cell lines

STEPHEN SULLIVAN[1], DIETER EGLI[1], HIDENORI AKUTSU[2], DOUGLAS A. MELTON[3], KEVIN EGGAN[1] AND CHAD A. COWAN[4]

[1]*Stowers Medical Institute and Harvard Stem Cell Institute, Department of Cellular and Molecular Biology, Harvard University, 7 Divinity Avenue, SF 457, Cambridge, MA 02138, USA*
[2]*National Research Institute for Child Health and Development Department of Reproductive Biology 2-10-1 Okura, Setagaya, Tokyo 157–8535, Japan*
[3]*Howard Hughes Medical Institute and Department of Molecular and Cellular Biology, Harvard University, 7 Divinity Avenue, Cambridge, MA 02138, USA*
[4]*Massachusetts General Hospital, Center for Regenerative Medicine, Cardiovascular Research Center, and Harvard Stem Cell Institute, Richard B. Simches Research Center, 185 Cambridge Street, Boston, MA 02114, USA*

Introduction

Human embryonic stem cells are culture artifacts derived from ICM cells placed in a culture system selecting for the capacity to proliferate. They are an important new experimental system for studying human development and modelling human disease (Trounson, 2005). There is, and will continue to be, a need for deriving new human ES cell lines for both basic research and clinical applications due to loss of genetic and epigenetic stability during human ES cell culture and manipulation (Cowan *et al.*, 2004; Ludwig *et al.*, 2006; Buzzard *et al.*, 2004; Maitra *et al.*, 2005) and the considerable variance within the human genome between individuals (Redon *et al.*, 2006).

It is still not known why specific lines have different levels of robustness and proclivities to differentiate into specific lineages, although variables in culture history and genetic background of the lines will likely play a large part (Allegrucci and Young, 2006). How such variables affect these innate characteristics of the lines is currently a focus of interest in the field. There is also interest in how ICM cells change in culture to become human ES cells and how these cells can change in culture. The more lines that are generated, the clearer an idea we will have as to what is happening.

Traditionally, human ES cells have been derived from day 5/day 6 blastocysts and propagated on mouse feeder layers. Using the method presented here, components of animal origin are present in the culture; however, this is likely only to be a concern for clinical applications. At present, this derivation method is still a robust method

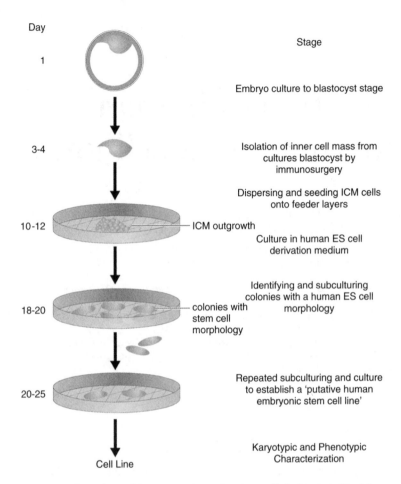

Figure 1 Overview of human embryonic stem cell derivation. Deriving new human ES cell lines requires more than 3 weeks uninterrupted and focused effort. Human embryos are cultured to blastocyst stage as described in **Chapter 3.** The ICM is removed from the blastocyst by immunosurgery and large clumps of ICM cells are transferred to a good quality feeder layer. After several days, ICM outgrowths will appear. Those with a human ES cell morphology (i.e. cells with prominent nucleoli tightly packed together, growing in a flat 2-D morphology) are isolated and subcloned. Successive culture and subcloning produce 'putative human ES cell lines'. These should be extensively characterized (see **Chapters 7, 8, 9 and 10** for details) to verify they are new human ES cell lines.

for deriving new human ES cell lines, requiring fewer embryos than for derivations under xeno-free conditions (Lee *et al.*, 2006). Given the scarcity of human embryos available to most laboratories worldwide, this method will likely be the most used protocol for some time to come. This process of derivation consists of several steps that are highlighted in **Figure 1**.

Overview of protocol

This chapter assumes the human embryos have been cultured to blastocyst stage as described in **Chapter 3**. The derivation process lasts typically between 3 and 6 weeks and involves preparation of mitotically-inactivated feeder layers, isolation of the inner cell mass (ICM) from cultured blastocysts, seeding the ICM clumps on feeders layers, culturing these explants in human ES cell derivation medium and subculturing resulting outgrowths bearing a human ES cell morphology.

Materials, reagents and equipment

Sterile phosphate buffered saline without Ca^{2+} and Mg^{2+} (PBS), pH 7.2 (INVITROGEN/GIBCO, Cat#1404-133)

Gelatin, Type A from porcine skin (SIGMA Cat#G1890)

10 and 15 cm tissue culture dishes (FALCON, Cat#35-3005 and Cat#35-3025 respectively)

Industrial single-edged razor blades (VWR, Cat#55411-050)

Knockout Dulbecco's modified Eagle medium (Knockout DMEM; INVITROGEN/GIBCO, Cat#10829)

KO-Serum Replacement (INVITROGEN/GIBCO, Cat#10828-018)

Plasmanate (BAYER, Cat#0026-0613-20)

Fetal bovine serum (HYCLONE, Cat#SH30070.03)

Glutamax-I (INVITROGEN/GIBCO, Cat#35050-061)

Non-essential amino acids (INVITROGEN/GIBCO, Cat#11140050)

50 units/mL penicillin and 50 µg/mL streptomycin (INVITROGEN/GIBCO, Cat#15070-063)

β-mercaptoethanol (INVITROGEN/GIBCO, Cat#21985-023)

12 ng/mL recombinant hLIF (CHEMICON INTERNATIONAL, Cat#LIF1010)

basic Fibroblast Growth Factor (bFGF, INVITROGEN/GIBCO, Cat#13256-029)

0.05% trypsin/EDTA (INVITROGEN/GIBCO, Cat# 25300-054).

Acid tyrodes (SPECIALTY MEDIA, Cat#MR004-D)

Derivation media (SIGMA, Cat#S1-1939)

Collagenase IV (INVITROGEN/GIBCO, Cat#21985-023)

Rabbit anti-human RBC antibodies (INTERCELL TECHNOLOGIES, Cat#AG28840)

Guinea pig sera complement (SIGMA, Cat#S1639)

Isopropanol

0.22 μm 500 mL steritop filter units (MILLIPORE Cat#SCGPS05RE)

Mitomycin C (Sigma Cat#M-0503)

Mycoalert Mycoplasma Detection Assay (Cambrex, Cat#LT07-218)

Equipment

37°C water bath

Flame pulled thin capillaries

Mouth-controlled suction device or Stripper micropipettor (MID-ATLANTIC DIAGNOSTICS Cat#MXL3-STR) with 175 μm capillary tubes (MID-ATLANTIC DIAGNOSTICS Cat#MXL3-175)

Dissection microscope

Inverted microscope

Heated microscope stage

Laminar air hood with a HEPA filter

Tissue culture incubator

CO_2 monitor

Thermometer

Hygrometer

Sterile door strips/ gloves/

Media/Solutions

MEF Medium (90% Dubelbecco's Modified Eagles Medium, 10% Fetal Bovine Serum, 50 units/mL penicillin and 50 μg/mL streptomycin).

Freezing Medium (90% Fetal Bovine Serum, 10% DMSO)

Human ES cell derivation medium (80% Knockout DMEM, 10% KO-Serum Replacement, 10% Plasmanate, 2 mM Glutamax-I, 1% non-essential amino acids, 50 units/mL penicillin and 50 μg/mL streptomycin, 0.055 mM beta-mercaptoethanol, and 5 ng/mL bFGF).

Human ES cell growth medium (80% Knockout DMEM, 20% Serum Replacement, 4 ng/mL bFGF, 2 mM Glutamax-I, 0.055 mM beta-mercaptoethanol)

Complement solution (dilute sera complement 1:10 in human ES cell derivation medium prior to use. Aliquot prior to use.)

0.1% Gelatin (1 g gelatin in 1 L MilliQ quality water, followed by sterile filtering)

Mitomycin C solution (1 mg mitomycin C powder is dissolved in 100 mL KO-DMEM and filter sterilized. Aliquot prior to use.)

Protocols

Derivation of mouse embryonic fibroblasts (MEFs)

Cull a pregnant ICR mouse (12.5–13.5 days post coitum) in accordance with local animal welfare guidelines. Use two methods of culling (e.g. CO_2 or avertin exposure followed by cervical dislocation) to verify the animal is dead prior to dissection.

1. Place the mouse facing upward on the bench and soak its fur with 70% ethanol. Soaking the fur minimizes the risk of contamination with mycoplasma and other organisms and also prevents fur getting in the way of the dissection.

2. Dissect the abdomen, unravelling and pushing the intestinal tract to one side to reveal the uterine horns. Extract the uterine horns by snipping above the genitalia and just below the ovaries as shown in **Figure 2(A)**.

3. Transfer the uterine horns to a fresh dish of 1× PBS supplemented with 1× Pen/Strep (PBS+P/S). Cut the uterine wall between individual embryos and using a forceps, tease the embryos into the PBS+P/S by tearing the mesometrium with two sets of sharp tipped forceps. Day 12.5–13.5 embryos should resemble those shown in **Figure 2(B)**.

4. Transfer the embryos to a fresh 10 cm Petri dish with 10 mL PBS+P/S. Hold the head of the embryo with one pair of blunt forceps, and carefully remove all the red viscera using sharp forceps with a pecking motion.

5. Cut off the limbs and tail of the embryo using the forceps. Finally, remove the head of the embryo and transfer the remaining trunk to a new Petri dish with PBS+P/S.

6. Mince the eviscerated embryo sections very well with two razor blades and transfer the pieces to several milliliters of prewarmed 1× trypsin EDTA (2 mL/embryo). Triturate the cells continuously until the trypsin solution becomes cloudy with liberated cells. Be careful not to over-trypsinize the cells as this will lead to lysis, clumping and loss of cells. About 5 minutes continuous vortexing of the embryo pieces should be sufficient.

7. Neutralize the protease with MEF medium (1 mL/embryo) (serum in the medium contains trypsin inhibitor and so stops the reaction).

8. Let the cell suspension sit for 5 minutes. Adipose tissue will generally float to the top of the supernatant, and cell clumps will fall down to form a pellet. Remove

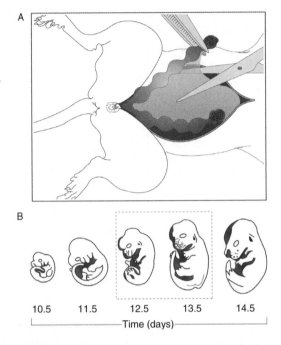

Figure 2 Murine embryo extraction for feeder preparation. (**A**) The uterus from a 12.5–13.5 dpc pregnant mouse is removed as follows: After a longitudinal section is made from above the diaphragm to the genitalia, the intestinal tract is unwound and pushed to one side, revealing the uterine horns behind. The uterine horns are held up with forceps and a sharp scissors is used to snip the mesometrium so they can be separated from the abdominal cavity. Subsequently, the embryos are removed from the uterine horns. (**B**) Typical embryos from day 12.5 to 13.5: the embryos should be similar in size and development to those shown.

the fat with a pipette, and carefully transfer the cell suspension to a fresh 15 mL centrifugation tube without disturbing the pellet.

9. Pellet the cells by centrifugation (300 $g/\sim$1,000 rpm for 5 min).

10. Resuspend the cells in MEF medium (1 mL/embryo). Measure cell number by counting an aliquot of the cells using a haemocytometer and seed the flask or dish to be used for generating feeder layers with 1×10^5 MEFs/cm^2.

11. Add additional MEF medium to the dishes/plates so the medium is at least 0.5 cm in depth.

12. Leave the cells to sit down overnight in the incubator (37°C, 5% CO_2) before expanding these cultures and mitotically inactivating them.

Note: Remember MEFs and other primary cell cultures are a common source of microbial contamination in the cell culture facility. To reduce the risk of contamination, always soak euthanized mice in 70% ethanol prior to embryo extraction, dip uterine horns in 70% ethanol before embryo removal, and culture the MEFs in medium supplemented with 1× Pen/Strep for 2 days after the MEFs are derived.

Note: MEFs should be specifically screened for mycoplasma (See **Chapters 1 and 2** for more on mycoplasma) before being used as feeders for human ES cell culture. We recommend an enzymatic assay [Mycoalert mycoplasma detection kit made by Cambrex www.Cambrex.com] as a quick and straightforward way to screen cultures for mycoplasma. To increase the sensitivity of this assay, MEFs should have been cultured without antibiotics for at least 48 hr prior to the mycoplasma screening.

Generation of mitotically-inactivated feeder layers

Proliferating MEFs will compete with human ES cells for media nutrients and space so feeder layers are mitotically-inactivated before human ES cells are seeded on them. There are two commonly-used ways to mitotically-inactivate MEFs: treatment with drug mitomycin C or exposure to γ irradiation.

Mitomycin C treatment

1. Culture the MEFs in 10 µg/mL mitomycin C for 2 hr at 37°C, 5% CO_2.

2. It is important to wash as much mitomycin C away after the MEFs have been inactivated to form feeder layers, otherwise the human ES cells may themselves be affected by the drug. Aspirate the mitomycin C solution and wash the feeders thoroughly four times with PBS.

3. Disaggregate feeders with warm 0.25% trypsin/EDTA solution. It is important to minimize the exposure of the feeders to trypsin/EDTA as the cells will begin to senesce or lyse. After about 1 min, the cells will 'round up'. Repeatedly tap the flask off the bench to help dislocate the feeders and closely monitor them under the microscope to identify when most have been dislodged. All in all this process should take less than 2 min.

4. Add 0.5–1 volumes of serum-containing MEF medium to stop the trypsinization reaction. Pellet the feeders down by centrifugation (300 g/∼1,000 rpm, 5 min).

5. Seed flasks, dishes or wells with 1–2 × 10^5 cells/cm^2. Wash the cells three times in PBS to remove serum that will cause differentiation of human ES cells. Grow the cells in human ES cell medium overnight.

6. If feeders are not immediately required or an excess has been generated, the cells should be frozen and stored in liquid nitrogen as described in **Chapter 5.**

γ irradiation

1. Culture freshly isolated MEFs to confluence and split 1:5 (generating passage 2 MEFs).

2. Once these passage 2 MEFs are confluent, trypsinize and resuspend them in a few milliliters MEF media, (we routinely resuspend a pellet generated from a confluent T175 flask of MEFs into 3 mL MEF medium). Irradiate the MEF pellet for 25 min at 200–250 Rad/min for a total exposure of 5,000–6,250 Rad.

3. Pellet the irradiated cells (300 g/~1,000 rpm, 5 min).

Seeding feeder layers

1. Treat the culture vessel with 0.1% gelatin and allow the gelatin to sit for at least 5 min before aspirating it off. This gelatine preparation prevents 'rolling-up' of confluent feeder layers.

2. Mitotically inactivated MEFs should be seeded onto gelatinized culture vessels with a density of ~50,000 cells/cm^2.

Isolation of the inner cell mass from cultured blastocysts

Figure 3 shows high and low quality blastocysts prior to immunosurgery. Derivation efficiency is at a maximum when high-quality blastocysts are used. Immunosurgery

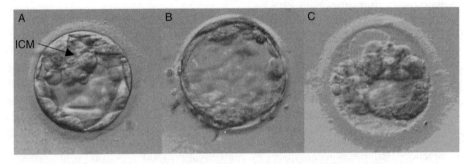

Figure 3   Day 1 Embryos cultures to blastocyst stage. (A) A good quality blastocyst (there is obvious cavitation, and large numbers of cells in the ICM, no graininess or lysis). (B) shows a blastocyst with few ICM cells (this can still produce a human ES cell line). (C) shows a necrotic embryo (extensive graininess and lysis).

allows the removal of the ICM from individual blastocysts and consists of three main steps:

(a) Dissolving the zona pellucida with acid tyrodes solution.

(b) Selectively killing the trophectoderm by complement-mediated lysis. The blastocyst is incubated with anti-human RBC antibody and subsequently complement. The cells of the trophectoderm layer have tight junctions between them providing a physical barrier which protects the ICM cells such damage.

(c) Removal of the ICM from the lysed trophecoderm cells.

Figure 4 shows an overview of the immunosurgery procedure and **Figure 5** shows blastocysts undergoing immunosurgery.

1. As the success of this procedure is dependent on exposing the blastocyst for the correct duration, and thus requires constant observation under the microscope, it is best to perform immunosurgery on one embryo at a time.

2. Prepare Petri dishes for embryo immunosurgery as follows: Make a series of 30 μL microdrops across three plates (see **Figure 4**): three of acid Tyrodes solution, three of derivation medium in the first dish (Plate 1); three of antibody solution, three of derivation medium, and three of Complement in the second (Plate 2), and six of derivation medium for the third (Plate 3). The drops are covered with embryo-tested mineral oil (to prevent evaporation) and while Plate 1 is left at room temperature, Plates 2 and 3 are placed in an incubator (5% CO_2, 37°C) for at least 1 hr and only removed directly prior to use.

3. Place a heating plate onto the stage of the dissection microscope and set the heating plate temperature to 37°C.

4. Prepare the mouth controlled suction device (assemble mouthpiece, rubber tubing, holder and glass pipette) and triturate ~200 μL FBS using the pipette. Coating the pipette with serum prevents the embryo sticking to the inside of the glass pipette during transfer. Avoid the aspiration of mineral oil into the glass pipette.

5. Remove dishes containing the embryo cultured to blastocyst stage, and the dish with microdrops from the incubator (See **Chapter 3** for details of human embryo culture). Locate the blastocyst under the microscope and transfer the embryo to the first microdrop of acid Tyrodes solution (AT) using a mouth pipette. Move the blastocyst from drop to drop as shown in **Figure 4**.The embryo should spend 1–2 s in each drop of AT. Watch the embryo carefully under the microscope as it is transferred from drop to drop. When the embryo has been transferred to the third drop of AT, the zona pellucida should have become thinner and be nearly digested. Transfer the embryo through the second row of microdrops containing derivation medium to neutralize the AT and then to the third row containing microdrops of the human antibody solution.

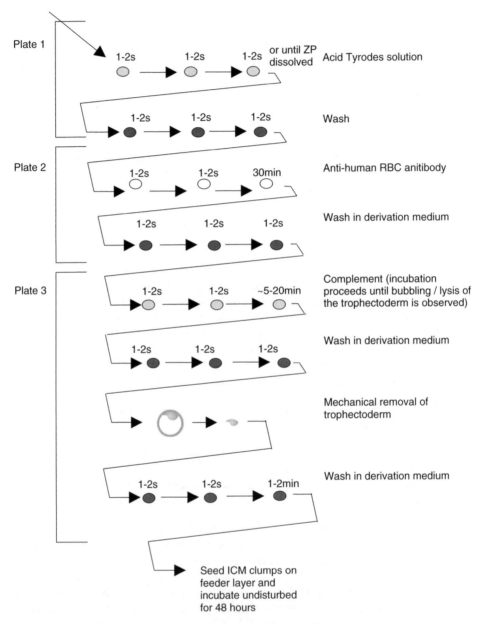

Figure 4 Extracting ICM by immunosurgery. This schematic shows the various steps and incubation times for extracting the ICM from the blastocyst. The small ovals represent microdrops in the dishes between which the embryo is transferred. Acid Tyrodes treatment and initial washing of the embryo is done in Plate 1, antibody incubation, second washing and complement incubation is carried out in Plate 2. Additional washing and trophectoderm removal is carried out in Plate 3 after which the ICM is seeded on a feeder layer.

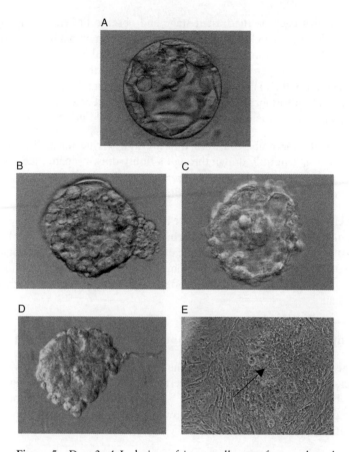

Figure 5 Day 3–4 Isolation of inner cell mass from cultured blastocyst by immunosurgery. (**A**) Embryo before incubation acid tyrodes (AT) solution. (**B**) Embryo after AT incubation, see the thinning of the zona pellucida. (**C**) Collapsed embryo after immunosurgery, note the trophectoderm 'bubbling'. (**D**) ICM after stripping of lysed trophectoderm layer. (**E**) Clump of ICM cells attached to feeder layer.

6. Transfer the embryo from the first to the third drop of antibody solution (1–2 s in the first and second microdrops). Once the embryo is in the third drop of antibody solution, the dish is placed back in the incubator for 30 min.

7. Remove the dish and place it back on the heated stage. Resume transferring the embryo from drop to drop using the mouth pipette initially through the fourth row of derivation medium (to wash off excess antibody) and then the fifth row of complement (1–2 s in the first and second drops of complement). Once the embryo has been transferred into the third drop of complement, place the dish back in the incubator. Leave for 15 min.

8. Place the dish back on the heated stage and observe the embryo for lysis of the trophectoderm cells and the collapse of the embryo in on itself. This Complement-mediated lysis is commonly described as trophectoderm 'bubbling' as the cells become large, transparent and bubble-like (See **Figure 5**). If no such 'bubbling' is evident, place the dish back in the incubator. Extend the incubation by intervals of 5 min until the embryo has collapsed and all trophectoderm cells are lysed. The incubation time for the embryo in complement should not be longer than 30 min.

9. After washing the embryo a third time by transferring through all drops of the sixth row, it is carefully drawn through a thinly drawn pipette, the bore of which is just smaller than the diameter of the embryo. The ICM should detach from the lysed trophectoderm cells with 1–2 draws up the pipette.

10. Transfer the detached ICM cell cluster to microdrops of human ES cell derivation medium and proceed as shown in **Figure 4.** Finally, transfer the ICM to a well of a four-well plate seeded with mitotically-inactivated feeders and containing pre-warmed human ES derivation medium. Place this plate back into the incubator and do not disturb for at least 48 hr, after which attachment and outgrowth should be evident.

Breaking up the ICM, growing explants and analyzing colony morphology

At this stage of the derivation process, the goal is to increase 'putative human ES cell' number, and not to be too much concerned with removing differentiated from the outgrowths. However, explants should not be allowed to become over-confluent as all cells will begin to differentiation. Here now are some general notes on maximizing derivation efficiency:

- During the initial dispersion of the ICM-derived outgrowth, a part of the initial outgrowth should be left intact, in case the reseeded clumps fail to attach and/or grow after transfer to new wells.

- Do not be tempted to expand the cultures too quickly, or to treat the cells too harshly, breaking them into clumps that are too small to sustain proliferation.

- Remember not to seed the feeder layers too densely (should seed at $>50,000$ cells/cm^2) - MEFs form cell clusters that can easily be mistaken for early human ES cell outgrowths.

- Partially change medium when the original medium on the cells becomes acidic (medium becomes light pink/yellow). If changed too often, medium will have less auto-stimulatory factors secreted from the human ES cells themselves that encourage their proliferation.

- Mechanical disaggregation of human ES cell colonies is described in detail in **Chapter 5 (Routine Protocols)**.

- Colonies displaying a human ES cell morphology, are most readily identifiable as such, by their prominent and numerous nucleoli and their flattened shape of

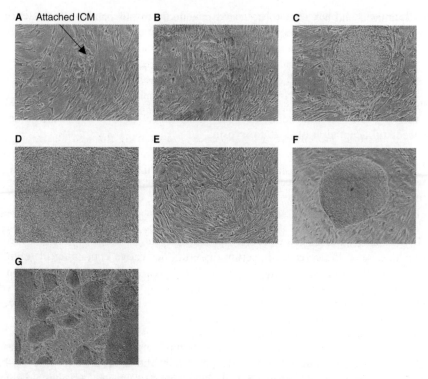

Figure 6 Identifying and subculturing colonies with a human cell morphology. (A–D) show the same colony with human ES cell morphology at days 5, 7, 10, 13. The colony is ready to mechanically disaggregate at day 13. (E) shows a P1 colony from this mechanical disaggregation. (F) shows a P2 colony. (G) shows a P5 culture successfully adapted to trypsinization.

their colonies, and the ability to distinguish individual cells within the colony (See **Figure 6**). Other morphologies that also arise from outgrowths include fibroblast-like cells (which are indistinguishable from feeders), spread out trophoblast cells, and epitheloid-like cells.

The protocol outlined using mouse feeder layers has been described in detail here as it as been independently used to repeatedly derive human ES cells in several laboratories over three years and so we consider it our most established protocol.

The use of animal feeders has some disadvantages, such as risks of pathogen transmission and viral infection (Richards *et al.*, 2002; Amit and Itskovitz-Eldor, 2003; Amit *et al.*, 2004; Rosler *et al.*, 2004). Also, foreign sugars may be transferred from feeders to the glycoproteins of the human ES cells (Martin *et al.*, 2005). These concerns, however, will probably only be significant if cells or their derivates are to be used therapeutically. For purposes of using human ES cells as a basic *in vitro* tool, this protocol has proved sufficient. While some laboratories are actively seeking alternatives to mouse feeders (human fibroblast feeders, human fallopian feeders, feeder conditioned medium, chemically defined medium), we feel these protocols are still being optimized and have yet to be tested repeatedly in our hands. Should you wish to try one of these alternative protocols, we direct you to the papers cited.

Adapting newly derived human ES cells to trypsin

Freshly derived lines can be adapted to trypsin as early as passage 3, when you have two wells of a six-well plate confluent with human ES cell colonies. Here are guidelines for successful adaptation of lines to trypsin:

- Initial splits with trypsin should be low i.e. 1:2–3. We trypsinize one of the two wells with trypsin and seed it into two or three new wells, while continuing to mechanically disaggregate cells the other well, in case there is a problem with enzymatic disaggregated cultures.

- Remember that cells are extremely prone of lysis by forces induced by rough pipetting when the cells are in serum or serum derivative containing medium. Do not make the mistake of scraping the cell cultures after the trypsin reaction has been stopped with serum or serum containing medium. Only scrape the wells when the trypsinization reaction is still proceeding, otherwise a large proportion of cells will lyse and seeding efficiencies will be very low.

- Ensure good plating efficiencies by gently pipetting the cells, and by carefully observing the cells under the microscope as they round up after trypsin has been added. As soon as all human ES cells have rounded up and become apparent as individual cells in the colony, wells should be scraped and the reaction stopped with serum or serum containing medium.

- We only trypsinize the cultures to clumps of ten cells or more during these early passages. If you have trouble with trypsinization, disaggregation with accutase or collagenase is recommended. When performing the second and third tryspinization on the culture, we gradually raise the split ratio to 1:2 and 1:4 respectively.

- Some have suggested that enzymatic disaggregation leads to a higher incidence of genetic abnormalities in lines of equivalent passage number [Mitalipova *et al.* 2005]. However, we point out that cultures can be expanded much more readily with enzymatic disaggregation and that comparisons between mechanically and enzymatically disaggregated cultures of equivalent passage is misleading (i.e. cells in trypsinized cultures will have undergone a substantially greater number of population doublings than mechanically disaggregated cultures of equivalent passage).

Protocol for trypsinization

Routine trypinization of human ES cells is covered in the Extended Protocols section of **Chapter 5**. For initial adaptation of human ES cell lines to trypsin we proceed as follows:

1. Warm aliquots of 0.05% trypsin EDTA solution and serum containing medium to 37°C in a waterbath.

2. Remove the medium from the well containing the culture to be disaggregated. Rinse the well twice with $1 \times$ PBS (1 mL per well of a four-well plate).

3. Add 1 mL warm 0.05% trypsin EDTA solution. Immediately transfer the plate to the heated plateform of an inverted microscope and watch carefully as the human ES cells' morphology changes and these cells become refractile and spherical. When about 90% of the human ES cells round up, scrape the well until cell clusters (>10 cells or more) are floating in the medium. Do not wait until the feeders become refractile before stopping the reaction. Add 0.5 mL medium to stop the trypsin reaction.

4. Carefully remove the cell suspension with a P1000 pipette and transfer to a 15 mL centrifuge tube.

5. Pellet the human ES cells by centrifugation (300 g/$\sim$1,000 rpm for 5 min).

6. Carefully resuspend the pellet and seed the newly prepared wells according to the appropriate split ratio.

7. Half-change the medium in the well with fresh medium 24 hr after trypsinization. The human ES cell colonies should have adhered to the feeder layers and show signs of proliferation. Some cell death may also be evident and this medium change should remove most of the floating dead cells.

Troubleshooting

When performing immunosurgery, the trophectoderm fails to exhibit bubbling and the embryo does not collapse? What has happened?

There is a problem with the antibody and/or the complement. Ensure the correct dilutions were made when making up the solutions and that fresh preparations of both solutions are used in the next derivation.

The ICM fails to attach or proliferate. Why?

This could be due to overexposure the blastocyst to complement, thereby damaging the ICM cells. It could also be due to use of poor quality feeder layers or medium.

After trituaration through a drawn glass capillary, some trophectoderm is still attached to the ICM. Should I triturate further at risk of damaging the ICM?

No, do not triturate further. Removal of all trophectoderm is not essential. In fact, having some trophectoderm still present can facilitate attachment of the ICM to the feeder layer and promote outgrowth.

I have an ICM outgrowth but after mechanical disaggregation it stopped growing. What happened?

This may be due to the cells being manipulated too roughly, breaking the colonies into two small clumps to support initial proliferation. Be more gentle, pipetting the cells more slowly next time.

None of my outgrowths produced a line displaying human ES cell morphology that continued to proliferate. What can I change to improve derivation frequencies in the next attempt.

You should probably change serum replacement/FBS batches. Screening serum batches for human ES cell growth is very important. Seed previously established human ES cell lines at different densities using media supplemented with different batches. Culture the cells for two weeks and assess the cultures morphologically and by immunoFACS (see **Chapter 9** for characterizing human ES cell cultures in this way). Use the serum batch that supports proliferation (but not differentiation or cell death) of human ES cells.

- Feeders provide a matrix and growth factors for your ICM outgrowths. Poor feeder support of the outgrowth can also be the reason the putative human ES cells cease proliferating, always ensure feeders are seeded at the appropriate density ($\sim$50,000 cells/cm^2) are of high quality (e.g. with a passage number $\leq$3, an upright, slender cell morphology, and no evidence of senescence).

- Poor embryo development and explant outgrowth can be due to microbial contamination. Filter and screen all solutions and media prior to using them for derivation of new lines. Microbial contamination and the avoidance of it are discussed in detail in **Chapters 1 and 2**. The media used to derive and culture human ES cells are extremely rich and so are susceptible to becoming contaminated with microbes. Routine use of antibiotics masks contamination which slows cell growth and as these cells are very sensitive the presence of antibiotics can be toxic to the cells. Remember feeders are a common source of contamination. Feeders should always be screened for microbes including mycoplasma as outlined in **Chapter 1**.

- The problem could also be due to physical changes in the cell culture, i.e. temperature, humidity, pH of medium. Make sure your incubator and heated plates are functioning and are set to the correct temperature (37°C). Make sure to keep the water tray in the incubator full of autoclaved water. Change the medium on the cells before it becomes too acidic (i.e. yellow).

- Remember organic solvents and other toxic aerosols are not introduced to the derivation facility. Other harsh chemicals like PFA also should not be used within the derivation facility, refrain from using perfume/aftershave. Human embryonic cultures are very sensitive to such aerosols and they can perturb cell proliferation if present in the atmosphere.

References

Allegrucci C and LE Young. (2006). Differences between human embryonic stem cell lines. *Hum Reprod Update* **13**(2): 103–20.

Amit M and J Itskovitz-Eldor (2003). Human feeder layers for human embryonic stem cells. *Biol Reprod* **68**: 2150–2156.

Amit M, C Shariki, V Margulets and J Itskovitz-Eldor. (2004). Feeder layer- and serum-free culture of human embryonic stem cells. *Biol Reprod* **70**: 837–845.

Buzzard JJ, NM Gough, JM Crook and A Colman. (2004). Karyotype of human ES cells during extended culture. *Nat Biotechnol* **22**: 381–382; author reply 382.

Cowan CA, I Klimanskaya, J McMahon, J Atienza, J Witmyer, JP Zucker, S Wang, CC Morton, AP McMahon, D Powers and DA Melton. (2004). Derivation of embryonic stem-cell lines from human blastocysts. *N Engl J Med* **350**: 1353–1356.

Lee JB, JE Lee, JH Park, SJ Kim, MK Kim, SI Roh and HS Yoon. (2005). Establishment and maintenance of human embryonic stem cell lines on human feeder cells derived from uterine endometrium under serum-free condition. *Biol Reprod* **72**: 42–49.

Ludwig TE, ME Levenstein, JM Jones, WT Berggren, ER Mitchen, JL Frane, LJ Crandall, CA Daigh, KR Conard, MS Piekarczyk, RA Llanas and JA Thomson. (2006). Derivation of human embryonic stem cells in defined conditions. *Nat Biotechnol* **24**: 185–187.

Maitra A, DE Arking, N Shivapurkar, M Ikeda, V Stastny, K Kassauei, G Sui, DJ Cutler, Y Liu, SN Brimble, K Noaksson, J Hyllner, TC Schulz, X Zeng, WJ Freed, J Crook, S Abraham, A Colman, P Sartipy, S Matsui, M Carpenter, AF Gazdar, M Rao and A Chakravarti. (2005). Genomic alterations in cultured human embryonic stem cells. *Nat Genet* **37**: 1099–1103.

Martin MJ, A Muotri, F Gage and A Varki. (2005). Human embryonic stem cells express an immunogenic nonhuman sialic acid. *Nat Med* **11**: 228–232.

Mitalipova MM, RR Rao, DM Hoyer, JA Johnson, LF Meisner, KL Jones, S Dalton and SL Stice. (2005). Preserving the genetic integrity of human embryonic stem cells. *Nat Biotechnol* **23**(1): 19–20.

Redon R *et al.* (2006). Global variation in copy number in the human genome. *Nature* **444**, 444–454.

Richards M, CY Fong, WK Chan, PC Wong and A Bongso. (2002). Human feeders support prolonged undifferentiated growth of human inner cell masses and embryonic stem cells. *Nat Biotechnol* **20**: 933–936.

Rosler ES, GJ Fisk, X Ares, J Irving, T Miura, MS Rao, MK Carpenter *et al.* (2004). Long-term culture of human embryonic stem cells in feeder-free conditions. *Dev Dyn* **229**: 259–274.

Trounson A. (2005). Human embryonic stem cell derivation and directed differentiation. *Ernst Schering Res Found Workshop* 27–44.

5

Standard culture of human embryonic stem cells

JEREMY M. CROOK, RACHEL HORNE AND ALAN COLMAN
ES Cell International Pte Ltd, 11 Biopolis Way, 05-06 Helios, Singapore 138667

Introduction

Since the first reports of successful isolation and culture of human embryonic stem (ES) cells by Ariff Bongso (1994) and *bona fide* cell lines from human blastocysts by James Thomson (1998), methods for working with human ES cells have steadily evolved. Although founded on protocols for generating mouse ES cells, the unique *in vitro* growth characteristics and desired applications of human ES cells have required a more conscientious approach to their maintenance and investigation. Thus, for the dual capacity of prolonged self-renewal while maintaining a stable development potential, the hall-mark characteristics of ES cells, together with ensuring safety and quality assurance, specialized methods to derive, culture, passage, expand, preserve, characterize and differentiate human ES cells are essential.

There are now published a plethora of approaches to their culture and cryopreservation. For example, variants of traditional human ES cell culture incorporate fibroblast feeder cells of either murine or human origin, combined with fetal bovine or more recently human serum or serum replacement with or without bovine or human serum albumin (Thomson *et al.*, 1998; Reubinoff *et al.*, 2000; Richards *et al.*, 2002, 2003; Amit *et al.*, 2003; Koivisto *et al.*, 2004; Ellerstrom *et al.*, 2006). Other methods utilize fibroblast-conditioned medium with feeder free growth matrices of Collagen IV, fibronectin, laminin, or Matrigel (Xu *et al.*, 2001; Noaksson *et al.*, 2005; Choo *et al.*, 2006). More contemporary feeder- and serum-independent methods seek to further define the physicochemical environment through supplementation with one or more cytokines, usually including basic fibroblast growth factor (bFGF), and defined growth substrates again including collagen, fibronectin and/or laminin (Ludwig *et al.*, 2006).

While the multitude of approaches reflects the need to further explore and develop new technologies in the field, there is an increasing need to discern suitable and reliable methods which can be widely employed across different laboratories as standard procedures for consistent and reproducible outcomes. Options should include standard protocols that satisfy the needs of basic science as well as methods compliant with the more stringent regulatory requirements of applied research and clinical product development. Consequently, this chapter describes a selection of standard and reliable protocols for human ES cell culture and preservation.

Overview of protocols

The protocol section of this chapter is extensive. To avoid overloading readers new to human ES cell culture, the protocol section has been divided into two parts: "routine protocols" and "extended protocols". The section **"Routine protocols"** includes culturing, freezing and thawing human ES cells on mouse and human feeders, testing for contamination, and disaggregation. Both FBS and KO-SR based media cultures are described in Routine Protocols. This in addition to work described in **Chapter 4** should be enough to allow a laboratory to become proficient at culturing most of the human ES cell lines available. The section **"Extended protocols"** includes vitrification, feeder free culture, trypsinization, culture on Matrigel, and controlled rate freezing.

Materials, reagents and equipment

(i) Cells

Fibroblast feeders

In-house derived mouse embryonic fibroblasts

Day 12.5–13.5 embryos of inbred mouse strains 129sv, CBA, CF1 or Balb/C mice are commonly used in many labs to generate mouse embryonic feeders (MEFs). In our experience, MEFs produced from outbred strains do not seem to adequately support human ES cell culture.

Commercially available mouse and human fibroblast feeders

Mouse and human fibroblasts (MEFs and HFs respectively) alternatively certified for human ES cell culture can be purchased from the American Type Culture Collection (ATCC; http://www.atcc.org) and Specialty Medium (http://www.specialtymedium.com). Unlike MEFs, we routinely employ commercially available HFs including embryonic, neonatal and adult fibroblasts, between passages 6 and 9. Importantly, significant variation can be expected between the efficacy of different fibroblast lines, which in turn can be human ES cell line-dependant. As such, we recommend selection of a most optimal/compatible feeder cell line through line comparison by the individual user.

Human ES cells

The protocols outlined in this chapter can be used to varying degrees of success with all lines specified in **Chapter 2**.

(ii) Medium reagents

Dulbecco's Modified Eagle Medium (DMEM; Invitrogen/Gibco Catalog # 11960–044)
L-Glutamine (200 mM; Invitrogen/Gibco Catalog # 25030–081)
Penicillin/streptomycin (Invitrogen/Gibco Catalog # 15070–063)
Non-essential amino acids (NEAA; Invitrogen/Gibco Catalog # 12383-014)
β-Mercaptoethanol (Invitrogen/Gibco Catalog # 21985-023)
Insulin–transferrin–selenium (ITS; Invitrogen/Gibco Catalog # 41400-045)
Human Recombinant Basic Fibroblast Growth Factor (bFGF; Invitrogen/Gibco Catalog # 13256-029)
Knockout™ Dulbecco's modified eagle medium (KO-DMEM; Invitrogen/Gibco Catalog # 10829-018)
Knockout™ serum replacement (KO-SR; Invitrogen/Gibco Catalog # 10828-028)
human ES cell qualified fetal bovine serum (FBS; Hyclone Catalog # SH30070-03)

(iii) Generic reagents

PBS+ (with Ca^{2+} and Mg^{2+}; Invitrogen/Gibco Catalog # 1404-133)
PBS- (without Ca^{2+} and Mg^{2+}; Invitrogen/Gibco Catalog # 14190-144)
Trypsin/EDTA (0.25%; Invitrogen/Gibco Catalog # 25200-056)
TrypLE™ Select (Invitrogen/Gibco Catalog # 12605-010)
HEPES 1M (Invitrogen/Gibco Catalog # 15630-106)
Collagenase IV (Invitrogen/Gibco Catalog # 17104-019)
Mitomycin-C (Sigma M-0503; note: carcinogen, cytotoxic, light sensitive)
Gelatin (Type A from porcine skin; Sigma Catalog # G-1890)
Ethylene glycol (EG; Sigma Catalog # G-1890)
Dimethyl sulfoxide (DMSO; Sigma Catalog # D-2650)
Sucrose (Sigma Catalog # S-7903)
Human plasma fibronectin (Sigma Catalog # F2006)
CryoStor™ CS10 (Biolife solutions Catalog # 99-610-DV)
Distilled water (dH_2O)
Isopropanol

(iv) Plastic-Ware

Vitrification straws (sterile; LEC instruments)
CBS high security straws (sterile; Cryo Bio System Catalog # 014651)
Centre well organ culture dishes (2.89 cm^2; Falcon Catalog # 35–3037)
75 cm^2 flasks (Falcon Catalog # 35–3136)

175 cm^2 flasks (Falcon Catalog # 35–3112)
15 mL tubes (Falcon Catalog # 35–2096)
50 mL tubes (Falcon Catalog # 35–2070)
Cell scrapers (Falcon Catalog # 35–3086)
Four-well plates (NUNC Catalog # 176740)
1.0 mL cryovials (NUNC Catalog # 377224)
4.5 mL cryovials (NUNC Catalog # 379146)
500 mL stericup receiver flasks (Millipore Catalog # SCOOB05RE)
250 mL 0.22 μm steritop filter units (Millipore Catalog # SCGPT02RE)
0.22 μm Luer lock syringe filters (Millipore Catalog # SLGVR25LS)
Cryo 1°C freezing container (Nalgene Catalog # 5100–0001)

(v) Specialized equipment

Class 2 biosafety cabinet
Dissection microscopes
Warm stages
Aspiration system
Planar biological freezer (Planar Kryo 360-1.7)
SYMS manual sealing unit for CBS straws (Cryo Bio System 007213)
Gamma-irradiator
Liquid nitrogen storage canister
Liquid nitrogen Dewar
Bunsen burner
Haemocytometer
Dissection scissors
Forceps
2.5 mL syringes
Aluminum cryocanes
12G and 18G needles
Pasteur pipettes
Instrument sleeves (Leica 11520145)
Glass capillaries (1.0 mm outer diameter; Harvard Apparatus Catalog # GC100T-15)

Routine protocols

Fibroblast feeders

(i) Solutions and media

1% gelatin stock solution

1. Dissolve 0.4 g of gelatin in 40 mL dH$_2$O in a 50 mL tube.

2. Autoclave at 121°C for 20 min.

3. Store 20 mL aliquots at −20°C.

0.1% gelatin working solution

1. Combine 20 mL 1% stock solution with 180 mL dH$_2$O.

2. Store for up to 4 weeks at 4°C.

Fibroblast feeder media

The media described in **Table 1** are routinely used in our laboratory for culturing MEFs and human adult dermal and foreskin fibroblasts. For other commercially available feeders we recommend preparing medium according to the manufacturer's instructions.

Combine components and filter medium through a Steritop filter flask into a Stericup receiver unit.

Store for up to 2 weeks at 4°C.

Table 1   Media for murine and human feeders

Materials/Reagents	Vendor	Catalog #	Volume
Murine feeder medium			
DMEM	Invitrogen/Gibco	11960-044	442.5 mL
L-Glutamine (200 mM)	Invitrogen/Gibco	25030-081	5 mL
FBS	Hyclone	SH30070-03	50 mL
Penicillin/streptomycin (optional)	Invitrogen/Gibco	15070-063	2.5 mL
Human feeder medium			
DMEM	Invitrogen/Gibco	11960-044	392 mL
L-Glutamine (200 mM)	Invitrogen/Gibco	25030-081	5 mL
FBS	Hyclone	SH30070-03	50 mL
Insulin-transferrin-selenium supplement (ITS)	Invitrogen/Gibco	41400-045	5 mL
Penicillin/streptomycin (optional)	Invitrogen/Gibco	15070-063	2.5 mL

Fibroblast feeder freezing medium for standard freezing (Table 2)

1. Combine components in **Table 2** below.

Table 2   Fibroblast feeder freezing medium

DMSO	Sigma	Cat # D-2650	10%
FBS	Hyclone	Cat # SH30070-03	90%

Fibroblast feeder thaw medium for standard frozen cells

1. **Murine feeder medium** and **Human feeder medium** for MEFs and HFs respectively.

Mitomycin-C solution

1. Dissolve 2 mg of Mitomycin-C in 1 mL of PBS and filter through a 0.22 μm syringe filter.

0.05% Trypsin/EDTA

1. Dilute 1 mL 0.25% Trypsin/EDTA in 4 mL PBS-.

2. Store for up to 1 week at 4°C.

(ii) Fibroblast feeder production and maintenance

Preparing gelatin coated plates (for MEF related cultures only)

1. Add 0.1% gelatin to the plates, ensuring there is enough to cover the entire surface.

2. Incubate for at least 1 hr at room temperature (RT), ensuring that the gelatin does not dry out.

(iii) Deriving and culturing murine embryonic fibroblasts

We derive and culture MEFs according to the protocol previously outlined in **Chapter 4**.

(iv) Subculturing fibroblast feeders

The following instructions are based on a split ratio of 1:2 using T175 flasks. However, although we routinely split MEFs and HFs 1:2–1:4, the optimal split ratio may range from 1:2 to 1:10 depending on the cell line and/or passage number. Irrespective, it is preferable to under-split rather than over-split. Proceed according to the suppliers instructions when commercially available cells are used.

1. Ensure the flask of cells is approximately 80% confluent.

2. Aspirate medium from flask.

3. Wash the flask with 20 mL PBS (without Ca^{2+} and Mg^{2+}).

4. Add 3 mL trypsin–EDTA (0.05%) to the flask for 1–2 min.

5. Tap the flask to remove cells.

6. Add 10 mL feeder medium to the flask and thoroughly mix with detached cells.

7. Transfer the cell suspension to a 50 mL tube.

8. Wash flask with 10 mL of feeder medium and combine with the cell suspension.

9. Check flask to ensure that less than 10% of cells remain attached.

10. Centrifuge cells at 600 g/~2,000 rpm for 2 min.

11. Aspirate supernatant from pelleted cells and resuspend cells in 10 mL of feeder medium. Mix well.

12.   Transfer 5 mL cell suspension to each of two labeled 175 cm^2 flasks containing 50 mL of feeder medium.

13.   Incubate flask for 48 hr at 37°C in a 5% CO_2 incubator.

14.   Change medium every 48 hr thereafter until cells reach about 80% confluence, and are ready for further passaging, harvesting for freezing or mitotic inactivation.

(v) Standard freezing of fibroblasts

We routinely freeze between 2×10^6 and 8×10^6 cells per milliliter of freezing medium in cryovials.

1.   Ensure the flask of cells is approximately 80% confluent.

2.   Aspirate feeder medium from each flask.

3.   Wash each flask three times with 20 mL PBS (without Ca^{2+} and Mg^{2+}).

4.   Add 3 mL Trypsin–EDTA (0.05%) to each flask for 1–2 min.

5.   Tap each flask to detach cells.

6.   Add 10 mL feeder medium to each flask and thoroughly mix with detached cells.

7.   Transfer the cell suspension to a 50 mL tube.

8.   Rinse each flask with 10 mL feeder medium and add to the cell suspension.

9.   Check each flask to ensure that less than 10% of cells remain attached.

10.   Perform a cell and viability count.

11.   Centrifuge cells at 600 $g/\sim$2,000 rpm for 2 min.

12.   Aspirate supernatant and resuspend cells in 2 mL of freezing solution. Mix well.

13.   Transfer cell suspension to cryovials, dispensing 1 mL per vial.

14.   Immediately place the cryovials into a suitable freezing container (lidded sty-rofoam box or a "Mr. Frosty" freezer box — do not use an unlidded freezing container as this will result in non-uniform freezing and cell death) and store at −80°C for 24 hr.

15.   Transfer the cryovials into liquid nitrogen (LN$_2$) storage.

(vi) Thawing fibroblasts

1.   Transport cryovial from LN$_2$ storage.

2.   Thaw at 37°C in water-bath, leaving an ice sliver.

3. Transfer the cell suspension from the cryovial to 10 mL feeder medium in a tube.

4. Rinse cryovial with 1 mL feeder medium and combine with cell suspension. Mix well.

5. Centrifuge cells at 600 g/~2,000 rpm for 2 min.

6. Aspirate supernatant from pelleted cells and resuspend cells in 3 mL feeder medium.

7. Perform a cell and viability count.

8. Transfer cells to labeled 175 cm^2 flasks containing 45 mL feeder medium.

9. Incubate flask for 48 hr at 37°C in a 5% CO_2 incubator.

10. Change medium every 48 hr thereafter until cells are used for human ES cell culture or until they reach about 80% confluence if mitotically active.

(vii) Mitotically inactivating fibroblasts

Mitomycin-C treatment

Fibroblasts need to be mitotically inactivated to be used as feeders to prevent their continued growth. We inactivate our feeders in batches using Mitomycin-C according to protocol seen in **Chapter 4**. We routinely seed feeders at 7×10^3 to 3×10^4 per cm^2 for subsequent inactivation. Optimal seeding may vary from one type of feeder to another. Mitotically inactivated cells can be frozen for later use or immediately plated for subsequent human ES cell culture.

Gamma irradiation

If the laboratory has access to a gamma irradiator, one can mitotically inactivate fibroblasts to generate feeder layers too. Similar to Mitomycin-C Treatment, we routinely seed feeders at 7×10^3 to 3×10^4 per cm^2 for subsequent inactivation. Optimal seeding may vary from one type of feeder to another. Mitotically-inactivated cells can be frozen for later use or immediately plated for subsequent human ES cell culture. If plating MEFs immediately after inactivation, prepare plates/flasks by coating with 0.1% gelatin (e.g. 1 mL for organ culture plates) for at least 1 hr prior to seeding. Aspirate the gelatin from the plates/flasks immediately prior to seeding. Finally, consider the gamma irradiation dose recommended below as a guide only. Optimal treatment should be determined by performing BrdU incorporation analysis to ensure >90% cells are mitotically inactive. We irradiate our feeders according to the protocol previously described in **Chapter 4**.

Human ES cells

(i) Solutions and media

FBS based human ES cell medium

Table 3   FBS based human ES cell medium

Materials/Reagents	Vendor	Catalog #	Volume
DMEM	Invitrogen/Gibco	Catalog # 11960-044	382 mL
L-Glutamine (200 mM)	Invitrogen/Gibco	Catalog # 25030-081	5 mL
FBS	Hyclone	Catalog # SH30070-03	100 mL
NEAA	Invitrogen/Gibco	Catalog # 12383-014	5 mL
ITS supplement	Invitrogen/Gibco	Catalog # 41400-045	5 mL
β-Mercaptoethanol	Invitrogen/Gibco	Catalog # 21985-023	915 μL
Penicillin/streptomycin (optional)	Invitrogen/Gibco	Catalog # 15070-063	2.5 mL

1. Combine DMEM (382 mL), 200 mM L-glutamine (5 mL), FBS (100 mL), NEAA (5 mL), ITS supplement (5 mL), β-mercaptoethanol (915 μL) and if necessary penicillin/streptomycin can also be added (2.5 mL). Refrain from routinely using antibiotic in cell cultures, as this can mask problems and allow selection for "super-bugs" that are resistant to the antibiotics.

2. Filter medium through using a 500 mL filter unit. Store for up to 2 weeks at 4°C.

3. Add the components in the order that they are listed as if, for example, FBS is added first and then NEAA, the acidic mix resulting can compromise components of the medium. This is true for all media listed in this chapter.

KO-SR based human ES cell medium

Table 4   KO-SR based human ES cell medium

Materials/Reagents	Vendor	Catalog #	Volume
KO-DMEM	Invitrogen/Gibco	Catalog # 10829-018	388 mL
KO-SR	Invitrogen/Gibco	Catalog # 10828-028	100 mL
L-Glutamine (200 mM)	Invitrogen/Gibco	Catalog # 25030-081	5 mL
NEAA	Invitrogen/Gibco	Catalog # 12383-014	5 mL
Penicillin/streptomycin (optional)	Invitrogen/Gibco	Catalog # 15070-063	2.5 mL

1. Combine KO-DMEM (388 mL), 200 mM L-glutamine (5 mL), KO-SR (100 mL), NEAA (5 mL) and optionally antibiotic (2.5 mL). Filter medium through a 500 ml filter unit.

2. Store for up to 2 weeks at 4°C.

(ii) Human ES cell maintenance

The most optimal human ES cell maintenance platform would be reproducible, economical, scaleable and comprise defined non-animal derived and non-adventitious components supporting a single homogenous differentiable stem cell population. Although contemporary methods more or less comprise combinations of the above specifications, to our knowledge no publicly disclosed protocols encompass all. For example, a more progressive method might omit serum and employ non-cellular matrices but require serum albumin and/or fail to be feeder-independent by inclusion of feeder conditioned medium. Hence, while certain aspects of conventional culture method (e.g. MEFs and β-mercaptoethanol) are likely unacceptable for regulatory and/or public approval, appropriately screened FBS and/or HFs pose less of a problem while ultimately being most beneficial to the bulk production of high quality human ES cells. Accordingly, in choosing a best platform the discerning operator should discriminate between research-based and clinically driven cell culture. Moreover, in striving for the latter, he/she must differentiate between the ultimate versus actual regulatory requirements towards a compliant and suitably robust method of cell production.

(iii) Subculturing human ES cells

Due to the ongoing controversy regarding the possible impact of enzymatic/bulk passaging on human ES cell karyotype (Draper *et al.*, 2004; Buzzard *et al.*, 2004; Mitalipova *et al.*, 2005), we recommend maintaining some human ES cells stocks through mechanical propagation. Although laborious, unsuitable for bulk expansion, and less amenable to cGLP/cGTP/cGMP, we and others have shown stable karyotype of mechanically propagated subcultures through >100 passages (Buzzard *et al.*, 2004; Mitalipova *et al.*, 2005). However, limited passaging by enzymatic disaggregation can be considered with concomitant regular karyotyping. Although the interval for serial passaging may vary between lines and culture conditions, as a rule of thumb it can be performed every 6–8 days of culture. Finally, the optimum split ratio when passaging human ES cells is dependant on the assessment of an experienced operator. However, suitable split ratios range from 1:1 to 1:8.

Producing cutting pipettes for microdissection ("cut and paste") subculture

1. Pass a glass capillary through the blue flame of a Bunsen burner for approximately 2 s until the glass softens (**Figure 1A**).

2. Remove the capillary from the flame and pull both ends horizontally, stretching the heated section.

3. Bend the capillary to break the stretched section.

4. Using the thick section of one half, knock off the thin section of the other half to create optimal cutting tips (**Figure 1B**).

5. Wrap the cutting pipettes in foil and sterilize by dry heat before use.

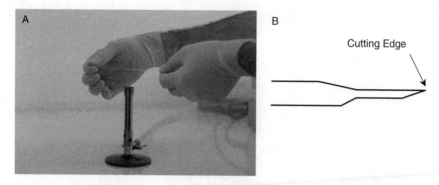

Figure 1 (**A**) Heating to soften a glass capillary for subsequent stretching and breaking to prepare cutting pipettes. (**B**) A schematic of a cutting pipette illustrating an optimal cutting tip and suitable edge.

Microdissection

Although widely employed, mechanical passaging by "microdissection" is labor intensive and unsuitable for bulk passaging. "Microdissection" requires easy access of a cell culture vessel to enable cutting. An advantage of mechanical passaging is the easy removal of differentiated and cystic material. Microdissection is best performed using either organ culture plates or 6 cm diameter culture plates, and is most compatible with feeder-based rather than feeder-free human ES cell culture. Finally, we recommend that plates be maintained on heated platforms at 37°C during passaging, while minimizing the time outside an incubator.

1. Working under a stereomicroscope, dissect the undifferentiated areas of colonies into several pieces using a cutting pipette. Exclude differentiated or cystic areas such as the central "button" region (**Figure 2A & B**). Although we routinely dissect pieces to approximately 2 mm^2, the optimal size may vary between human ES cell lines. Initiate dissection by scoring around the perimeter of a colony and the central "button" if present. Cutting from the centre to the periphery of the colony, dissect segments as illustrated by **Figure 2B**.

2. Using a 20 μL pipette tip, gently lift each dissected piece of undifferentiated colony from its substratum, allowing it to float freely within the medium.

3. Using a pipette, carefully aspirate the healthy human ES cell colony pieces and transfer to a new fibroblast coated culture plate. As a guide, 5–8 healthy colony pieces can be transferred to a fresh organ culture plate or 30–60 pieces per 6 cm diameter culture plate.

4. Ensure that the colony pieces are evenly distributed across each plate and transfer to a humidified incubator, equilibrated to 37°C and 5% CO_2.

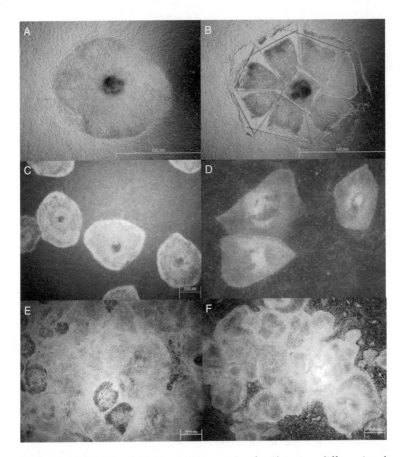

Figure 2 (**A**) Bright field photomicrograph of a day 7 undifferentiated human ES cell colony cultured on MEFs through 61 passages. The colony is suitable for passaging, being an acceptable size and comprising high density human ES cell growth with well defined edges and uniform morphology. (**B**) Bright field photomicrograph of the same colony shown in (**A**) following microdissection by scoring with a cutting pipette. The differentiated or cystic central "button" region would be excluded from passaging. (**C–F**) Bright field photomicrographs of human ES cell colony colonies cultured on MEFs, HFs or non-cellular matrices human plasma fibronectin and Matrigel respectively. For HF based cultures, fibroblasts represent a cGMP and clinically compliant human foreskin cell line, FDA approved for cell therapy (**D**).

Cell "scraping"

In our experience, cell "scraping" is useful for passaging human ES cells grown on HF or feeder-free platforms. However, it is less amendable to MEF based human ES cell culture. Compared to microdissection, it allows less discrimination between human ES cells and feeders during harvesting and serial transfer. However, due to their inactivation and eventual loss of viability, contamination by feeders of transferred material does not represent a problem. For feeder based culture we recommend plating dissociated human ES cells to feeders seeded at a lower density (25–30%) than recommended for microdissected cultures. Moreover, when subculturing from microdissected cultures, allow 2–4 weeks for cultures to adapt. Although potentially useful for bulk-culture, similar to microdissection, this method requires direct access to culture vessels for mechanical dispersion. For example, it is not compatible with closed cell factories.

1) Remove differentiated/cystic material from the cultures by dissecting and aspirating.

2) Aspirate spent human ES cell culture medium, removing any debris, and add fresh human ES cell culture medium.

3) Dissociate the cultures by scoring back and forth quickly and evenly with a P200 pipette tip or a glass Pasteur pipette, to obtain 0.5–1 mm^2 fragments.

4) Lift the fragments from the substratum using a cell scraper.

5) Use a pipette to triturate any large clumps.

6) Split the disaggregated cultures 1:3–1:6. Pipette to transfer to fresh culture vessels.

7) Ensure even dispersion of fragments across each culture vessel and transfer to a humidified incubator, equilibrated to 37°C and 5% CO_2.

Enzymatic passaging

While a variety of methods for enzymatic bulk transfer have been published, we recommend a collagenase-based approach. Although other laboratories employ Trypsin/EDTA based transfer, in our experience human ES cells are more sensitive to Trypsin/EDTA compared to collagenase. This is likely in part due to the release of cells as a single-cell suspension rather than clumps respectively, the latter seemingly preferred by human ES cells when initiating subculture. A caveat for collagenase digestion is that it is most compatible with harvesting and serial passaging of HF and feeder–free based human ES cells. Moreover, similar to other methods for bulk transfer, it allows less discrimination between human ES cells and feeders during harvesting and serial transfer. To reiterate from above, due to their inactivation and eventual loss of viability, contamination by feeders of transferred material does not represent a problem.

The protocol detailed below combines Collagenase IV digestion with cell scraping, reducing the time of harvest. Alternatively, spontaneous detachment of colonies without scraping can be achieved through extended exposure of cultures to collagenase for 30 min. The latter variant is compatible with closed culture platforms such as cell factories.

1) Equilibrate PBS with Ca^{2+} and Mg^{2+} and human ES cell culture medium to 37°C and 5% CO_2.

2) Dilute Collagenase IV stock solution 1:10 with PBS (with Ca^{2+} and Mg^{2+}) to make 1 mg/mL working solution.

3) Remove differentiated/cystic material from the cultures by dissecting and aspirating.

4) Aspirate medium from culture vessels and rinse twice with PBS with Ca^{2+} and Mg^{2+}.

5) Add Collagenase IV working solution (e.g. 0.5 mL per organ culture dish, 1 mL per 25 cm^2 flask, 3 mL per 75 cm^2 flask) and incubate at 37°C for 8 min.

6) When harvesting from organ culture dishes, use a P1000 pipette to lift and triturate the colonies. When harvesting from flasks, use a cell scraper to gently abrade the colonies.

7) Transfer cell suspension to a suitable Falcon tube and further dissociate colony fragments using a pipetteman (i.e. generate cell fragments $1/10^{th}$–$1/20^{th}$ the size of a mature colony).

8) Add three volumes of warm medium (i.e. 3× the volume of Collagenase IV working solution).

9) Centrifuge at 300 g/∼1,000 rpm for 2 min, remove supernatant, and resuspend cells in human ES cell culture medium.

10) Transfer human ES cell suspension to freshly prepared culture vessels. For inactivated MEF (6×10^4 feeders per cm^2) based culture, split the human ES cell suspension 1:8–1:10. For inactivated HF (5–6×10^4 feeders per cm^2) based culture, split cells 1:3–1:6.

Extended protocols

Fibroblast feeders

(i) Solutions and media

Fibroblast Feeder freezing medium for programmable control rate freezing (Table 5)

Table 5   Fibroblast feeder freezing medium

CryoStor™ CS10	Biolife Solutions	Catalog # 99-610-DV

Fibroblast feeder thaw medium for programmable control rate frozen cells

For MEFs use mouse feeder medium. If HFs are being thawed, use human feeder medium. Both media are shown in **Table 1**.

Human ES cells

(i) Solutions and media

Human feeder conditioned culture medium

1. Plate mitotically inactivated HFs in feeder culture medium and incubate overnight (e.g. 8×10^6 cells in 50 mL medium per 175 cm^2 flask).

2. The following day change the medium to fresh KO-SR based culture medium supplemented with 10 ng/mL bFGF.

3. Every 24 hr thereafter collect and sterile filter the conditioned medium.

4. Store conditioned medium for up to 1 week at 4°C.

5. Repeat steps 1 to 4 for up to 10 days re-using the same feeder preparation.

6. For human ES cell culture, supplement conditioned medium with 25–50 ng/mL bFGF and 5 mM L-glutamine immediately prior to use.

Human ES cell vitrification medium base

1. Combine DMEM (15.6 mL), FBS (4 mL), and 1M HEPES (0.4 mL).

2. Filter medium through a syringe filter into a 50 mL tube.

3. Store up to 1 week at 4°C.

Human ES cell vitrification medium: 1 M sucrose medium

1. Add 3.42 g sucrose to 6 mL Medium Base in a 15 mL tube.

2. Dissolve sucrose at 37°C.

3. Bring volume to 8 mL with additional medium base.

4. Add 2 mL FBS.

5. Filter medium through a syringe filter into a fresh 15 mL tube.

6. Store up to 1 week at 4°C.

Human ES cell vitrification medium: 10% vitrification solution

1. Combine medium base (2 mL) with ethylene glycol (0.25 mL) and DMSO (0.25 mL).

2. Store up to 1 week at 4°C.

Human ES cell vitrification medium: 20% vitrification solution

1. Combine medium base (0.75 mL) with 1 M sucrose medium (0.75 mL), ethylene glycol (0.5 mL) and DMSO (0.5 mL).

2. Store up to 1 week at 4°C.

Human ES cell freezing medium for programmable control rate freezing

1. CryoStor™ CS10

Human ES cell thaw medium for vitrified cells: medium base

1. See p. 67

Human ES cell thaw medium for vitrified cells: 1 M sucrose medium

1. See p. 67

Human ES cell thaw medium for vitrified cells: 0.2 M sucrose medium

1. Combine medium base (4 mL) and 1 M sucrose medium (1 mL).

2. Store up to 1 week at 4°C.

Human ES cell thaw medium for vitrified cells: 0.1 M sucrose medium

1. Combine medium base (4.5 mL) and 1 M sucrose medium (0.5 mL).

2. Store up to 1 week at 4°C.

Human ES cell thaw medium for programmable control rate frozen cells

1. Use FBS or KO-SR based human ES cell culture medium respectively.

Matrigel stock solution

1. Thaw Matrigel to 4°C overnight.

2. Using cold tubes and cold pipettes combine 1 mL of 4°C Matrigel with 15 mL of 4°C KO-DMEM.

3. Aliquot and store at −20°C until required.

Fibronectin stock solution

1. Dissolve 1 mg human plasma fibronectin in 1 mL dH_2O equilibrated to 37°C.

2. Aliquot and store at -20°C until required.

Collagenase IV stock solution

1. Dissolve 20 mg Collagenase IV in 1 mL dH_2O.

2. Aliquot and store at -20°C until required.

Protocols

(i) Programmable control rate freezing of fibroblast feeders

We routinely freeze between 2×10^6 and 8×10^6 cells per milliliter of freezing medium in cryovials.

1. Switch on and program the biological control rate freezer according to **Table 6.**

2. Ensure the flask of cells is approximately 80% confluent.

3. Aspirate feeder medium from each flask.

4. Wash each flask three times with 20 mL PBS (without Ca^{2+} and Mg^{2+}).

5. Add 3 mL trypsin–EDTA (0.05%) to each flask for 1–2 min.

6. Tap each flask to detach cells.

7. Add 10 mL feeder medium to each flask and thoroughly mix with detached cells.

8. Transfer cell suspension to a 50 mL tube.

Table 6 Control rate freezing program for fibroblast feeders

Step (Ramp)	From (°C)	To (°C)	Rate (°C/min)	Time (min)	Action
1	4	4	0	5	Hold
2	4	−4	−1.0	—	Cold activation
3	−4	−40	−28.0	—	Auto seeding
4	−40	−12	+25.0	—	—
5	−12	−40	−1.0	—	Dehydration
6	−40	−60	−5.0	—	—
7	−60	−90	−10.0	—	Rapid cool
8	−	−90	—	—	Hold

9.   Rinse each flask with 10 mL feeder medium and add to the cell suspension.

10.   Check each flask to ensure that less than 10% of cells are still attached.

11.   Perform a cell and viability count.

12.   Centrifuge cells at 600 g/~2,000 rpm for 2 min.

13.   Aspirate supernatant and resuspend cells in 2 mL of freezing solution. Mix well.

14.   Transfer cell suspension to cryovials, dispensing 1 mL per vial.

15.   Ensuring the biological control rate freezer has reached 4°C.

16.   Load the cryovials into the freezing chamber.

17.   Select Run from the main menu to commence freezing.

18.   With completion of a run (i.e. cooled to −180°C), quickly transfer the cryovials by plunging into LN_2.

19.   Store until further required.

(ii) Fibronectin based feeder-free human ES cell culture

The method described below supports human ES cell expansion on a fibronectin matrix with HF conditioned medium in the absence of a pre-plated feeder cell layer. A characteristic of this system and other feeder-free platforms is the spontaneous formation of fibroblast-like cells from human ES cells. Spontaneous differentiation can be observed at the periphery of human ES cell colonies and should be minimally harvested when passaging for human ES cell subculture. The level of fibroblast induction tends to be increased with lower initial plating density of human ES cell fragments during subculture.

1.   Dilute stock human plasma fibronectin (section **"Solutions and media"**, p. 69) to 50 µg/mL using PBS-.

2.   Coat a suitable culture vessel with 100 µL per cm^2 working solution and incubate for 2–4 hr at RT.

3.   Following coating, aspirate excess fibronectin solution and immediately plate human ES cells in HF conditioned culture medium (see p. 67).

4.   As a general rule, split cultures 1:2–1:3 for subculture.

5.   Expand to confluence over 6–8 days.

(iii) Matrigel based feeder-free human ES cell culture

The protocol described below is a variant of method originally described by Xu et al. (2001). Although in our experience Matrigel shows greater efficacy for extended hESC culture than other commonly employed non-cellular matrices including fibronectin, as a murine derived reagent it is deemed suitable for research application only.

1. Thaw stock Matrigel solution (see p. 68) overnight at 4°C.

2. Maintaining stock on ice, dilute 1:2 (final 1:30) with cold DMEM using cold tubes and cold pipettes.

3. Coat a suitable culture vessel (e.g. 1 mL/well for a six-well culture plate) and incubate for 1 hr at RT.

4. Following coating, aspirate excess Matrigel solution and immediately plate human ES cells in HF conditioned culture medium (see p. 67) at the same split ratio used for regular passaging (1:3–1:6).

5. Expand to confluence over 6–8 days.

(iv) Open straw vitrification of human ES cells

Vitrification is a conventional approach to cryopreserving human ES cells (Reubinoff *et al.*, 2001) derived from methods developed for freezing blastocysts (Vajta *et al.*, 1998). Although widely used, it is generally associated with poor post-thaw cell viability, typically yielding approximately 1–10% survival. In addition, as a relatively protracted procedure including open straw containment potentially exposing cells to contaminating microbial and viral pathogens before, during and after LN_2 storage (Rall, 2003), this method is recommended for research grade cell stocking only. Moreover, it is generally used in conjunction with mechanical harvesting by microdissection (section **"Routine protocols"**). Alternatively, our more contemporary and efficacious closed straw freezing method (see p. 73) developed primarily for cGLP/cGTP/cGMP application towards clinical compliance (Crook and Kravets, 2005) may be employed.

1. Prewarm a four-well plate ("vitrification plate") to 37°C comprising medium base (p. 67) in well 1, 10% vitrification solution (p. 67) in well 2, and 20% vitrification solution (p. 68) in well 3 (**Figure 3**).

2. Label 5 mL cryovials for storage and identification of straws.

3. Prepare a rinsing plate with fresh human ES cell medium equilibrated to 37°C and 5% CO_2 (e.g. 1 mL medium per organ culture dish).

4. Punch a hole through the side of each cryovial using an 18G needle heated in a Bunsen burner flame. This avoids the build up of LN_2 vapor pressure within a cryovial during the process of thawing.

5. Load the cryovials to cryocanes and immerse to LN_2 for pre-cooling, without submerging the top of the cryovial.

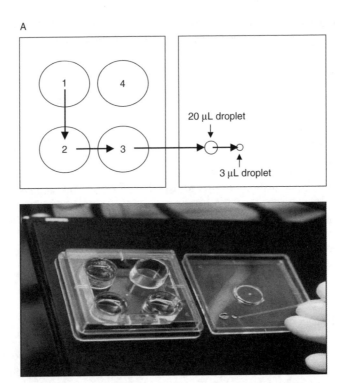

Figure 3 (**A**) Diagram of a "vitrification plate" illustrating the arrangement of solutions in order of application. Well 1 holds medium base, well 2 holds 10% vitrification solution, well 3 holds 20% vitrification solution, and well 4 is empty. (**B**) Photograph of vitrification plate, mirroring the accompanying diagram.

6. To ensure suitable post-thaw cell viability, the following steps must be completed within 3 min for each straw.

7. Prepare human ES cells for freezing according to steps 1–2 of microdissection (see p. 63).

8. Transfer colony fragments to the equilibrated rinsing plate.

9. Using a P20 pipettor, transfer the colony fragments to well 1 of the vitrification plate.

10. For each straw, prepare a fresh 20 µL drop of 20% vitrification solution on the underside of the vitrification plate lid (**Figure 3**).

11. Transfer 6 to 9 colony fragments from well 1 to well 2 of the vitrification plate (**Figure 3**). Leave for 1 min.

12. Transfer the colony fragments from well 2 to well 3 of the vitrification plate (**Figure 3**), avoiding contact of the fragments with the walls of the well. Leave for 25 s.

13. Transfer the colony fragments in the smallest possible volume to the 20 µL drop of 20% vitrification solution.

14. Using a P10 pipettor, aspirate the fragments in a 3 µL volume from the 20 µL drop and deposit as separate small, peaked droplet on the lid.

15. Immediately abut the narrow end of a vitrification straw to the side of the droplet at about a 30° angle to the plane of the dish. The droplet should be drawn into the straw by capillary action, forming a 1 mm long medium column. In the event that a few pieces are not drawn up, do not attempt to recover them.

16. Promptly plunge the straw into LN_2.

17. Quickly transfer the straw to an appropriately labeled cryovial in the LN_2, avoiding thawing of contained cells.

18. Store until further required.

(v) Closed straw programmable control rate freezing of human ES cells

Although routinely used in our laboratory for research grade human ES cell stocking, the method described below was primarily developed for cGLP/cGTP/cGMP application towards clinical compliance (Crook and Kravets, 2005). Accordingly, we have employed it for cGMP production of clinically-compliant human ES cell lines and related production of cGMP master cell banks as highlighted in *Nature*: Clinical-use stem cells made in Singapore — Lines designed for safe use in humans make their debut, 26 July 2006, doi:10.1038/news060724-8. The method includes automated slow rate freezing, a protein free FDA approved cryomedium, and sterile hermetically sealed FDA approved straws for closed cell containment. Moreover, it affords high post-thaw cell viability, typically yielding approximately 80–90% survival, and is amendable to high-throughput processing for master or working cell bank generation. Importantly, a programmable control rate freezer permitting automated or manual ice nucleation is a preferred requirement.

1. Switch on and program the biological control rate freezer according to **Table 7**.

2. Pre-warm PBS+ to 37°C and equilibrate human ES cell culture medium to 37°C and 5% CO_2.

3. Dilute collagenase stock solution (Extended Protocols section **"Solutions and media"**, p. 69) 1:20 in PBS+ to make a working solution of 1 mg/mL.

Table 7 Control rate freezing program for human ES cells

Step (Ramp)	From (°C)	To (°C)	Rate (°C/min)	Time (min)	Action
1	4	4	0	5	Hold
2	4	−8	−1.0	—	Cold activation
3	−8	−8	0	10	Soak
4	−8	−8	0	—	Manual seeding
5	−8	−8	0	5	Hold
6	−8	−38	−0.8	—	Dehydration
7	−38	−100	−10.0	—	Rapid cool
8	−100	−180	−35.0	—	Plunge
9	—	−180	—	—	Hold

4. Aspirate human ES cell culture medium from culture vessels and rinse three times with PBS with Ca^{2+} and Mg^{2+} (e.g. 1 mL per organ culture dish or 5 mL per 25 cm^2 flask).

5. Add collagenase working solution to culture vessels (e.g. 0.5 mL per organ culture dish or 1 mL per 25 cm^2 flask) and incubate at 37°C and 5% CO_2 for 8–10 min.

6. If using organ culture dishes, use a P200 pipette to flush and detach colony fragments and cells from the substratum. If using a flask, use a policeman to gently scrape the colonies from the flask combined with gentle flushing with 1 mL of human ES cell culture medium.

7. Transfer colony fragments to a 15 mL tube with 6 mL 37°C human ES cell culture medium and further disaggregate using a 5 mL pipette; aim to generate cell clumps about 1/10th–1/20th the size of a mature human ES cell colony.

8. Add at least five volumes of 37°C human ES cell culture medium (i.e. 5× the volume of collagenase working solution).

9. Centrifuge at 300 g/~1,000 rpm for 3 min, remove supernatant, and resuspend cells in CryoStor™ CS10 solution (cryo-medium). The volume of cryo-medium depends on the number of straws to be prepared, with 0.2 mL required per straw.

10. Maintain cells in cryo-medium at 4°C (e.g. on ice) for 10 min.

11. Transfer 0.2 mL of cells in cryo-medium to a 0.5 mL CBS™ high security straw using a P200 pipette. The straw must be held at the top, where the safety stopper ensures sterility is maintained throughout the filling procedure. The open end of the straw must be protected from any contamination.

12. The open ends of the straw should be placed in the jaws of a sealing unit, and each sealed hermetically. If necessary label each straw for future identification.

13. Maintain the straws at 4°C (e.g. on ice) until all have been sealed.

14. Ensuring the biological control rate freezer has reached 4°C, load the straws into the freezing chamber.

15. Select Run from the main menu to commence freezing.

16. For manual ice nucleation, seed at −8°C, after soaking the cells at −8°C for 5 min. Ice-nucleate the cells by contacting LN_2 cooled forceps to the straws in the region of the medium/cell column.

17. Press enter to resume the profile.

18. With completion of a run (−180°C), quickly remove the straws avoiding thawing of contained cells and transfer into LN_2.

19. Store until required.

(vi) Thawing vitrified human ES cells

1. Prewarm a four-well plate ("thawing plate") to 37°C comprising 0.2 M sucrose medium (Extended Protocols/Human ES cells section **"Solutions and media"**) in well 1, 0.1 M sucrose medium (Extended Protocols/Human ES cells section **"Solutions and media"**) in well 2, and medium base (section **"Solutions and media"**) in wells 3 and 4 (**Figure 4**).

2. Quickly transfer cryovial(s) containing frozen straw(s) from LN_2 storage canisters to a holding container filled with LN_2. Avoid thawing of frozen cells.

3. Remove a straw using forceps from the cryovial, again avoiding thawing of the cells.

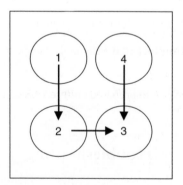

Figure 4 Diagram of a thawing plate illustrating the arrangement of solutions in order of application. Well 1 holds 0.2 M sucrose medium, well 2 holds 0.1 M sucrose medium, and wells 3 and 4 hold medium base.

4. Working quickly, hold the straw between thumb and middle finger and at a slight angle submerge the narrow end to well 1 of the thawing plate.

5. Immediately following melting of the cell/medium column, place a finger on the top of the straw.

6. The cells/medium column will be forced from the straw with gas expansion. Expel any remnant liquid from the straw by placing the tip of a P200 pipette in the top end of the straw and ejecting.

7. After 1 min, transfer the colony fragments to well 2.

8. After 5 min transfer the colony fragments to well 3.

9. After 5 min transfer the colony fragments to well 4.

10. After 5 min, plate the fragments to prepared culture vessels for recovery and expansion.

(vii) Thawing programmable control rate frozen human ES cells

1. Quickly transfer frozen straw(s) from LN_2 storage canisters to a holding container filled with LN_2. Avoid thawing of frozen cells.

2. Similarly cut off the other end of the straw proximal to the seal.

3. Using a P100 pipette, empty the straw into a 14 mL conical centrifuge tube with 6 mL warm human ES cell culture medium. Expel the straw contents by placing the pipette tip in the top end of the straw and ejecting.

4. Rinse the straw with additional 0.5 mL human ES cell culture medium.

5. Centrifuge cells at 300 g/~1,000 rpm for 3 min.

6. Aspirate supernatant from pelleted cells and resuspend in 0.5 mL human ES cell culture medium.

7. Transfer the cell suspension to prepared culture vessels for recovery and expansion.

Troubleshooting

The feeders are growing very slowly. What can I do?

This could be due to several causes: overexposure to trypsin, seeding the feeders at too low a density, problems with the medium or serum, sub-optimal freezing or thawing of the feeder stock, disaggregating the cells at the wrong time (either when they are too sparse or too confluent, or contamination of the feeder culture.

If none of these is immediately the obvious cause, make up new feeder medium, and resuscitate a new vial from feeder stocks. Feeders should be seeded at a density greater than 20,000 cells/cm^2. Avoid disaggregating the cells too early or too late (i.e. after the cells have reached confluence and become quiescent).

Why are the MEFs/mitotically-inactivated feeders peeling off the plate?

Either the cells have been seeded at too high a density or the cells were not success-fully mitotically inactivated. Seed at a lower density, making sure the cells have been mitotically inactivated. Peeling of cell layers from the dish or flask is less likely to happen if the surface has been gelatinized prior to seeding the cells.

Why are the human ES cells growing very slowly?

The most common causes of this include plating the human ES cells at too low a density, disaggregating the cells aggressively prior to seeding, poor quality feeders that do not support the human ES cells sufficiently, or there is something wrong with the medium (osmolarity, pH, nutrient deficiency, contamination).

Maintain the human ES cell cultures at a high confluence, splitting flasks 1:4 or less.

Always use good quality, low passage feeders, properly mitotically inactivated, seeded at the correct density. Feeders, being a primary line, are a likely source of contamination. Even if every thing looks okay, contaminated feeders drastically reduce the proliferation rate of human ES cells. Feeders should be grown without any antibiotic for at least 48 hr prior to testing with Mycoalert® mycoplasma detection kit, to verify that they are suitable for human ES cell culture.

Always batch test serum/serum replacement to ensure it supports human ES cells before purchasing a large batch. This is one of the best ways of ensuring consistent proliferation of human ES cells in culture and reducing variability between experiments. If you are beginning human ES culture for the first time, use a serum batch tested by others, and compare it to the serum batch you have in house.

If the problem is ongoing and none of the proceeding explanations appears a likely cause, measure the osmolarity of the medium, also verify the incubator has the appropriate CO_2 levels and humidity.

Another common error is to leave the cells out of the incubator for too long, where they are exposed to sub-optimal CO_2 levels and temperatures. When looking at human ES cell cultures, take only one or two plates or dishes out at a time to minimize exposure to these sub-optimal conditions. Consider investing in a heated stage for your microscope.

Human ES cell colonies display columnar "vertical" rather than lateral colony expansion and begin to differentiate. Why?

Inadequate feeder layer support. Use better quality feeders.

During routine expansion, human ES cell colonies are differentiating spontaneously. What can I do to prevent this?

Avoid over-seeding and change the culture media every 1–2 days. Add bFGF to the medium immediately before use, to ensure that there is sufficient inhibition of differentiation. This factor is very labile and will lose its activity if left at RT for too long.

Do not allow the cultures to become too confluent, maintain sufficient fresh culture medium on the cells. Remove differentiated cells (e.g. central "button" region of large colonies) by microdissection and aspiration.

The pipette breaks during mechanical disaggregation.

The glass capillary has been over-stretched during the production of the pipette. Ensure the glass capillary is pulled to ~0.25–1.0 mm diameter.

The cutting pipette tears, not cuts colonies.

Cutting tip is too blunt. Ensure the glass capillary is pulled to ~0.25–1.0 mm diameter. Ensure the cutting tip is sharp and without jagged edges.

During enzymatic disaggregation, cells/colonies fail to separate from substratum, even with combined scraping.

This happens due to sub-optimal activity of the disaggregating enzyme (e.g. collagenase, trypsin). Increase the enzyme concentration or extend the time of incubation.

During enzymatic disaggregation, a large white clump forms in the suspension and seeding efficiency is very low.

The cells lyse when over-exposed to disaggregating enzymes like trypsin. When cells lyse they become sticky, causing the formation of a large cellular aggregate.

References

Amit M, V Margulets, H Segev, K Shariki, I Laevsky, R Coleman and J Itskovitz-Eldor. (2003). Human feeder layers for human embryonic stem cells. *Biol Reprod* **68**: 2150–2156.

Bongso A, CY Fong, SC Ng and S Ratnam. (1994). Isolation and culture of inner cell mass cells from human blastocysts. *Human Reprod* **9**: 2110–2117.

Choo A, J Padmanabhan, A Chin, WJ Fong and SKW Oh. (2006) Immortalized feeders for the scale-up of human embryonic stem cells in feeder and feeder-free conditions. *J Biotech* **122**: 130–141.

Crook JM and L Kravets (2005) Cell preservation method. Patent International Publication Number WO2005/118785 A1.

Ellerstrom C, R Strehl, K Moya, K Andersson, C Bergh, K Lundin, J Hyllner and H Semb (2006) Derivation of xeno-free human ES cell line. *Stem Cells Express*. June Epub ahead of print.

Koivisto H, M Hyvarinen, AM Stromberg, J Inzunza, E Matilainen, M Mikkola, O Hovatta and H Teerijoki. (2004) Cultures of human embryonic stem cells: serum replacement medium or serum-containing medium and the effect of basic fibroblast growth factor. *Reprod Biomed Online* **9**: 330–337.

Ludwig TE, ME Levenstein, JM Jones, WT Berggren, ER Mitchen, JL Frane, LJ Crandall, CA Daigh, KR Conard, MS Piekarczyk, RA Llanas and JA Thompson. (2006) Derivation of human embryonic stem cells in defined conditions. *Nat Biotechnol* **24**: 185–187.

Noaksson K, N Zoric, X Zeng, MS Rao, J Hyllner, H Semb, M Kubista and P Sartipy. (2005) Monitoring differentiation of human embryonic stem cells using real-time PCR. *Stem Cells* **23**: 1460–1467.

Rall WF (2003) Avoidance of microbial cross-contamination of cryopreserved gametes, embryos, cells and tissues during storage in liquid nitrogen. *The Embryologists'* **6**: 4–15.

Reubinoff BE, MF Pera, G Vajta and A Trounson. (2001) Effective cryopreservation of human embryonic stem cells by the open pulled straw vitrification method. *Human Reprod* **16**: 2187–2194.

Reubinoff BE, MF Pera, CY Fong, A Trounson and A Bongso. (2000) Embryonic stem cell lines from human blastocysts: somatic differentiation in vitro. *Nat Biotechnol* **18**: 399–404.

Richards M, Tan S, Fong CY, Biswas A, Chan WK, Bongso A. (2003) Comparative evaluation of various human feeders for prolonged undifferentiated growth of human embryonic stem cells. *Stem Cells* **21**: 546–556.

Richards M, CY Fong, WK Chan, PC Wong, A Bongso. (2002) Human feeders support prolonged undifferentiated growth of human inner cell masses and embryonic stem cells. *Nat Biotechnol* **20**: 882–883.

Thomson JA, J Itskovitz-Eldor, SS Shapiro, MA Waknitz, JJ Swiergiel, VS Marshall and JM Jones. (1998) Embryonic stem cell lines derived from human blastocysts. *Science* **282**: 1145–1147.

Vajta G, P Holm, M Kuwayama, PJ Booth, H Jacobsen, T Greve, and H Callesen. (1998) Open pulled straw (OPS) vitrification: a new way to reduce cryoinjuries of bovine ova and embryos. *Mol Reprod Dev* **51**: 53–58.

Xu C, MS Inokuma, J Denham, K Golds, P Kundu, JD Gold and MK Carpenter. (2001) Feeder-free growth of undifferentiated human embryonic stem cells. *Nat Biotechnol* **19**: 971–974.

6

Chemically-defined culture of human embryonic stem cells

JULIE HSU CLARK AND SHENG DING

Department of Chemistry and Cell Biology, The Scripps Research Institute, 10550 North Torrey Pines, La Jolla, CA 92037, USA

Introduction

Conventional human ES cell culture conditions rely on serum cocktails and/or feeder cells to maintain an undifferentiated human ES cell phenotype. Although these undefined conditions are widely used to expand human ES cell cultures, they are not optimal for self-renewal due to their tendencies to also contain factors with negative activities (i.e. inducing differentiation or cell death)[1]. The presence of undesired factors in undefined conditions results in batch-to-batch culture variability, requiring addition of certain factors to counteract their effects (Wang *et al.*, 2005; Xu *et al.*, 2005). Consequently, the establishment of a well-defined culture condition would facilitate practical applications of human ES cells, and allow the study and control of signaling inputs that regulate self-renewal or differentiation of human ES cells.

The chemically-defined conditions described in this chapter are largely based on the notion that a basal medium composite free of differentiation-inducing factors (mainly from serum) with additional proliferation-promoting and/or differentiation-inhibiting factors would be sufficient for self-renewal of human ES cells. BMP-like activity induces differentiation in most commonly used human ES cell culture medias (Xu *et al.*, 2002, 2005). The proprietary Knockout™ Serum Replacement (Invitrogen), formulated for undifferentiated ES cell culture, directly replaces FBS in existing protocols has recently been shown to contain strong BMP-like activity (Xu *et al.*, 2005). We have

---

[1] The reliance on non-human support systems also diminishes the therapeutic application of human ES cells due to the risk of contamination of animal pathogens or immunoincompatible animal products (Martin *et al.*, 2005).

found that the omission of serum and serum replacement cocktails reduces BMP-like activity. The remaining differentiation activities can be inhibited with 20 ng/mL bFGF allowing expansion of undifferentiated human ES cells. Two independent human ES cell lines (H1, HSF6), grown on a Matrigel-coated surface[2] and serially passaged under these chemically defined conditions[3], maintain expression of pluripotency markers (See **Chapter 7**). They also retain a characteristic human ES cell morphology, possess a normal karyotype; and form complex teratomas 4–5 weeks after inoculation into immunocompromised 'nude' mice.

It should be noted that several other different chemically-defined conditions have also been developed, in which factors such as Wnt3a (or Wnt/β-catenin pathway agonists), TGFβ, Activin A or Nodal are also used in addition to bFGF[4,5,6]. While all of these conditions have been shown to maintain/propagate the human ES cell phenotype, the gene expression profiles of human ES cells under these conditions are not likely to be identical and their differentiation tendency toward different lineages may also vary. It can be expected that further development of culture conditions would allow derivation and propagation of new human ES cell lines while maintaining superior self-renewal and differentiation capacity.

Most differentiation protocols involve undefined conditions that result in the generation of heterogeneous cell types. Starting from human ES cell colonies, addition of specific growth factors (e.g. noggin, Activin A, BMP)[7] to our CDM can induce

---

[2] It should be noted that additional efforts are needed to further define the optimal extra cellular matrix (ECM) and albumin supplement for self-renewal as well as improve the culture condition for efficient clonal expansion. Both currently used Matrigel and BSA are not chemically-defined and may have batch to batch variation (e.g. some BSA still contains BMP-like activity). Defining the optimal ECM and albumin supplement will be crucial in making the culture condition completely free of animal products.

[3] This chemically-defined medium (CDM) has been shown to support H1 and HSF6 human ES cells for extended culture (over 10 months). HUES 1,6,8,9 and H9 human ES cell lines have displayed good proliferation and morphology, for greater than 10 passages, under this CDM condition as well.

[4] Vallier's RPMI-based CDM requires bFGF (12 ng/mL) and Activin (10 ng/mL)/Nodal (100 ng/mL), the use of FBS-derived ECM and a relatively higher concentration of BSA for supporting long-term self-renewal of human ES cells (James, 2005; Vallier, 2005) . The discrepancy of Vallier's observation, that the use of bFGF as the only growth factor supplement is not sufficient to support prolonged self-renewal of human ES cells, but the addition of Activin/Nodal is also required in addition to bFGF, versus the chemically defined conditions described here may be explained by the following possibilities: our use of 20 ng/mL bFGF vs. 12 ng/mL in Vallier's CDM, growth factor activity is often concentration-dependent; Vallier's use of much higher concentration of BSA; Matrigel has been shown to contain TGFβ (Kleinman et al., 1982; Vukicevic et al., 1992) (which, however, could not substitute Activin/Nodal in Vallier's CDM); FBS-derived ECM may contain other undefined factors; and certain components in the N2 and B27 supplements may provide some self-renewal promoting activity.

[5] Ludwig's TeSR1 contains bFGF ($\sim$100 ng/mL), TGFβ ($\sim$0.6 ng/mL), LiCl ($\sim$1 mM), GABA and pipecolic acid and has been tested for long-term self-renewal and derivation of human ES cells (Ludwig et al., 2006). The addition of GABA in Ludwig's TeSR1 is based on the microarray data which shows an increase in GABA-A receptor β-3 subunits in human ES cells. Pipecolic acid enhances GABA-A receptor response. LiCl inhibits GSK3 and is a known activator of the Wnt/β-catenin pathway.

[6] Lu's chemically defined human ES cell cocktail (HESCO) consists of bFGF (4 ng/mL), insulin (160 μg/mL), transferrin (88 μg/mL), Wnt3a (100 ng/mL), April (a proliferation-inducing ligand) (100 ng/mL) or BAFF (B cell-activating factor belonging to the TNF family) (100 ng/mL), albumin (2.5 mg/mL), and cholesterol lipid supplement (2.5×) (Lu et al., 2006).

[7] Noggin, Activin A and Wnt3a have been shown to have different activities for self-renewal or differentiation depending on the context (e.g. addition of 100 ng/mL Nodal or 20 ng/mL Activin A to the

selective differentiation toward neural, definitive endoderm/pancreatic and early cardiac muscle cells in monolayer (Yao *et al.*, 2006). These chemically-defined conditions provide a more robust platform for the further study of human ES cell self-renewal and directed differentiation. These conditions will also enable the use of approaches (e.g. genomic, proteomic and high throughput screening studies) that had been compromised in the past by complex undefined media.

Overview of protocol

Chemically-defined culture involves adapting human ES cells from serum containing medium on feeder layers to chemically defined conditions (including growth on Matrigel). Also included are our protocols for thawing, culturing and freezing human ES cells under chemically defined conditions.

Materials, reagents and equipment (Table 1)

Table 1 Materials, reagents and equipment

Material/Reagent	Vendor	Catalog #
1 mM non-essential amino acids	Invitrogen GIBCO	11140-035
2 mL cryogenic vials	Corning Incorporated	430659
5 mL disposable serological pipets	Fisher Scientific	13-676-10H
Six-well cell culture plates	Corning Incorporated	3506
55 mM β-mercaptoethanol	Invitrogen GIBCO	21985-023
200 mM Glutamax	Invitrogen GIBCO	35050-061
B27 Supplement-without vitamin A 50×	Invitrogen GIBCO	12587-010
bFGF	Invitrogen GIBCO	13256-029
Bovine albumin (BSA) Faction V 7.5%	Invitrogen GIBCO	15260-037
DMEM/F-12	Invitrogen GIBCO	11330-032
Dimethyl sulfoxide (DMSO)	Sigma	D2650
Dispase	Invitrogen GIBCO	17205-041
ES cell qualified fetal bovine serum	Invitrogen GIBCO	10439-024
Matrigel basement membrane matrix (growth factor reduced)	Becton, Dickinson and Company	CB-40230
N2 supplement 100×	Invitrogen GIBCO	17502-048
Phosphate buffered saline 1×	Invitrogen GIBCO	14190-144
Stericup 0.22 μm	Millipore Corporation	SCGPU05RE
Stratacooler/Cell freezing box	Stratagene Cloning Systems	400005

CDM described here clearly induce differentiation of human ES cells); these discrepancies may be largely explained by their dosage effects and existence of other factors in the media (e.g. in the presence of their differentiation activity counteracting factors, their proliferation and/or survival promoting activity might be seemingly dominant).

Protocol

A. Preparation of solutions and culture media (Tables 2–4)

Table 2 Chemically defined medium (CDM)

Reagent	Vendor	Catalog #	Quantity
1× DMEM/F12	Invitrogen GIBCO # 11330-032		460 mL
7.5% Bovine albumin (BSA) Faction V (150×)	Invitrogen GIBCO # 15260-037		3.33 mL
β-Mercaptoethanol 55 mM (500×)	Invitrogen GIBCO # 21985-023		1 mL
B27 supplement—without vitamin A (50×)[1]	Invitrogen GIBCO # 12587-010		10 mL
bFGF	Invitrogen GIBCO # 13256-029		10 μg
Glutamax 200 mM (100×)	Invitrogen GIBCO # 35050-061		5 mL
N2 supplement (100×)[2]	Invitrogen GIBCO # 17502-048		5 mL
Non-essential amino acids 1 mM (100×)	Invitrogen GIBCO # 21985-023		5 mL

CDM should be stored in the dark at 4°C and is good for up to 2 weeks. Only warm up the amount of medium needed for your application since bFGF is very labile.

[1]The B27 Supplements (GIBCO) contain D-biotin, BSA (fatty acid free, Fraction V), Catalase, L-Carnitine HCl, Corticosterone, Ethanolamine HCl, D-Galactose (Anhyd.), Glutathione (Reduced), Insulin (Human, Recombinant), Linoleic Acid, Linolenic Acid, Progesterone, Putrescine.2HCl, Sodium Selenite, Superoxide Dismutase, T-3/Albumin Complex, DL Alpha-Tocopherol, DL Alpha Tocopherol Acetate, Transferrin (Human, Iron-Poor) (information provided by GIBCO). It should be noted while having slower growth rate, N2-based CDM is sufficient for long-term self-renewal of human ES cells and may represent a more basal condition. In addition, B27 alone or 0.5× N2 plus 0.5× B27 can also be used as the supplements for CDM condition.

[2]The N2 Supplements (GIBCO) contain 1 mM human Transferrin (Holo), 8.61 μM recombinant human Insulin, 1 mM Progesterone, 1 mM Putrescine and 1 mM Selenite (information provided by GIBCO).

Table 3 Dispase working solution

Reagent	Vendor	Catalog #	Quantity
1× DMEM/F12	Invitrogen GIBCO # 11330-032		10 mL
Dispase	Invitrogen GIBCO # 17205-041		10 mg

Make up a fresh solution of 1 mg/mL Dispase in DMEM/F12 media when needed.

Table 4 2× freezing media

Reagent	Vendor	Catalog #	Quantity
CDM (see **Table 2**)			20%
Dimethyl sulfoxide (DMSO)	Sigma # D2650		20%
ES cell qualified fetal bovine serum[1]	Invitrogen GIBCO # 10439-024		60%

Bottles should be sprayed with 70% ethanol solution before placing in tissue culture hood. All solutions must be filter sterilized with 0.22 μm pore size nitrocellulose filters before use (Millipore Corporation # SCGPU05RE). All waste disposals should be carried out in accordance with the biohazard guidelines of your particular institution.

[1]For optimal freeze down and thaw of CDM culture, undefined ES cell qualified fetal bovine serum is included in the freezing media, but is removed from cells immediately upon thaw.

B. Matrigel coating protocol

1.   Thaw Matrigel solution on ice (a whole bottle takes typically 2–3 hr to thaw completely from −80°C).

2. Chill pipette tips, media and any other transfer equipment that may come into contact with the Matrigel solution.

3. Dilute Matrigel stock solution to 80 μg/mL with prechilled DMEM/F12. Mix solution until uniform. Matrigel will polymerize at room temperature and will not easily go into solution if media is not pre-chilled.

4. Dispense enough diluted Matrigel solution to cover the surface of your vessel. (For routine culture a six-well culture dish is standard (see Table 1), but CDM cultures can also be easily grown in 15 cm dishes.)

5. Incubate at room temperature for 2 hr or overnight at 4°C. If properly sealed Matrigel-coated plates can be stored at 4°C for up to 1 month.

6. There is no need to wash coated surfaces before use, simply remove Matrigel solution. Do not allow Matrigel-coated surface to dry out, if necessary aspirate Matrigel from wells and dispense a small amount of medium to each well.

C. Adapting human ES cells from standard to chemically defined conditions

If done properly all traces of irradiated feeder cells should be gone after 2–3 passages. Day 1:

1. Gently lift your human ES cell colonies off the feeder layer with regular splitting techniques (e.g. mechanical or enzymatic detachment) in chemically-defined conditions. It is important that the human ES cell colonies be as large and intact as possible since human ES cells will be separated from single cell feeders by gravity precipitation.

2. Place detached human ES cells in a 15 mL conical tube and allow colonies to slowly fall to the bottom (~5 min, depending on the size of the colonies).

3. Remove media gently without disturbing the loose cluster of cells and resuspend in 1 mL of CDM (see **Table 2**).

4. Break colonies up by pipetting up and down 2–3 times with a P1000.

5. Add an appropriate amount of CDM to cell solution to achieve a split ratio of 1:3 and mix.

6. Aspirate Matrigel solution from new plates and dispense human ES cell clusters to Matrigel-coated plates.

7. Incubate cells at 37°C, 5% CO_2.

Day 3:

8. Medium on cells will be changed for the first time on day 3. At day 2 you should see good attachment of your colonies as well as some cell debris. Henceforth, change medium daily and split when the culture is almost confluent.

D. Routine culture and passaging with dispase

1. Change medium every day until culture is 80% confluent or colonies are large enough for passaging (~6–8 days).

2. Remove differentiated colonies mechanically either with a glass pipette tip or via vacuum aspiration. (see **Figure 1**)

3. Make up a fresh solution of 1 mg/mL dispase (see **Table 3**) in DMEM/F12, warm solution to 37°C and filter sterilize.

4. Aspirate CDM medium and wash cells with PBS.

5. Aspirate PBS and dispense enough dispase solution to cover cells.

6. Incubate at 37°C until the edges of colonies begin to lift off the Matrigel (~5 min). Watch culture carefully during enzymatic digestion.

7. Aspirate dispase solution and carefully wash culture with PBS to remove residual dispase.

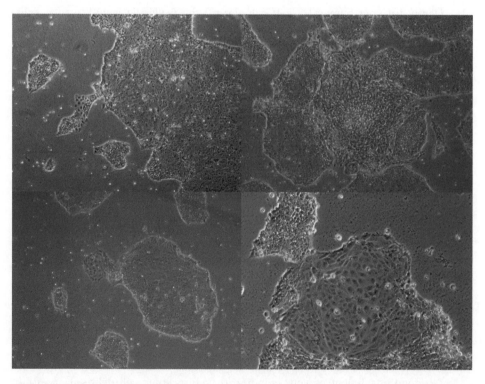

Figure 1 Examples of differentiated human ES cell colonies grown in chemically defined conditions.

8. Using a sterile 5 mL serological pipette, lift off colonies in a scraping motion while slowly dispensing CDM to facilitate removal.

9. Collect colonies in a 15 mL conical tube and centrifuge at 300 g/~1,000 rpm for 4 min.

10. Aspirate media from tube without disturbing the pellet of cells.

11. Tap tube gently to loosen human ES cell pellet and resuspend in 1 mL of CDM.

12. Break up the colonies into small clusters of 6–10 cells with a P1000 pipette by tituration. (see **Figure 2**)

13. Add an appropriate amount of CDM to the cell solution to achieve a spilt ratio of 1:3 and gently invert tube to mix.

14. Transfer cell suspension to a fresh Matrigel-coated plate.

E. Freezing human ES cells

1. Prepare a fresh solution of 2× freezing media (see **Table 4**) and keep on ice.

2. Remove differentiated colonies mechanically either with a glass pipette tip or via vacuum aspiration.

3. Make up a fresh solution of 1 mg/mL dispase in DMEM/F12, warm to 37°C and filter sterilize before use.

4. Aspirate CDM media and wash cells with PBS.

5. Aspirate PBS and dispense enough dispase to cover cells.

6. Incubate at 37°C until edges of colonies begin to lift off Matrigel (~5 min). Watch culture carefully during enzymatic digestion.

Figure 2 Progression of human ES cell culture under chemically defined conditions. (**A**) Human ES cells are seeded as small clusters of 4–10 cells. (**B**) Small colonies are seen after overnight attachment. (**C**) In approximately 6 days, small cell clusters will develop into large human ES colonies with characteristic compact morphology.

7.   Aspirate dispase solution and carefully wash culture with PBS to remove residual dispase.

8.   Using a sterile 5 mL serological pipette, lift off colonies in a scraping motion while slowly dispensing CDM to facilitate their removal.

9.   Collect colonies in a 15 mL conical tube and centrifuge at 300 g/~1,000 rpm for 4 min.

10. Aspirate media from tube without disturbing the human ES cell pellet.

11. Tap tube gently to loosen human ES cell pellet and resuspend them with CDM at 1 mL per dish. Do not break up colonies.

12. Slowly add an equal volume of 2× freezing medium while agitating tube to allow for the gradual addition of DMSO.

13. Dispense your cells in 1 mL aliquots into pre-labeled cryogenic vials.

14. Place vials in cell freezing boxes and store in a −80°C freezer overnight.

15. Transfer frozen cell stocks into a liquid nitrogen freezer for long term storage.

F. Thawing human ES cells

1.   Thaw your vial of human ES cells in a 37°C water bath with gentle agitation until only a small ice pellet remains.

2.   Gently transfer your thawed human ES cells into a 15 mL conical tube.

3.   Slowly add 10 mL of prewarmed CDM to your cells while agitating tube to allow for the gradual dilution of DMSO.

4.   Centrifuge cells at 300 g/~1,000 rpm for 4 min.

5.   Aspirate media gently without disturbing the cell pellet and resuspend colonies in 12 mL of CDM per vial.

6.   Dispense 2 mL of cell solution to each well in a fresh Matrigel-coated six-well plate.

7.   Incubate cells at 37°C, 5% CO_2.

8.   Inspect cultures daily. Attachment of previously frozen colonies should be observed by day two.

9.   Change medium for the first time on day 3 and daily thereafter, split before the culture becomes confluent.

Troubleshooting

Problems generally arise if the human ES cells are seeded at too low a density (keep above 60% confluence and only carry out low dilution <1:3 splits), or too many of the aspects of culture are changed at once. Avoid the temptation to change medium too often (especially at lower confluences).

After human embryonic stem cells are transferred to Matrigel for the first time, there are some feeder cells still remaining. Will persistence of these feeder cells negatively affect the human ES cells?

As long as most of the feeders have been removed human ES cell culture should be fine. If the mouse feeder layer has been properly irradiated/inactivated their presence will be negligible after 2–3 passages. Even if they have not, serial passaging will select against them and other differentiated cells.

The colonies have excellent human ES cell colony morphology but grow slowly and are not ready to be split within 7–8 days. Is anything wrong?

Most human ES cells under CDM conditions grow slower than feeder/serum dependent cultures likely due to absence of proliferation-promoting factors from feeders/serum as well as smaller initial colony size. Ways to increase the growth rate are to seed larger colonies and/or seed colonies at higher densities. If a slow growth rate is noticed, increased differentiation or microbial contamination may be to blame (See **Chapter 5** for more details on recognizing differentiation and testing for contamination).

The frequency of differentiation in CDM culture has gradually increased over a several passages. Can the culture be rescued?

Cells grown under CDM conditions generally exhibit less differentiation (<2%) than those in feeder/serum cultures. However, differentiated cells if not regularly removed can propagate and dominate the culture. Prevention is the best course of action, daily monitoring of cultures and mechanically remove 'suspicious' cells regularly. If your culture starts to differentiate you may choose to mechanically collect good colonies and start a fresh culture. Continued differentiation despite daily pruning is most likely due to a reduction in bFGF activity in the medium. Different vendors carry bFGF with varying activities and you may need to batch test before use. Furthermore, bFGF is very labile. CDM containing bFGF should be stored in the dark at 4°C and used within 2 weeks. To minimize loss of bFGF activity, aliquot the bottle into 50 mL aliquots and warm up only the volume of CDM needed that day.

Under chemically defined conditions cells in my culture have a human ES cell morphology and exhibit a normal rate of proliferation, but I have noticed some cell debris. Should I be concerned?

Do not be alarmed! All is well as long as your culture retains correct ES cell morphology and adequate growth rate. Once human ES cell colonies in chemically defined conditions reach a certain size their growth rate increases and a greater amount of cell debris is observed with a net increase in live healthy human ES cells. If the debris increases rapidly and the proliferation rate of the cells decreases substantially conduct standard tests for contamination as outlined in **Chapter 1**.

Despite careful thawing and culture of frozen human ES cells from a cryovial, the survival and proliferation rates of the cells appear very low. What can I do?

Thawing survival will primarily depend on correct freezing. If cells are not surviving thaws, one might try increasing the seeding density per well or freezing larger colonies at a higher density. If there are other human ES cell cultures growing that have not been frozen, use these to verify that the culture conditions (including media) are not impeding survival or proliferation.

References

James D, AJ Levine, D Besser and A Hemmati-Brivanlou. (2005). TGFbeta/activin/nodal signaling is necessary for the maintenance of pluripotency in human embryonic stem cells. *Development* **132**(6): 1273–1282.

Kleinman HK, ML McGarvey, LA Liotta, PG Robey, K Tryggvason and GR Martin. (1982). Isolation and characterization of type IV procollagen, laminin, and heparan sulfate proteoglycan from the EHS sarcoma. *Biochemistry* **21**: 6188–6193.

Lu J, R Hou, CJ Booth, SH Yang and M Snyder. (2006). Defined culture conditions of human embryonic stem cells. *Proc Natl Acad Sci USA* **103**: 5688–5693.

Ludwig TE, ME Levenstein, JM Jones, WT Berggren, ER Mitchen, JL Frane, LJ Crandall, CA Daigh, KR Conard, MS Piekarczyk, RA Llanas and JA Thomson. (2006). Derivation of human embryonic stem cells in defined conditions. *Nat Biotechnol* **24**: 185–187.

Martin MJ, A Muotri, F Gage and A Varki. (2005). Human embryonic stem cells express an immunogenic nonhuman sialic acid. *Nat Med* **11**: 228–232.

Vallier L, M Alexander and RA Pedersen. (2005). Activin/Nodal and FGF pathways cooperate to maintain pluripotency of human embryonic stem cells. *J Cell Sci* **118**: 4495–4509

Vukicevic S, HK Kleinman, FP Luyten, AB Roberts, NS Roche and AH Reddi. (1992). Identification of multiple active growth factors in basement membrane Matrigel suggests caution in interpretation of cellular activity related to extracellular matrix components. *Exp Cell Res* **202**: 1–8.

Wang G, H Zhang, Y Zhao, J Li, J Cai, P Wang, S Meng, J Feng, C Miao, M Ding, D Li and H Deng. (2005). Noggin and bFGF cooperate to maintain the pluripotency of human embryonic stem cells in the absence of feeder layers. *Biochem Biophys Res Commun* **330**: 934–942.

Xu RH, X Chen, DS Li, R Li, GC Addicks, C Glennon, TP Zwaka and JA Thomson. (2002). BMP4 initiates human embryonic stem cell differentiation to trophoblast. *Nat Biotechnol* **20**: 1261–1264.

Xu RH, RM Peck, DS Li, X Feng, T Ludwig and JA Thomson. (2005). Basic FGF and suppression of BMP signaling sustain undifferentiated proliferation of human ES cells. *Nat Methods* **2**: 185–190.

Yao S, S Chen, J Clark, E Hao, GM Beattie, A Hayek and S Ding. (2006). Long-term self-renewal and directed differentiation of human embryonic stem cells in chemically defined conditions. *Proc Natl Acad Sci USA*. **103**: 6907–6912.

Section 2

Characterization of Human Embryonic Stem Cells

7

Phenotypic analysis of human embryonic stem cells

JONATHAN S. DRAPER[1,2,3], CHERYLE A. SÉGUIN[1] AND
PETER W. ANDREWS[3]

[1]*Department of Developmental Biology, Hospital for Sick Children, Toronto, Canada*
[2]*Samuel Lunenfeld Research Institute, Mount Sinai Hospital, Toronto, Canada*
[3]*Centre for Stem Cell Biology, University of Sheffield, Sheffield, UK*

Introduction

The phenotypic properties of a cell type represent a unique signature for identifying and distinguishing it from the range of other tissues present during the development and adult life of an organism. Unfortunately, different cell types can share single or sometimes multiple characteristics, for example, many genes are reused in distinct spatial and temporal points during development. For this reason utilizing several assays concomitantly to measure a panel of phenotypic characteristics can provide greater confidence when identifying cell types.

Numerous publications detail phenotypic traits common to a range of different human ES cell lines, including morphology, gene and protein expression, proliferation, karyotype and pluripotency. While not all of the phenotypic characteristics of human ES cells are practical to assay on a regular basis, many of these traits can easily be measured. These can often be performed without destroying the human ES cells in the process thereby making routine analysis of human ES cultures feasible. The monitoring of stock human ES cell cultures is especially important given that these cells generally exist as a heterogeneous population of undifferentiated and differentiating cells. To this end, this chapter will concentrate on describing a range of RNA, DNA and protein markers that are indicative of the human ES cell phenotype. It will also provide protocols that utilize these traits to enable the differentiation status and pluripotency of human ES cell cultures to be classified.

Overview of protocols

The most readily accessible of human ES cell phenotypic traits is cellular morphology which can and should be analyzed on a daily basis. Human ES cells have a high nuclear to cytoplasmic ratio with several prominent nucleoli visible. Many laboratories grow human ES cells on layers of mitotically inactivated feeders which can interfere with morphological and marker analysis of the spontaneous differentiation often observed in human ES cell cultures. The morphology of human ES cells varies according to the substrate upon which they are cultured. For instance, growth on Matrigel, a mixture of basement membrane components, leads to a flattened morphology. Whereas culturing human ES cells on feeders promotes cells to squeeze together (Ginis *et al.*, 2004; Xu *et al.*, 2001).

Many laboratories growing human ES cells on feeders typically use media containing Serum Replacement (Invitrogen) and bFGF (Amit *et al.*, 2000). Growth on substrates other than feeders can require either additional steps, for instance the conditioning of the media by co-culture with feeders (Xu *et al.*, 2001), or completely different culture media (e.g. **Chapter 6** gives details of chemically defined human ES cell medium). Although it can be expensive, feeder free culture is becoming an attractive alternative to feeder dependent culture. Moreover, there is increasing evidence that some feeder free culture systems can reduce spontaneous differentiation compared to human ES cell culture with feeders. This may be due to factors like BMPs that induce differentiation (Beattie *et al.*, 2005; Rosler *et al.*, 2004; Xu *et al.*, 2005).

Cell surface markers that had initially been used to characterize human embryonal carcinoma cell lines (the pluripotent stem cells derived from germ cell tumors) can also be used to classify human ES cells (Reubinoff *et al.*, 2000; Thomson *et al.*, 1998). Although some of these cell surface markers are proteins (Thy1 and alkaline phosphatase), others are glycolipids (SSEA3 and SSEA4), or keratin sulfates (TRA-1-60, TRA-1-81 and GCTM). SSEA3 is a sensitive marker of the undifferentiated state of human ES cells (Draper *et al.*, 2002; Henderson *et al.*, 2002). Other useful cell surface markers, including SSEA1, A2B5, VIN-IS-53 and VIN-IS-56, are expressed on subsets of differentiating human ES cells (Andrews *et al.*, 1996; Draper *et al.*, 2002). The pan human cell surface marker TRA-1-85 is a useful tool for distinguishing human from non-human cells, including murine feeders (Williams *et al.*, 1988) (see **Table 1**). When analyzed by flow cytometry, surface markers are a powerful tool for profiling cultures of human ES cells on a single cell basis. This gives an accurate picture of spontaneous differentiation within a culture. Moreover, due to the non-invasive nature of staining human ES cells with surface markers, it is also possible to isolate cells based upon their surface expression profile and then reclaim discrete population of live cells by fluorescence activated cell sorting (FACS) for further analysis or culture (Sidhu and Tuch, 2006). With FACS, gating of cells that are positive or negative for surface antigen expression is fairly stringent. Only a small percentage of the starting population are retrieved, so starting with many cells ($>10^7$ cells) is recommended. In addition to their application in flow cytometry, cell surface markers can also be used to stain both fixed and live colonies of human ES cells *in situ*, providing additional information on their phenotypic state.

The antibodies that recognize the cell surface markers described in this chapter are now readily available from a range of commercial and non-profit organizations (see **Table 2**). Laboratories with heavy consumption of some cell surface markers may consider purchasing the hybridoma cell lines and producing their own concentrated supernatant, which we find to be the equal of many commercial antibody preparations (see Developmental Studies Hybridoma Bank (DSHB), www.uiowa.edu/~dshbwww/).

Table 1 Antibodies for flow cytometry and immunohistochemistry

Markers of human embryonic stem cells

Antigen	Antibody (Species and isotype)	Cellular location/Comments/References/Supplier ($N = Nuclear$; $CS = Cellsurface$; $C = Cytoplasmic$)
SSEA3	MC631 (Rat IgM)	CS/Globoseries Glycolipid/(Andrews et al., 1982b; Draper et al., 2002; Fenderson et al., 1987; Henderson et al., 2002; Shevinsky et al., 1982)/DSHB & Chemicon
SSEA4	MC813-70 (Mouse IgG)	CS/Globoseries Glycolipid/(Draper et al., 2002; Fenderson et al., 1987; Henderson et al., 2002; Kannagi et al., 1983)/DSHB & Chemicon
TRA-1-60	TRA-1-60 (Mouse IgM)	CS/Keratin sulfate/(Andrews et al., 1984a; Badcock et al., 1999; Draper et al., 2002; Henderson et al., 2002)/DSHB & Chemicon
TRA-1-81	TRA-1-81 (Mouse IgM)	CS/Keratin sulfate/(Andrews et al., 1984a; Badcock et al., 1999; Draper et al., 2002; Henderson et al., 2002)/DSHB & Chemicon
Alkaline Phosphatase	TRA-2-54 (Mouse IgG)	CS/Liver–Bone–Kidney Alkaline Phosphatase/(Andrews et al., 1984b; Draper et al., 2002; Henderson et al., 2002)/DSHB & Chemicon
Alkaline Phosphatase	TRA-2-49 (Mouse IgG)	CS/Liver–Bone–Kidney Alkaline Phosphatase/(Andrews et al., 1984b)/DSHB & Chemicon
Thy-1	CD90 (Mouse IgG)	CS/immunoglobulin supergene family/(Andrews et al., 1983; Andrews et al., 1987; Draper et al., 2002)/Abcam # ab23894
GCTM-2	GCTM-2 (Mouse IgM)	CS/Keratin sulfate/ (Pera et al., 1988; Reubinoff et al., 2000)/Dr M. Pera
Oct-4	Anti-Oct-4 (Mouse IgG)	N/Pluripotency associated transcription factor/(Nichols et al., 1998)/Santa Cruz # sc-5279
Nanog	Anti-Nanog (Rabbit poly)	N/Pluripotency associated transcription factor/(Chambers et al., 2003)/Cosmo Bio. Co. #RCAB0002P-F

Markers of differentiating human ES cells and human specific antigens

See also **Chapters 11–14** for more on specific markers of extraembryonic, endodermal, mesodermal and ectodermal cell types.

Antigen	Antibody (Species and Isotype)	Cellular location/Comments/References/Supplier ($N = Nuclear$; $CS = Cellsurface$; $C = Cytoplasmic$)
Ok[a]	TRA-1-85 (Mouse IgG)	CS/Pan human marker/(Draper et al., 2002; Henderson et al., 2002; Williams et al., 1988)/DSHB
SSEA1 (le[x])	MC480 (Mouse IgM)	CS/Lactoseries Glycolipid; on differentiating hES & hEC/(Draper et al., 2002; Fenderson et al., 1987; Henderson et al., 2002; Solter and Knowles, 1978)/DSHB & Chemicon

(continued overleaf)

Table 1 (*continued*)

Antigen	Antibody (Species and Isotype)	Cellular location/Comments/References/Supplier ($N = Nuclear; CS = Cellsurface; C = Cytoplasmic$)
CD30	CD30 (Mouse IgG)	CS/TNF receptor superfamily; upregulated on aneuploid hES cells/(Herszfeld *et al.*, 2006)/Dako # Ber-H2
GT_3	A2B5 (Mouse IgM)	CS/Neural/(Draper *et al.*, 2002; Eisenbarth *et al.*, 1979; Fenderson *et al.*, 1987)/R&D # MAB1416
GD_3	VINI556 (Mouse IgM)	CS/Neural/(Andrews *et al.*, 1990; Draper *et al.*, 2002)/DSHB
GD_2	VIN2PB22 (Mouse IgM)	CS/Neural/(Andrews *et al.*, 1990; Draper *et al.*, 2002)/DSHB
Desmin	Desmin (Mouse IgG)	C/Muscle specific actin/(Li *et al.*, 1989)/Dako # D33
AFP	AFP (Mouse IgG)	C/Def. and ext.emb. endoderm/(Andrews *et al.*, 1982a)/R&D MAB1368
FoxA2	FoxA2 (Goat Poly)	N/(AKA: HNF3b) Endoderm/(Kaestner *et al.*, 1994)/Santa Cruz # sc-6554
Cdx2	Cdx2 (Mouse IgG)	N/Endoderm and trophect./(Beck *et al.*, 1995)/Biogenex # MU392-UC
Brachyury	Brachyury (Rab Poly)	N/Mesoderm/(Wilkinson *et al.*, 1990)/Abcam # ab20680

Table 2 Human primers for RT-PCR
There are several methods of extracting total RNA from human ES cells. We favor Tri Reagent (Sigma) or the RNAeasy kit (Qiagen), both of which can provide good yields of RNA. We perform reverse transcription using the Superscript II kit (Invitrogen) and then use Taq polymerase for the PCR reaction. Consult the manufacturer's protocols for further details.

Markers of human embryonic stem cells

Gene	Primer sequence	Product size	PCR conditions	Reference
OCT-4	F: CGACCATCTGCCGCTTTGAG R: CCCCCTGTCCCCCATTCCTA	577 bp	94°C 1'; 58°C 1'; 72°C 1'26 cycles	(Nichols *et al.*, 1998)
Nanog	F: CAGAAGGCCTCAGCACCT R: CTGTTCCAGGCCTGATTGTT	216 bp	94°C 1'; 55°C 1'; 72°C 1'; 26 cycles	(Chambers *et al.*, 2003)
SOX2	F: CCCCCGGCGGCAATAGCA R: TCGGCGCCGGGGAGATACAT	448 bp	94°C 1'; 58°C 1'; 72°C 1'; 26 cycles	(Avilion *et al.*, 2003)
FOXD3	F: ACGACGGGCTGGAAGAGAAGGA R: CGGCGCCCAGGCTATTGAG	957 bp	94°C 1'; 65°C 1'; 72°C 1'; 30 cycles	(Hanna *et al.* 2002)
UTF1	F: ACCAGCTGCTGACCTTGAAC R: TTGAACGTACCCAAGAACGA	210 bp	94°C 1'; 58°C 1'; 72°C 1'; 30 cycles	(Okuda *et al.*, 1998)
Rex1	F: CGTACGCAAATTAAAGTCCAGA R: CAGCATCCTAAACAGCTCGCAGAAT	306 bp	94°C 1'; 56°C 1'; 72°C 1'; 30 cycles	(Henderson *et al.*, 2002)

Table 2 (*continued*)

Markers of human embryonic cell differentiation (common spontaneous differentiation)

Gene	Primer sequence	Product size	PCR conditions	Reference
Markers of Endoderm Differentiation				
Gata4	F: ACCACAAGATGAACGGCATCAACC R:GGAAGAGGGAAGATTACGCAGTGA	612 bp	94°C 1'; 56°C 1'; 72°C 1' 30 cycles	(Kuo *et al.*, 1997)
Gata6	F: TCTACAGCAAGATGAACGGCCTCA R: AGGTGGAAGTTGGAGTCATGGGAA	312 bp	94°C 1'; 56°C 1'; 72°C 1' 30 cycles	(Morrisey *et al.*, 1998)
HNF4α	F: TGTGTGTGAGTCCATGAAGGAGCA R: AGTCATACTGGCGGTCGTTGATGT	431 bp	94°C 1'; 56°C 1'; 72°C 1' 28 cycles	(Duncan *et al.*, 1997)
AFP	F: AGAACCTGTCACAAGCTGTG R: GACAGCAAGCTGAGGATGTC	676 bp	94°C 1'; 57°C 1'; 72°C 1' 30 cycles	(Andrews *et al.*, 1982a)
Markers of Trophectoderm Differentiation				
Cdx2	F:CCTCCGCTGGGCTTCATTCC R:TGGGGGTTCTGCAGTCTTTGGTC	295 bp	94°C 1'; 60°C 1'; 72°C 1' 30 cycles	(Beck *et al.*, 1995)
Eomes	F: TCACCCCAACAGAGCGAAGAGG R: AGAGATTTGATGGAAGGGGGTGTC	374 bp	94°C 1'; 58°C 1'; 72°C 1' 30 cycles	(Ciruna and Rossant, 1999)
HCG	F: CAGGGGACGCACCAAGGATG R: GTGGGAGGATCGGGGTGTCC	510 bp	94°C 1'; 62°C 1'; 72°C 1' 30 cycles	(Muyan and Boime, 1997)
Markers of Ectoderm Differentiation				
NeuroD1	F: AAGCCATGAACGCAGAGGAGGACT R: AGCTGTCCATGGTACCGTAA	579 bp	94°C 1'; 66°C 1'; 72°C 1' 30 cycles	(Lee *et al.*, 1995)
SOX1	F: CTCACTTTCCTCCGCGTTGCTTCC R: TGCCCTGGTCTTTGTCCTTCATCC	848 bp	94°C 1'; 58°C 1'; 72°C 1' 30 cycles	(Pevny *et al.*, 1998)
Markers of Mesoderm Differentiation				
Brachyury	F: GGTCTCGGCGCCCTCTTCCTC R: GGGCCAACTGCATCATCTCCACA	402 bp	94°C 1'; 63°C 1'; 72°C 1' 30 cycles	(Herrmann, 1991)
HBZ	F: CTGACCAAGACTGAGAGGAC R: ATGTCGTCGATGCTCTTCAC	224 bp	94°C 1'; 61°C 1'; 72°C 1' 30 cycles	(Peschle *et al.*, 1985)

Global gene expression analysis (e.g. transcriptional microarrays and SAGE) allows assessment of the transcriptional status of many genes simultaneously. Numerous attempts have now been made to characterize the transcriptome of human ES cells (Brandenberger *et al.*, 2004; Dvash *et al.*, 2004; Liu *et al.*, 2006; Sato *et al.*, 2003; Sperger *et al.*, 2003). Some differences exist between the expression profiles of human ES cell lines, probably due to line-to-line variation and differences in cell culture.

However, many more similarities are evident suggesting that a molecular blueprint for the pluripotent state exists. It is worth noting that a transcriptome profiling paper analyzing human ES cells on a single cell level has yet to be reported; instead reports have been performed on whole populations of cells, which likely include contamination with differentiated cells. We have performed detailed transcriptome profiling on populations of undifferentiated human ES cells enriched by sorting for SSEA3 positive cells. This approach appears to increase the fidelity of the profile generated, however, profiles for both the sorted and non-sorted populations are not that different (Enver et al., 2005). Although the lists of genes that are developmentally regulated in human ES cells are expansive, little or nothing is known about the function of the majority of these genes in maintenance of human ES cell self-renewal or pluripotency. For the time being, many of these genes are merely potentially useful markers of the human ES cell phenotype. Fortunately, by capitalizing on work performed in murine ES cells and embryos, it is becoming clear that several genes that have been characterized as essential regulators of self-renewal and pluripotency in the mouse also appear to perform the same roles in human ES cells. Most notable among these are pluripotency-associated transcriptional regulators Oct-4, Sox2 and Nanog (Boyer et al., 2005; Chambers et al., 2003; Hyslop et al., 2005; Matin et al., 2004; Niwa et al., 2000). The involvement of these transcription factors in the maintenance of ES cell phenotype often makes them useful as markers for the undifferentiated state (see **Chapter 8** for more on genetic analysis of human ES cell cultures). Many other genes are now recognized to be highly expressed in human ES cells and downregulated as differentiation occurs, for instance, TDGF1 and EBAF (Enver et al., 2005).

For routine analysis of the differentiation status of human ES cultures, a more limited analysis of gene expression patterns by PCR based approaches is often more feasible. These include techniques such as RT-PCR and quantitative real time PCR (qRT-PCR). **Table 2** describes a number of useful primer sequences for assessing the phenotype of human ES cell cultures using these techniques.

It is now well appreciated that the majority of developmentally-regulated markers are expressed at several discrete time points throughout development and again later in the adult organism, and this should be kept in mind. Oct-4, Sox2 and Nanog show no exception to this rule: both Oct-4 and Nanog are expressed later in development in germ cells, while Sox2 is expressed in neural tissues (Chambers et al., 2003; Ellis et al., 2004; Pesce et al., 1998). Thus, assaying for expression of a panel of markers can confirm the undifferentiated state of cells with a high degree of certainty.

Assessing pluripotency in human ES cells is comparatively more difficult than in murine ES cells for ethical reasons. Currently, the most commonly used tests of pluripotency in laboratories are in vitro differentiation and teratoma formation (see **Chapters 10 and 9** respectively). In vitro differentiation typically involves removing the cells from an environment that promotes self-renewal and placing them into conditions that favor differentiation. There are numerous ways of achieving differentiation in human ES cells, including embryoid body formation, the addition of signaling ligands (e.g. BMP4, Activin-A) or chemical agents (e.g. 10^{-5} M retinoic acid, 3 mM hexamethylene bisacetamide or 1% dimethylsulfoxide). **Table 2** describes several markers that are useful for assessing the differentiated status of human ES cells.

Materials, preparation of materials, reagents and equipment

Materials 1: Flow cytometry and FACS buffers

1. PBS without Ca^{2+} and Mg^{2+} (Invitrogen, cat # 14190-094).

2. 0.05% Trypsin–EDTA (Invitrogen 25300-054)

3. Wash buffer (5% fetal bovine serum in PBS)

4. DMEM/FBS (10% fetal bovine serum in DMEM Gibco, cat # 1196-069)

5. 0.2 µM syringe filter (Nalgene, cat # 195–2520)

Materials 2: Immunohistochemistry

1. PBS without Ca^{2+} and Mg^{2+} (Invitrogen, cat # 14190-094).

2. 4% paraformaldehyde

3. Blocking solution (PBS + 5% FBS + 0.1% Triton X-100)

4. PBST (PBS + 0.1% Triton X-100)

Materials 3: RT-PCR

1. Trizol (Invitrogen # 15596-026) or RNAeasy Kit (Qiagen # 74104) for RNA purification

2. Superscript II reverse transcriptase (Invitrogen # 18064-022)

3. Primers (see **Table 2**)

4. Taq polymerase

5. dNTPs

Protocols

Preparation of samples for FACS analysis

1. Aspirate medium from cells, and wash once with PBS (without Ca^{2+}/Mg^{2+})

2. Add trypsin–EDTA and incubate at 37°C for 5 min, tapping the culture vessel occasionally to help dislodge cells.

3. Inactivate trypsin by adding DMEM/FBS (i.e. 9 mL/T25 flask) and triturate to break up clumps and form a single cell suspension. Remove an aliquot of cells to count using a hemocytometer.

4. Pellet cells by centrifugation at 300 g/~1,000 rpm for 5 min.

5. Remove the supernatant and gently flick the tube to disperse the cell pellet.

6. Resuspend cell pellet to a concentration of 10^7 cells per milliliter in wash buffer.

7. Immunostaining can be carried out in individual 1.5 mL tubes, or in round bottom 96 well plates with self-sealing covers for more high-throughput analysis. Dilute primary antibodies in wash buffer for a total of 50 μL/sample.

8. Mix cell suspension with antibodies by adding 50 μL of cells (i.e. 5×10^5 cells) to each 50 μL of antibody dilution.

9. Seal samples, and incubate for 1 hr at 4°C with gentle shaking.

10. Pellet cells (1.5 mL tubes centrifuge 300 g/~1,000 rpm for 5 min; 96 well plate centrifuge 300 g/~1,000 rpm for 3 min) and remove supernatant.

11. Wash cells by adding 100 μL of wash buffer per sample, seal, and agitate by flicking tube to resuspend cells. Spin down as above.

12. Remove supernatant, and repeat for a total of three washes.

13. Remove supernatant from final wash, and add 50 μL of fluorochrome-labeled secondary antibody diluted in wash buffer to each sample.

14. Seal tubes/plate and incubate 1 hr at 4°C with gentle shaking.

15. Spin down cells and perform three washes.

16. Resuspend the cells at about 5×10^5 cells/mL in wash buffer, and proceed to analysis by flow cytometry.

Preparation of samples for FACS retrieval

1. Prepare single cell suspension from human ES cultures by following Steps 1 through 5 as described in flow cytometry protocol above.

2. Dilute primary antibodies in wash buffer and add 200 μL of diluted antibody per 10^7 cells in 15 mL conical tube. Incubate for 15 min at 4°C with gentle shaking.

3. Wash the cells by adding 10 mL of wash buffer, and pellet cells by centrifugation at 200 g for 5 min. Remove supernatant and gently flick the tube to disperse the cell pellet.

4. Repeat wash step once.

5. Following second wash, resuspend cell pellet in fluorochrome-labeled secondary antibody diluted in wash buffer (200 μL per 10^7 cells). Incubate for 15 min at 4°C with gentle shaking.

6. Wash twice as before.

7. Remove supernatant and gently flick tube to disperse the pellet.

8. Resuspend cells in wash buffer at 10^7 cells/mL and proceed to FACS. If cells are to be cultured after sorting, sort into human ES culture medium supplemented with antibiotics.

Immunostaining human ES cells in monolayer culture

1. In order to minimize amount of antibody required, we suggest culturing human ES cells in four-well dishes (1.5 cm well diameter) for immunostaining; volumes presented below reflect this but can be adjusted for culture dishes used.

2. Remove culture media, rinse cells twice with PBS (without Ca^{2+}/Mg^{2+}), and fix with freshly prepared 4% paraformaldehyde (250 μL/well) for 15 min at room temperature.

3. Aspirate fixative, rinse twice with PBS. At this stage, fixed cells can be stored in PBS at 4°C for several days, or can be used directly for immunostaining.

4. Incubate in blocking solution for 1 hr at room temperature.

5. Rinse briefly with PBST.

6. Dilute primary antibodies according to manufacturer's instructions in blocking buffer (minimum volume of 200 μL). Incubate either 1 hr at room temperature or overnight at 4°C with gentle rocking.

7. Wash twice briefly in PBST.

8. Dilute fluorochrome-labeled secondary antibodies in blocking buffer (minimum volume of 200 μL), and incubate 1 hr at room temperature in dark.

9. Wash twice briefly in PBST.

10. Dilute nuclear stain 1/400 in PBST (YOYO, Invitrogen cat # Y3606; DRAQ5, Alexis Biochemicals cat # 1305-889-001-R200, DAPI) and incubate for 20 min at room temperature in the dark. Alternatively, nuclear stain can be included along with the secondary antibody precluding the need for an additional step.

11. Wash twice briefly in PBS, cover cell layer completely with PBS, and proceed with imaging. Dishes should be stored in dark at 4°C before imaging.

Troubleshooting

At what concentration should I use the antibodies?

For concentrated hybridoma supernatants we typically use a dilution between 1:2 and 1:10. For monoclonal antibodies the typical range of dilutions for ascites we use is between 1:100 and 1:1,000. Polyclonal antibodies should be tested in the range of 1:50 to 1:500. We strongly recommend that the dilution should be empirically determined by prior titration to produce maximal binding for a given cell type.

How can I optimize the FACS process to retrieve better quality live cells?

Consultation with an experienced FACS machine operator may also be required in order to optimize nozzle size and sheath pressure used by the FACS machine to minimize damage to the human ES cells. In addition, whenever an experiment requires reclamation of live cells through cell surface marker selection it is strongly advised that the primary and secondary antibodies be filter sterilized prior to use. Collection of cells into antibiotic-containing medium can minimize post-sort contamination.

What negative controls should I use for flow cytometry?

We typically use the antibody produced by the parent myeloma cell line P3X63Ag8 as a negative control. Include a negative control for each secondary antibody if several are being used. These P3X negative controls should be used in addition to the secondary only negative controls.

Can I assay intracellular antigens by flow cytometry?

Yes, we use the Intraprep kit (Beckman Coulter # IM2389) with good results for markers like Oct-4.

How do I increase throughput while minimizing the amount of reagents used during antibody staining of human ES cells?

Staining cells in four-well plates requires about 100 μL solution per well. When antibody supplies are limited or when staining the cells with multiple antibodies, we us eight-well chamber slides (Nunc LabTek Cat #177402) which require less than 50 μL of solution per well. When staining cells in either four-well plates or eight-well chamber slides we advise keeping the vessels in a sealed box containing damp tissues for overnight incubation to avoid evaporation of staining solution.

References

Amit M, MK Carpenter, MS Inokuma, CP Chiu, CP Harris, MA Waknitz, J Itskovitz-Eldor and JA Thomson. (2000). Clonally derived human embryonic stem cell lines maintain pluripotency and proliferative potential for prolonged periods of culture. *Dev Biol* **227**: 271–278.

Andrews GK, M Dziadek and T Tamaoki. (1982a). Expression and methylation of the mouse alpha-fetoprotein gene in embryonic, adult, and neoplastic tissues. *J Biol Chem* **257**: 5148–153.

Andrews PW, G Banting, I Damjanov, D Arnaud and P Avner. (1984a). Three monoclonal antibodies defining distinct differentiation antigens associated with different high molecular weight polypeptides on the surface of human embryonal carcinoma cells. *Hybridoma* **3**: 347–361.

Andrews PW, J Casper, I Damjanov, M Duggan-Keen, A Giwercman, J-I Hata, A Von Keitz, LHJ Looijenga, JW Oosterhuis, M Pera, M Sawada, H-J Schmoll, NE Skakkebæk, W Van Putten and P Stern. (1996). Comparative analysis of cell surface antigens expressed by cell lines derived from human germ cell tumours. *Int J Cancer* **66**: 806–816.

Andrews PW, PN Goodfellow and DL Bronson. (1983). Cell-surface characteristics and other markers of differentiation of human teratocarcinoma cells in culture. In: *Teratocarcinoma Stem Cells*. Silver, Martin and Strickland, eds. Cold Spring Harbour Press, New York, pp 579–590.

Andrews PW, PN Goodfellow, LH Shevinsky, DL Bronson and BB Knowles. (1982b). Cell-surface antigens of a clonal human embryonal carcinoma cell line: morphological and antigenic differentiation in culture. *Int J Cancer* **29**: 523–531.

Andrews PW, LJ Meyer, KL Bednarz and H Harris. (1984b). Two monoclonal antibodies recognizing determinants on human embryonal carcinoma cells react specifically with the liver isozyme of human alkaline phosphatase. *Hybridoma* **3**: 33–39.

Andrews PW, E Nudelman, S Hakomori and BA Fenderson. (1990). Different patterns of glycolipid antigens are expressed following differentiation of TERA-2 human embryonal carcinoma cells induced by retinoic acid, hexamethylene bisacetamide (HMBA) or bromodeoxyuridine (BUdR). *Differentiation* **43**: 131–138.

Andrews PW, JW Oosterhuis and I Damjanov. (1987). Cell lines from human germ cell tumours. In: *Teratocarcinoma and Embryonic Stem Cells: a Practical Approach*. E. Robertson, ed. IRL Press, Oxford, pp 207–248.

Avilion AA, SK Nicolis, LH Pevny, L Perez, N Vivian and R Lovell-Badge. (2003). Multipotent cell lineages in early mouse development depend on SOX2 function. *Genes Dev* **17**: 126–140.

Badcock G, C Pigott, J Goepel and PW Andrews. (1999). The human embryonal carcinoma marker antigen TRA-1-60 is a sialylated keratan sulfate proteoglycan. *Cancer Res* **59**: 4715–4719.

Beattie GM, AD Lopez, N Bucay, A Hinton, MT Firpo, CC King and A Hayek. (2005). Activin A maintains pluripotency of human embryonic stem cells in the absence of feeder layers. *Stem Cells* **23**: 489–495.

Beck F, T Erler, A Russell and R James. (1995). Expression of Cdx-2 in the mouse embryo and placenta: possible role in patterning of the extra-embryonic membranes. *Dev Dyn* **204**: 219–227.

Boyer LA, TI Lee, MF Cole, SE Johnstone, SS Levine, JP Zucker, MG Guenther, RM Kumar, HL Murray, RG Jenner, DA Melton, DK Gifford, R Jaenisch and RA Young. (2005). Core transcriptional regulatory circuitry in human embryonic stem cells. *Cell* **122**: 947–956.

Brandenberger R, H Wei, S Zhang, S Lei, J Murage, GJ Fisk, Y Li, C Xu, R Fang, K Guegler, MS Rao, R Mandalam, J Lebkowski and LW Stanton. (2004). Transcriptome characterization elucidates signaling networks that control human ES cell growth and differentiation. *Nat Biotechnol* **22**: 707–716.

Chambers I, D Colby, M Robertson, J Nichols, S Lee, S Tweedie and A Smith. (2003). Functional expression cloning of Nanog, a pluripotency sustaining factor in embryonic stem cells. *Cell* **113**: 643–655.

Ciruna BG and J Rossant. (1999). Expression of the T-box gene Eomesodermin during early mouse development. *Mech Dev* **81**: 199–203.

Draper JS, C Pigott, JA Thomson and PW Andrews. (2002). Surface antigens of human embryonic stem cells: changes upon differentiation in culture. *J Anat* **200**: 249–258.

Duncan SA, A Nagy and W Chan. (1997). Murine gastrulation requires HNF-4 regulated gene expression in the visceral endoderm: tetraploid rescue of Hnf-4(−/−) embryos. *Development* **124**: 279–287.

Dvash T, Y Mayshar, H Darr, M McElhaney, D Barker, O Yanuka, KJ Kotkow, LL Rubin, N Benvenisty and R Eiges. (2004). Temporal gene expression during differentiation of human embryonic stem cells and embryoid bodies. *Hum Reprod* **19**: 2875–2883.

Eisenbarth GS, FS Walsh and M Nirenberg. (1979). Monoclonal antibody to a plasma membrane antigen of neurons. *Proc Natl Acad Sci USA* **76**: 4913–4917.

Ellis P, BM Fagan, ST Magness, S Hutton, O Taranova, S Hayashi, A McMahon, M Rao and L Pevny. (2004). SOX2, a persistent marker for multipotential neural stem cells derived from embryonic stem cells, the embryo or the adult. *Dev Neurosci* **26**: 148–165.

Enver T, S Soneji, C Joshi, J Brown, F Iborra, T Orntoft, T Thykjaer, E Maltby, K Smith, RA Dawud, M Jones, M Matin, P Gokhale, J Draper and PW Andrews (2005). Cellular differentiation hierarchies in normal and culture-adapted human embryonic stem cells. *Hum Mol Genet* **14**: 3129–3140.

Fenderson BA, PW Andrews, E Nudelman, H Clausen and S Hakomori. (1987). Glycolipid core structure switching from globo- to lacto- and ganglio-series during retinoic acid-induced differentiation of TERA-2-derived human embryonal carcinoma cells. *Dev Biol* **122**: 21–34.

Ginis I, Y Luo, T Miura, S Thies, R Brandenberger, S Gerecht-Nir, M Amit, A Hoke, MK Carpenter, J Itskovitz-Eldor and MS Rao. (2004). Differences between human and mouse embryonic stem cells. *Dev Biol* **269**: 360–380.

Henderson JK, JS Draper, HS Baillie, S Fishel, JA Thomson, H Moore and PW Andrews. (2002). Preimplantation human embryos and embryonic stem cells show comparable expression of stage-specific embryonic antigens. *Stem Cells* **20**: 329–337.

Herrmann BG. (1991). Expression pattern of the Brachyury gene in whole-mount TWis/TWis mutant embryos. *Development* **113**: 913–917.

Herszfeld D, E Wolvetang, E Langton-Bunker, TL Chung, AA Filipczyk, S Houssami, P Jamshidi, K Koh, AL Laslett, A Michalska, L Nguyen, BE Reubinoff, I Tellis, JM Auerbach, CJ Ording, LH Looijenga and MF Pera. (2006). CD30 is a survival factor and a biomarker for transformed human pluripotent stem cells. *Nat Biotechnol* **24**: 351–357.

Hyslop L, M Stojkovic, L Armstrong, T Walter, P Stojkovic, S Przyborski, M Herbert, A Murdoch, T Strachan and M Lako (2005). Downregulation of NANOG induces differentiation of human embryonic stem cells to extraembryonic lineages. *Stem Cells* **23**: 1035–1043.

Inzunza J, S Sahlen, K Holmberg, AM Stromberg, H Teerijoki, E Blennow, O Hovatta and H Malmgren. (2004). Comparative genomic hybridization and karyotyping of human embryonic stem cells reveals the occurrence of an isodicentric X chromosome after long-term cultivation. *Mol Hum Reprod* **10**: 461–466.

Kaestner KH, H Hiemisch, B Luckow and G Schutz. (1994). The HNF-3 gene family of transcription factors in mice: gene structure, cDNA sequence, and mRNA distribution. *Genomics* **20**: 377–385.

Kannagi R, NA Cochran, F Ishigami, S Hakomori, PW Andrews, BB Knowles and D Solter. (1983). Stage-specific embryonic antigens (SSEA-3 and -4) are epitopes of a unique globo-series ganglioside isolated from human teratocarcinoma cells. *Embo J* **2**: 2355–2361.

Kuo CT, EE Morrisey, R Anandappa, K Sigrist, MM Lu, MS Parmacek, C Soudais and JM Leiden. (1997). GATA4 transcription factor is required for ventral morphogenesis and heart tube formation. *Genes Dev* **11**: 1048–1060.

Lee JE, SM Hollenberg, L Snider, DL Turner, N Lipnick and H Weintraub. (1995). Conversion of Xenopus ectoderm into neurons by NeuroD, a basic helix-loop-helix protein. *Science* **268**: 836–844.

Li ZL, A Lilienbaum, G Butler-Browne and D Paulin. (1989). Human desmin-coding gene: complete nucleotide sequence, characterization and regulation of expression during myogenesis and development. *Gene* **78**: 243–254.

Liu Y, S Shin, X Zeng, M Zhan, R Gonzalez, FJ Mueller, CM Schwartz, H Xue, H Li, SC Baker, E Chudin, DL Barker, TK McDaniel, S Oeser, JF Loring, MP Mattson and MS Rao. (2006). Genome wide profiling of human embryonic stem cells (hESCs), their derivatives and embryonal carcinoma cells to develop base profiles of U.S. Federal government approved hESC lines. *BMC Dev Biol* **6**: 20.

Matin MM, JR Walsh, PJ Gokhale, JS Draper, AR Bahrami, I Morton, HD Moore and PW Andrews. (2004). Specific knockdown of Oct4 and beta2-microglobulin expression by RNA interference in human embryonic stem cells and embryonic carcinoma cells. *Stem Cells* **22**: 659–668.

Morrisey EE, Z Tang, K Sigrist, MM Lu, F Jiang, HS Ip and MS Parmacek. (1998). GATA6 regulates HNF4 and is required for differentiation of visceral endoderm in the mouse embryo. *Genes Dev* **12**: 3579–3590.

Muyan M and I Boime. (1997). Secretion of chorionic gonadotropin from human trophoblasts. *Placenta* **18**: 237–241.

Nichols J, B Zevnik, K Anastassiadis, H Niwa, D Klewe-Nebenius, I Chambers, H Scholer and A Smith. (1998). Formation of pluripotent stem cells in the mammalian embryo depends on the POU transcription factor Oct4. *Cell* **95**: 379–391.

Niwa H, J Miyazaki and AG Smith. (2000). Quantitative expression of Oct-3/4 defines differentiation, dedifferentiation or self-renewal of ES cells. *Nat Genet* **24**: 372–376.

Okuda A, A Fukushima, M Nishimoto, A Orimo, T Yamagishi, Y Nabeshima, M Kuro-o, Y Nabeshima, K Boon, M Keaveney (1998). UTF1, a novel transcriptional coactivator expressed in pluripotent embryonic stem cells and extra-embryonic cells. *Embo J* **17**: 2019–2032.

Pera MF, MJ Blasco-Lafita, S Cooper, M Mason, J Mills and P Monaghan. (1988). Analysis of cell-differentiation lineage in human teratomas using new monoclonal antibodies to cytostructural antigens of embryonal carcinoma cells. *Differentiation* **39**: 139–149.

Pesce M, X Wang, DJ Wolgemuth and H Scholer. (1998). Differential expression of the Oct-4 transcription factor during mouse germ cell differentiation. *Mech Dev* **71**: 89–98.

Peschle C, F Mavilio, A Care, G Migliaccio, AR Migliaccio, G Salvo, P Samoggia, S Petti, R Guerriero, M Marinucci (1985). Haemoglobin switching in human embryos: asynchrony of zeta–alpha and epsilon–gamma-globin switches in primitive and definite erythropoietic lineage. *Nature* **313**, 235–8.

Pevny LH, S Sockanathan, M Placzek and R Lovell-Badge. (1998). A role for SOX1 in neural determination. *Development* **125**: 1967–1978.

Reubinoff BE, MF Pera, CY Fong, A Trounson and A Bongso. (2000). Embryonic stem cell lines from human blastocysts: somatic differentiation in vitro. *Nat Biotechnol* **18**: 399–404.

Rosler ES, GJ Fisk, X Ares, J Irving, T Miura, MS Rao and MK Carpenter. (2004). Long-term culture of human embryonic stem cells in feeder-free conditions. *Dev Dyn* **229**: 259–274.

Sato N, IM Sanjuan, M Heke, M Uchida, F Naef and AH Brivanlou. (2003). Molecular signature of human embryonic stem cells and its comparison with the mouse. *Dev Biol* **260**: 404–413.

Shevinsky LH, BB Knowles, I Damjanov and D Solter. (1982). Monoclonal antibody to murine embryos defines a stage-specific embryonic antigen expressed on mouse embryos and human teratocarcinoma cells. *Cell* **30**: 697–705.

Sidhu KS and BE Tuch. (2006). Derivation of three clones from human embryonic stem cell lines by FACS sorting and their characterization. *Stem Cells Dev* **15**: 61–69.

Solter D and BB Knowles. (1978). Monoclonal antibody defining a stage-specific mouse embryonic antigen (SSEA-1). *Proc Natl Acad Sci USA* **75**: 5565–5569.

Sperger JM, X Chen, JS Draper, JE Antosiewicz, CH Chon, SB Jones, JD Brooks, PW Andrews, PO Brown and JA Thomson. (2003). Gene expression patterns in human embryonic stem cells and human pluripotent germ cell tumors. *Proc Natl Acad Sci USA* **100**: 13350–13355.

Thomson JA, J Itskovitz-Eldor, SS Shapiro, MA Waknitz, JJ Swiergiel, VS Marshall and JM Jones. (1998). Embryonic stem cell lines derived from human blastocysts. *Science* **282**: 1145–1147.

Wilkinson DG, S Bhatt and BG Herrmann. (1990). Expression pattern of the mouse T gene and its role in mesoderm formation. *Nature* **343**: 657–659.

Williams BP, GL Daniels, B Pym, D Sheer, S Povey, Y Okubo, PW Andrews and PN Goodfellow. (1988). Biochemical and genetic analysis of the OKa blood group antigen. *Immunogenetics* **27**: 322–329.

Xu C, MS Inokuma, J Denham, K Golds, P Kundu, JD Gold and MK Carpenter. (2001). Feeder-free growth of undifferentiated human embryonic stem cells. *Nat Biotechnol* **19**: 971–974.

Xu RH, RM Peck, DS Li, X Feng, T Ludwig and JA Thomson. (2005). Basic FGF and suppression of BMP signaling sustain undifferentiated proliferation of human ES cells. *Nat Methods* **2**: 185–190.

8

Genetic and epigenetic analysis of human embryonic stem cells

LAURIE A. BOYER[1], RUDOLF JAENISCH[1,2] AND MAISAM MITALIPOVA[1]

[1] Whitehead Institute for Biomedical Research, Cambridge, MA 02142, USA
[2] Massachusetts Institute of Technology, Cambridge, MA 02142, USA

Introduction

Human embryonic stem (ES) cells have been derived from blastocyst-stage embryos (Thomson *et al.*, 1998) and are thought to hold great promise for human medicine because of their unlimited potential to become any one of the ~200 cell types in the body. A number of issues remain to be resolved, however, before human ES cells can be used effectively and routinely in cell-based therapies. One such obstacle is the lack of standardization in techniques employed for derivation and maintenance of human ES cell lines. Recent evidence indicates that some methods for culture of human ES cells can influence their genetic integrity and differentiation potential. For example, human ES cells that are typically propagated in clumps by mechanical dissociation are more likely to maintain a normal chromosomal constitution (Mitalipova *et al.*, 2005; Buzzard *et al.*, 2004). In contrast, the faster and easier means of disaggregating human ES cells by enzymatic treatment such as trypsin digestion promotes chromosomal aneuploidy, particularly trisomy 12 and/or 17, in conjunction with aberrant gene expression (Draper *et al.*, 2004; Mitalipova *et al.*, 2005). Interestingly, such karyotypic abnormalities are also associated with germ cell tumors (Skotheim *et al.*, 2002) and neuroblastomas (Westermann and Schwab, 2002) underscoring the need to better understand the cause and consequence of these particular anomalies.

Mammalian preimplantation development is characterized by dynamic changes in DNA methylation patterns (Morgan *et al.*, 2005). In the mouse, it has been shown that culture can affect the methylation state of the genome and the integrity of genomic imprinting both in blastocysts and in ES cells (Humpherys *et al.*, 2001; Dean *et al.*, 1998; Mann *et al.*, 2004). These imprinting errors can lead to developmental defects and are also commonly associated with a variety of tumor phenotypes (Holm *et al.*, 2005).

In humans, it has long been known that aberrant DNA methylation is a hallmark of many tumors (Herman and Baylin, 2003) and that disruption of allele-specific methylation patterns at imprinted loci during gametogenesis is responsible for parental-specific inheritance of human disorders such as Prader–Willi/Angelman syndrome (Arnaud and Feil, 2005). Moreover, the incidence of Beckwith–Wiedeman Syndrome, a devastating imprinting disorder, is significantly increased in children born as a result of *in vitro* fertilization (IVF) treatment (Niemitz and Feinberg, 2004; DeBaun *et al.*, 2003). Studies indicated that imprinted gene expression can also be disrupted during the isolation or propagation of human ES cells. For example, abnormal DNA methylation patterns at CpG islands, which include regions associated with imprinting centers as well as those associated with known tumor suppressor genes, have been observed in human ES cells (Maitra *et al.*, 2005) and during human ES cell differentiation (Shen *et al.*, 2006). It has also been reported, however, that human ES cell lines can maintain stable DNA methylation patterns at imprinted loci under certain conditions (Rugg-Gunn *et al.*, 2005). It is not surprising then that individual human ES cell lines vary widely in the genetic and epigenetic state of their genomes (Ware *et al.*, 2006). Thus, embryo cultivation and cell propagation methods may influence the genetic and epigenetic status and ultimately the utility of human ES cells for research and medicine. Ultimately, human ES cell lines isolated and cultured under different conditions need to be compared to establish a robust correlation between culture history of a given ES cell line and its overall genetic and epigenetic state. Here, we describe experimental methods for the analysis of karyotype stability and for monitoring the allele-specific expression of imprinted loci as a surrogate assay for DNA methylation status in human ES cells.

Materials, reagents, and equipment

I. Karyotype analysis

A. Stock reagents

Ethidium bromide stock solution

Sigma Chemical Company, catalog# E7637

Stock concentration: 10 µg/mL prepared in 1× Hanks Balanced Salt Solution without calcium and magnesium

Acetic acid, glacial

Fisher Scientific, catalog# A38-212, 2.5L

Methyl alcohol anhydrous

Fisher Scientific, catalog# A412-4, 4L

Potassium chloride (KCl)

Fisher Scientific, catalog# P217-500, 500 g

B. Working reagents

KaryoMAX® Colcemid solution

Invitrogen, catalog# 15210-040

Working concentration: 10 μg/mL

Ethidium bromide solution

Working concentration: 1 μg/mL

Add 10 mL ethidium bromide stock solution to 90 mL sterile distilled water.

Filter with a 0.2 μm filter and store protected from light at −20°C. A 5 mL aliquot can be stored at 4°C for 1 month.

1× Hanks balanced salt solution without calcium and magnesium

Invitrogen, catalog# 14170

Store at room temperature.

1× Trypsin–EDTA

Invitrogen, catalog# 25300

Pipette into 5 mL aliquots and store frozen at −20°C.

Prewarm to 37°C before use.

Potassium chloride hypotonic solution

Working concentration: 0.075M KCl

Dissolve 5.6 g KCl in 1L of sterile distilled water.

Prewarm to 37°C before use.

Fixative

3:1 ratio of methanol: acetic acid

Prepare fresh before each use.

II. Epigenetic analysis

A. Stock reagents

SuperScript TMIII first–strand synthesis system for reverse-transcription polymerase chain reaction (RT-PCR)

Invitrogen, catalog# 18080-051

PUREGENE DNA purification kit

Gentra, catalog# D-5010A

PureLink micro-to-midi total RNA purification system

Invitrogen, catalog# 12183-018

QiAquick gel extraction kit

Qiagen, catalog# 28706

Qiaquick PCR purification kit

Qiagen, catalog# 28106

Taq PCR core kit

Qiagen, catalog# 201223

Topo® TA cloning system

Invitrogen, catalog# K2500-20

B. Working reagents

Collagenase, type IV

Invitrogen, catalog# 17104-019, 1 *g*

Working concentration: 1 mg/ml

Dissolve collagenase in DMEM F12 without supplements (base media for human ES cell culture; Invitrogen catalog# 11330-032). **Note:** Avoid using serum-containing media as it will inhibit the activity of collagenase.

Equipment

- Sorvall RT Legend tabletop centrifuge (or similar)

- Microfuge

- PTC-100 Programmable Thermal Controller (MJ Research, Inc.) or similar thermocycler

Protocols

I. Karyotype analysis

Chromosomal abnormalities can lead to developmental defects and altered growth properties, which is a major consideration when using human ES cell lines for cell-based therapy. It has been reported that prolonged passage using enzymatic dissociation of human ES cell colonies can lead to chromosomal abnormalities including partial trisomy (Draper *et al.*, 2004; Mitalipova *et al.*, 2005). Therefore, a complete cytogenetic analysis including fluorescence *in situ* hybridization (FISH) and Q-banding should be considered in order to identify the chromosomal origin of the trisomy. If more than one cell line is cultured in the lab, it may also be advisable to periodically request DNA fingerprint analysis to rule out any cross-contamination among these lines.

Procedure for preparation of human ES cells for karyotyping

This protocol is designed for preparation of cells from a T-25 Flask (effective growth area $\sim$25 cm^2), but volumes given can be adjusted to accommodate any culture dish or flask.

1. The medium should be changed 24 hr prior to performing the protocol.

   This stimulates division and increases the number of dividing human ES cells. **Note:** This protocol is compatible with any passage technique or cell culture condition including the use of feeder cells.

2. Add 60 μL of *ethidium bromide* (working solution) to each flask containing 5 mL of media and gently swirl to mix (final concentration 0.012 μg/mL)

3. Incubate the culture at 37°C with 5% CO_2 for 40 min.

4. Add 60 μL of *colcemid solution* to the medium previously treated with ethidium bromide in **Step 2** (final concentration 0.12 μg/mL).

5. Incubate at 37°C with 5% CO_2 for 20 min.

6. Collect medium and save in labeled centrifuge tube.

7. Rinse the flask with 1–2 mL of Hanks balanced salt solution (1× without calcium and magnesium). Add the Hanks rinse solution to the labeled centrifuge tube from **Step 6**.

8. Add 1.5–2 mL of pre-warmed trypsin–EDTA (1× without calcium and magnesium) to the flask and incubate at 37°C. Cells will begin to detach from the surface in approximately 1–3 min.

9. Tighten the cap of the flask and tap for 3–5 s to dislodge cells. If human ES cells are growing on cell culture dishes, gently tap the dish to dislodge the cells.

10. Add 2 mL of Hanks balanced salt solution (1× without calcium and magnesium) to the flask or plate and triturate the cell suspension gently with a large bore disposable seralogical pipette to produce a single cell suspension. **Note:** Do not use micropipette tips to disrupt the cell aggregate because the small opening will damage the cells.

11. Transfer the cell suspension to the centrifuge tube saved in **Step 6**.

    Centrifuge the tube for 8 min at 250 g (~900 rpm) at room temperature.

12. Carefully aspirate and discard the supernatant without disturbing the cell pellet. Gently break up the cell pellet by flicking the outside of the tube with your finger.

13. Resuspend the cell pellet in 4–6 mL of hypotonic solution (0.075M KCl) pre-warmed to 37°C. Incubate the tube at 37°C for 20 min.

14. Carefully add 8–10 drops of freshly prepared working fixative solution to the hypotonic solution, mix gently, and allow to stand at room temperature for 5 min. **Note:** The cells are very delicate at this stage and require very gentle handling to prevent cell lysis.

15. Centrifuge the tube, discard the supernatant, and gently resuspend the cell pellet by flicking the tube with your finger. Slowly add 5 mL of fresh fixative (working solution) down the side of the centrifuge tube and incubate for 30 min at room temperature.

16. Repeat **Step 15** twice without any incubation period.

17. Spin the sample down and add about 1.5 mL of fresh fixative. Transfer the cell suspension to a 1.5 or 2.0 mL cryovial freezer tube. The cell pellet can be stored for 1–2 months at 4°C or at −10 to −20°C for long-term storage.

    The sample should then be sent to a specialist center for karyotypic analysis. A sample karyotype is shown in **Figure 1** (see below; color version in color plate section).

II. Epigenetic analysis

Genomic imprinting is an epigenetic mechanism for restricting the expression of a small subset of genes in a parent-of-origin specific manner and appears to be linked to

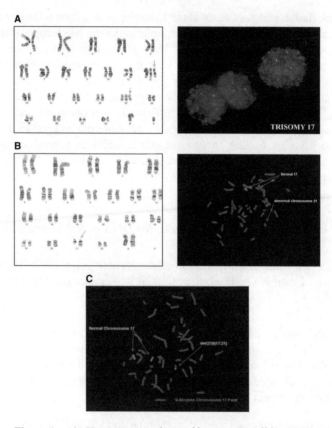

Figure 1 (A) Karyotype analysis of human ES cell line BG02 (Bresagen) that has been passaged by mechanical dissociation and collagenase treatment. This analysis showed an abnormal karyotype resulting in full trisomy of chromosome 12. (B) Karyotype analysis of human ES cell line H9 (NIH code WA09) passaged over 20 times (passage 40) with trypsin-EDTA. This analysis revealed an abnormal female karyotype with partial duplication of the long arm of chromosome 17 resulting in partial trisomy of chromosome 17 and partial deletion of the long arm of chromosome 21. FISH analysis on metaphase chromosomes using a cocktail probe that is specific for the HER-2/NEU and TOPO genes on chromosome 17q11.2-q12 together with a chromosome 17 centromere control probe was used to identify additional material present on the long arm of chromosome 21. Initially, the control cell line and the H9 (p40) human ES cell line showed a normal signal pattern, with no additional signals present on the long arm of the derivative chromosome 21. (C) A whole chromosome probe specific for chromosome 17 was then used to establish the position of the break point. The chromosome paint probe showed a fluorescent signal on the long arm of the derivative chromosome 21, confirming that the translocation was derived from chromosome 17 with the breakpoint distal to HER-2/NEU locus. This image is best seen in color; please refer to plate section to see the color image.

the DNA methylation status of each allele (Morgan *et al.*, 2005). Genomic imprinting plays a critical role in development and disease and may be disrupted during *in vitro* culture of blastocysts or human ES cells. This protocol describes a surrogate method in which the allele-specific expression of imprinted genes may be evaluated to assess the DNA methylation status of a given human ES cell line. This method first requires the identification of single nucleotide polymorphisms (SNPs) in a transcribed region of the gene of interest. SNPs are genetic differences that occur on average 1 in every 1200 base pairs and can be used to distinguish the maternal and paternal alleles. Here, we present an example for the analysis of IGF2, which is normally expressed from the paternal allele. However, this general protocol is amenable for analysis of any imprinted loci.

Preparation of human ES cells for DNA or RNA isolation

A confluent 35 mm tissue culture dish of human ES cells can yield ~50 µg of genomic DNA or ~1 µg of total RNA. For best results, it is recommended that DNA and RNA samples be isolated at the same time.

1. Separate the human ES cells from feeder cells by incubation with the working stock of collagenase for 1–2 min followed by pipetting to dislodge the human ES cell colonies. Use 0.5 mL for confluent 35 mm dish or adjust volumes accordingly. If feeders are not used for culture, the human ES cells can be removed from the culture dish by trypsin digestion.

2. Collect the supernatant in 15 mL conical tube and harvest the human ES cells in a Sorvall table top centrifuge at 300 *g* (~1,000 rpm) for 5 min at room temperature.

3. Remove supernatant and wash cell pellet once with 1 × PBS followed by centrifugation as above. **Note:** The cell pellet upon re-suspension in 1 × PBS can be split into multiple tubes for simultaneous isolation of DNA and RNA.

4. Use cell pellet for immediate genomic **DNA isolation** using the PUREGENE DNA Purification Kit or other suitable DNA isolation method. For total **RNA isolation,** use PuroLink Micro-to-Midi Total RNA Purification System or other suitable RNA isolation method. The integrity and purity of RNA should be checked on a Northern blot using conventional methods or by using the Agilent 2100 Bioanalyzer System (www.agilent.com). Alternatively, the cell pellet can be immediately frozen in liquid nitrogen and stored at −80°C for future use.

PCR amplification of IGF2 genomic DNA

The example given employs oligonucleotide primers designed to amplify exon 9 of the IGF2 locus (see **Table 1**).

Table 1 Oligonucleotide primers for PCR analysis of IGF2 gene expression in human embryonic stem cells

Name	Target region	Sequence	Product size (base pairs)		Tm
			DNA	**RNA**	(°C)
IGF2-2 Forward	Acc. X07868	5′-CTTGGACTTTGAGTCAAATTG–3′			51
			293	293	
IGF2-2 Reverse	Exon 9	5′-GGTCGTGCCAATTACATTTCA–3′			55

5. Set up PCR reactions as follows using Qiagen Taq PCR Core Kit components or similar PCR reagents.

   5 µL 10× PCR reaction buffer

   2 µL 25 mM $MgCl_2$ solution

   6 µL of each 2 µM primer

   1 µL of dNTP mix (10 mM of each dNTP)

   0.25 µL of AmpliTaq Polymerase

   0.2–0.5 µg of DNA template

   Total volume to 50 µL with distilled H_2O

6. The following PCR cycle parameters have been designed for use with the oligonucleotide primers listed in **Table 1**.

   Step 1: 94°C for 3 min

   Step 2: 94°C for 30 s

   Step 3: 55°C for 30 s

   Step 4: 72°C for 1 min

   Step 5: 30 cycles of steps 2–4

   Step 6: 72°C for 10 min

   Step 7: Store at 4°C

7. Analyze PCR products by electrophoresis using a 1% agarose gel with ethidium bromide to check for DNA fragments of the appropriate size (see **Table 1**). Purify PCR product using Qiagen PCR isolation kit or similar method and quantify the DNA. Purified DNA may be stored long-term at — 20°C.

Analysis of IGF2 allele sequence and SNP identification

It is recommended to clone the PCR product obtained in **step 7** prior to sequencing the PCR product using the Topo TA cloning system according to the manufacturer's instructions. In the following series of steps DNA sequences representing each allele will be cloned individually and each allele can thus be isolated as a homogeneous population. This will minimize contaminating PCR products and will ensure the most reliable sequencing results.

8. A minimum of ten colonies should be selected from the TOPO TA cloning reaction. Inoculate a tube containing 2 mL of Luria Broth (LB) plus ampicillin (50 ug/mL) for each individual colony and grow overnight at 37°C. It is best to pick a number of colonies to ensure adequate representation of both alleles for sequencing.

9. Isolate plasmid DNA using PureLink Quick Miniprep Kit or similar method. Purified DNA may be stored long-term at −20°C.

10. Use appropriate restriction enzymes (consult TOPO TA vector map supplied by the manufacturer) to digest the plasmid DNA isolated in **step 9**. For example, EcoRI can be used to analyze the IGF2 exon 9 clones. Analyze restriction digests by electrophoresis on 1% agarose gel with ethidium bromide for correct fragment sizes.

11. The cloned PCR products are now ready to be sequenced. Generally, 0.2–0.5 μg of the purified DNA and 3.2 pmol of either the M13 reverse or forward primer (primers are supplied in Topo TA cloning kit) are used for standard sequencing reactions. It is important to sequence cloned PCR products in the forward and reverse direction to distinguish between true SNPs and possible errors introduced during the sequencing process.

12. Align and compare all the sequencing results obtained from the same genomic region to identify SNPs. Since the goal is to identify allele-specific DNA sequence differences within the same cell line, it is only necessary to find single nucleotide differences among the aligned sequences. Any number of sequence analysis software packages such as Sequencher (Gene Codes Corporation) can be used for this application. **Note:** It is very important to verify the SNP results by independent multiple PCR reactions, sequencing multiple clones and/or by direct sequencing of the PCR products excised and purified from the agarose gel.

Once a SNP or set of SNPs has been identified, it is now reasonable to sequence the PCR products directly from **steps 5–7** during subsequent analyses (e.g. later passage) of the same human ES cell line.

Analysis of allele-specific expression of imprinted genes

The identification of SNPs (e.g. in exon 9 of the IGF2 locus) now allows for analysis of allele specific expression patterns. The following series of steps requires the generation of cDNA from total RNA isolated from the same human ES cells in which the SNP was identified, followed by PCR amplification of IGF2 exon 9, Topo TA cloning, and sequencing. It is important to note that dysregulation of imprinted gene expression is not an all or none phenomenon as it occurs over multiple cell divisions and prolonged passage. Thus it may be difficult to detect the expression of both alleles if such changes have occurred in only a small population of the total cells used for the analysis. Therefore, it is advisable to sequence a large number of clones (>25) during early passage of human ES cells or for the first analysis as it is important to institute a reference so that a robust correlation can be established between culture history and the integrity of imprinted gene expression.

12. Total RNA isolated in **step 4** was treated with DNAase I followed by cDNA synthesis using specific SuperScript TMIII First-Strand Synthesis System for Reverse-transcription (RT-PCR) according to the manufacturer's instructions.

13. IGF2 exon 9 DNA (or any sequence of interest) can be amplified using the same primers used for initial PCR reactions (listed in **Table 1**) as well as the PCR parameters as outlined in **steps 5–7**.

14. Clone the purified genomic PCR products using the Topo TA cloning system and analyze as in **steps 8–10**. A minimum of 15–20 clones should be used for sequence analysis. Perform sequencing reactions as described in **step 11**. The expression of each allele can be detected by the presence of the SNP in a proportion of the sequenced clones.

The disruption of normal imprinted gene expression patterns in IGF2 or other imprinted loci suggests that such dysregulation may have resulted from improper maintenance of allele-specific DNA methylation patterns. Further analysis of the DNA methylation status of each allele can be carried out with bisulfite modification followed by genomic sequencing or other methods. There are currently a number of detailed protocols describing such methods (Wong, 2006; Wong *et al.*, 2006).

Troubleshooting

I. Karyotype analysis

What is the minimal number of cells needed for statistical analysis of a culture's karyotype?

Generally, a minimum of 20 metaphase spreads is required for G-banding as not all cells in a population will harbor chromosomal abnormalities especially during early passage. This will require the use of approximately 1×10^6 human ES cells. FISH analysis should be performed on a minimum of 300 interphase nuclei to determine the frequency of the abnormality in a particular cell line.

What kind of karyotype analysis should be performed on the sample?

This depends entirely on the investigator and the intended use of the cells. Generally it is sufficient to perform the G-banding first and to proceed with a full cytogenetic analysis if a karyotype abnormality is detected.

How often should I evaluate my human ES cell cultures?

This is somewhat dependent on culture conditions and the methods used to passage the human ES cells. It has been observed that mechanical dissociation results in a stable karyotype over many generations, whereas enzymatic treatment such as trypsin digestion may lead to karyotypic abnormalities. The appearance of such abnormalities is highly variable and depends on culture and passage techniques as well as the person manipulating the cells. It is recommended that cells should be characterized every 15–20 passages.

Are there any indicators that the human ES cells have karyotypic abnormalities, i.e. growth rate and morphology differences?

The effects of genetic abnormalities in human ES cell cultures may not always be apparent as often times the cells maintain the expression of typical pluripotency markers. In some cases, there is a noticeable change in the rate of cell division. If this happens, the cells should be analyzed for karyotype abnormalities.

I do not have very many cells for karyotyping. What has happened?

It is possible that most of the cells were lysed during the procedure. The cell membrane is extremely fragile as a result of the procedure. Repeat the protocol taking care to gently manipulate the cells.

II. Epigenetic analysis

1. I cannot find a polymorphism in the sequence for my gene of interest.

It is predicted that SNPs occur only about 1 in 1,200 nucleotides so not all loci will contain SNPs. Multiple primer sets that encompass transcribed regions should be designed to maximize the potential for finding a SNP at a given locus. Also consult public databases (e.g. www.ncbi.nlm.nih.gov/SNP) for the location of previously mapped SNPs.

2. The RT-PCR amplification of cDNA to determine allele-specific expression is not working.

The most critical parameters are the primer design and annealing temperature. Repeat the PCR with a slight decrease in annealing temperature (usually 0.5–1°C steps). If this fails, examine the primers to ensure they are amplifying the predicted length fragment, that they do not contain intron or non-coding sequences, that they face in the appropriate 5′ to 3′ direction to allow amplification, and that they have a mutually compatible melting temperature. If all optimization efforts fail, design a new set of PCR primers.

References

Arnaud P and R Feil. (2005). *Birth Defects Res C Embryo Today* **75**: 81–97.

Buzzard JJ, NM Gough, JM Crook and A Colman. (2004). *Nat Biotechnol* **22**: 381–382; author reply 382.

Dean W, L Bowden, A Aitchison, J Klose, T Moore, JJ Meneses, W Reik and R Feil. (1998). *Development* **125**: 2273–2282.

DeBaun MR, EL Niemitz and AP Feinberg. (2003). *Am J Hum Genet* **72**: 156–160.

Draper JS, K Smith, P Gokhale, HD Moore, E Maltby, J Johnson, L Meisner, T Zwaka, JA Thomson and PW Andrews. (2004). *Nat Biotechnol* **22**: 53–54.

Herman JG and SB Baylin. (2003). *N Engl J Med* **349**: 2042–2054.

Holm TM, L Jackson-Grusby, T Brambrink, Y Yamada, WM Rideout 3rd and R Jaenisch. (2005). *Cancer Cell* **8**: 275–285.

Humpherys D, K Eggan, H Akutsu, K Hochedlinger, WM Rideout 3rd, D Biniszkiewicz, R Yanagimachi and R Jaenisch. (2001). *Science* **293**: 95–97.

Maitra A, DE Arking, N Shivapurkar, M Ikeda, V Stastny, K Kassauei, G Sui, D Cutler, Y Liu, SN Brimble, K Noaksson, J Hyllner, TC Schulz, X Zeng, WJ Freed, J Crook, S Abraham, A Colman, P Sartipy, S Matsui, M Carpenter, AF Gazdar, M Rao and A Chakravarti. (2005). *Nat Genet* **37**: 1099–1103.

Mann MR, SS Lee, AS Doherty, RI Verona, LD Nolen, RM Schultz and MS Bartolomei. (2004). *Development* **131**: 3727–3735.

Mitalipova MM, RR Rao, DM Hoyer, JA Johnson, LF Meisner, KL Jones, S Dalton and SL Stice. (2005). *Nat Biotechnol* **23**: 19–20.

Morgan HD, F Santos, K Green, W Dean and W Reik. (2005). *Hum Mol Genet* **14** Spec No 1: R47–R58.

Niemitz EL and AP Feinberg. (2004). *Am J Hum Genet* **74**: 599–609.

Rugg-Gunn PJ, AC Ferguson-Smith and RA Pedersen. (2005). *Nat Genet* **37**: 585–587.

Shen Y, J Chow, Z Wang and G Fan. (2006). *Hum Mol Genet* **15**: 2623–26235.

Skotheim RI, O Monni, S Mousses, SD Fossa, OP Kallioniemi, RA Lothe and A Kallioniemi. (2002). *Cancer Res* **62**: 2359–2364.

Thomson JA, J Itskovitz-Eldor, SS Shapiro, MA Waknitz, JJ Swiergiel, VS Marshall and JM Jones. (1998). *Science* **282**: 1145–1147.

Ware CB, AM Nelson and CA Blau. (2006). *Stem Cells* **24**: 2677–2684.

Westermann F and M Schwab. (2002). *Cancer Lett* **184**: 127–147.

Wong IH. (2006). *Methods Mol Biol* **336**: 33–43.

Wong HL, HM Byun, JM Kwan, M Campan, SA Ingles, PW Laird, AS Yang. (2006). *Biotechniques* **41**: 734–739.

Section 3
Manipulation of Human Embryonic Stem Cells

9

In vivo differentiation of human embryonic stem cells

SCOTT A. NOGGLE, FRANCESCA M. SPAGNOLI AND ALI H. BRIVANLOU
Laboratory of Molecular Vertebrate Embryology, Rockefeller University, 1230 York Avenue, New York, NY 10021, USA

Introduction

Embryonic stem cells (ES) are self-renewing pluripotent cells that are derived from the inner cell mass (ICM) of mammalian blastocysts and are able to differentiate into all adult cell types (Brivanlou *et al.*, 2003).

The broad developmental capacity of ES cells was first appreciated in the mouse (mES), and primarily used as a vehicle for genetic manipulation. The recent isolation of human ES cells, endowed with similar properties as murine ES cells, has opened a window to the study of human embryogenesis. Finally, the human ES cell model represents an unprecedented opportunity to address how much of the information obtained from the study of development in classic embryology systems, such as frog and chick, is relevant in the context of human development.

Lineage-specific differentiation of human ES cells has been studied principally by three assays, two *in vitro* and one *in vivo*. *In vitro* assays include differentiation of ES cell monolayers and the formation of embryoid bodies (EBs). The only *in vivo* assay used so far for human ES cells is their ability to generate teratomas (Keller, 2005; Spagnoli and Brivanlou, 2006) in immuno-compromised mice. While these methods are used as diagnostics for *bona fide* embryonic stem cells, they are far from the natural environment from which they originated and thus from an embryological standpoint, they do not necessarily mimic the cues that these cells are exposed to in an embryo. Two-dimensional *in vitro* cell culture is under the influence of extrinsic factors, which may or may not be relevant to the endogenous situation. EBs have the advantage of providing three-dimensional structures, but may lack the proper juxtaposition of cells ultimately required for the induction and patterning of specific organs. Finally, though human ES cells can give rise to derivatives of the three germ layers within teratomas, the surrounding adult environment in which they differentiate is far from the embryonic milieu. In order to study the behavior of human ES cells in a more authentic embryonic setting, we therefore used traditional experimental embryological techniques of engraftment and cell mixing to introduce human ES cells into mouse blastocysts to begin to address their potentials for lineage-specific differentiation.

Recently, reports from our group and others showed that human ES cells can be directly engrafted into an *in vivo* context and respond appropriately to inductive signals of both mouse and chicken embryos and achieve functional differentiation (Goldstein *et al.*, 2002; Muotri *et al.*, 2005; Tabar *et al.*, 2005; James *et al.*, 2006). Procedures for engraftment of human ES cells in the chick have been described elsewhere (Goldstein, 2006). In this chapter, we describe the assay that we have designed to engraft human ES cells into the mouse. The procedures are based on those described for the incorporation of murine ES cells into mouse embryos (Bradley, 1987; Nagy and Rossant, 1999). The pluripotent cells are incorporated directly into the mouse embryo at the blastocyst stage which, in developmental time, represents the closest related *in vivo* embryonic environment for the human ES cells. This provides a tool for understanding the basic embryological properties of human ES cells and can be extended in the future to study the basis of diseases. As teratoma formation is still the most established model for assessing differentiation, we also include a range of protocols for intramuscular, subcutaneous, intraperitoneal and kidney capsule injection of human ES cells into mice. Taken together these provide an informative assay with which to assess the developmental potential of human ES cells.

Overview of protocols

Two protocols for the introduction of human ES cells into pre-implantation mouse embryos are detailed in this chapter. A diagram of the procedure is shown in **Figure 1**. Both protocols require: (i) superovulation of female mice and timed matings to produce host embryos for injections or aggregations, (ii) preparation of pseudo-pregnant foster mothers to receive the manipulated embryos, (iii) preparation of human ES cells for aggregation or blastocyst injection, and (iv) harvesting embryos after the appropriate number of days of gestation for aggregation or injection. For blastocyst injection, embryos are harvested at the early balstocyst stage in the morning of embryonic day 3.5. For aggregations, the embryos are harvested at the eight-cell stage in the morning of embryonic day 2.5. The embryos are injected or aggregated with human ES cells and cultured *in vitro* to the expanded blastocyst stage on the afternoon corresponding to embryonic day 3.5. They are either transferred to the pseudo-pregnant foster mothers and allowed to develop *in vivo* or further grown *in vitro* as differentiating outgrowths. The resulting embryos can then be processed using common assays for immunofluoresence, RT-PCR, or *in situ* hybridization. As an additional diagnostic tool for the ability of human ES cells to differentiate *in vivo*, protocols are also provided for the induction of teratoma from human ES cells in adult severe combined immunodeficient (SCID) mice.

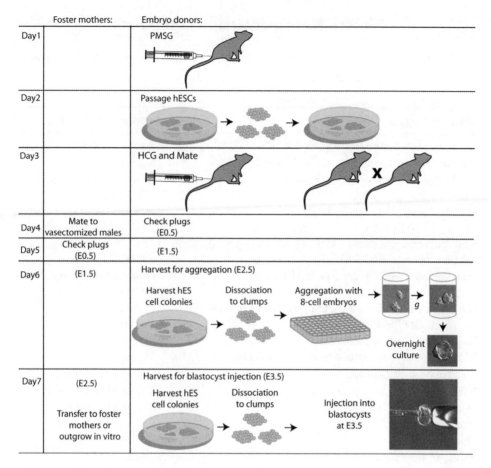

Figure 1 Diagram of mouse procedural timing, aggregations, and blastocyst injections. Timing of superovulation, mating and developmental progression for mice used in the procedures. Day 1: Mice for embryo donors are primed with PMSG. Day 2: human ES cells are passaged. Day 3: Embryo donor mice are primed with HCG and mated to fertile males. Day 4: Foster mothers are mated to vasectomized males and plugs are checked on embryo donor mice. Day 5: check plugs on foster mothers. Day 6: Embryo donors can be harvested at E2.5 for both protocols. For blastocyst injections, the E2.5 embryos are cultured overnight to Day 7 for injection at the equivalent of E3.5. Human ES cells are harvested either by microdissection or with enzyme immediately before the aggregation or injection procedures on Day 6 and 7, respectively. They are further dissociated into clumps of 10–15 cells. At E2.5, the clumps are combined with eight-cell pre-compacted mouse embryos, harvested on Day 6, in the conical bottomed wells of a 96-well plate, centrifuged and cultured overnight. At E3.5 (Day 7), the clumps are injected into blastocysts. In the afternoon of E3.5 (Day 7), the manipulated embryos from either protocol are transferred to pseudo-pregnant foster mothers or further cultured *in vitro* as embryonic outgrowths.

Materials, reagents, and equipment

A. Mice, reagents and surgical tools for producing host embryos, recipient foster mothers and teratomas by kidney capsule surgery

Material/reagent	Vendor Catalog #
Pregnant mare serum gonadotrophin (PMSG), 50 IU/mL	(Sigma #G4877)
Human chorionic gonadotrophin (HCG) 50 IU/mL	(Sigma #C0684)
Blastocyst donor mice: C57Bl6/6J	(Jackson Laboratory, Strain #000664)
Morula donor mice: B6CBAF1/J	(Jackson Laboratory, Strain #100011)
Stud male mice of proven fertility for mating to produce host embryos: B6CBAF1/J	(Jackson Laboratory, Strain #100011)
Vasectomized male mice: B6CBAF1/J	(Jackson Laboratory, Strain #100011)
Female mice to produce pseudo-pregnant recipient fosters: B6CBAF1/J	(Jackson Laboratory, Strain #100011)
10 g of 2,2,2-tribromoethanol	(Sigma #T48402)
10 mL 2-methyl-2-butanol (*tert*-amyl alcohol)	(Sigma #240486)
Alcohol prep pad, sterile	(Fisher #06-669-62)
Micro-dissecting forceps, curved	(Roboz #RS-5100)
Micro-dissecting forceps, Tip 0.10×0.06 mm	(Roboz #RS-4976)
Fine scissors	(Roboz #RS-6702)
Clamp	(Roboz #RS-7440)
Aspirator tube assemblies	(Sigma # A5177)
Reflex Autoclip	(Roboz #RS9260)
Wound clips	(Roboz #RS9962)
Calibrated glass capillary pipette, pulled over a flame	(VWR #53432921)
Tween-20	(Sigma #P1379)
M2 medium (with Hepes)	(Sigma #M17167)
35 mm tissue culture dishes	
4-0 gut suture attached to 3/8 circle C-13 needle	

B. Human ES cell culture reagents and tools

Suppliers and recipes for human ES cell culture media are found in **Table 1**.

Bunsen burner	(VWR #17928-027)
9 inch borosilicate glass Pasteur Pipettes, autoclaved	(Fisher #13-678-20D)
18 G 1.5 inch needle	(Becton Dickinson #305196)

Feeders: Mouse Embryonic Fibroblast (MEF) feeder cells from two different sources are used: MEFs prepared from E13 ICR embryos (strain CD-1 from Charles River

Table 1 Reagents and media preparation for human ES cell culture

Material/reagent	Catalogue no.	Final concentration
HUESM		
Knockout-SR	10828	20%
GlutaMAX	35050	2 mM
MEM non-essential amino acids	11140-050	0.1 mM
Penicillin–streptomycin	15140-122	100U/mL–0.1 mg/mL
2-Mercaptoethanol	21985-023	0.1 mM
B27 Supplement (without Vitamin A)	12587-010	1×
DMEM	11995-065	75%
H1-Medium		
Knockout-SR	10828	20%
L-Glutamine	25030-081	2 mM
MEM non-essential amino acids	11140-050	0.1 mM
Penicillin–streptomycin	15140-122	100U/mL–0.1 mg/mL
2-Mercaptoethanol	21985-023	0.110 mM
DMEM/F12	11330-032	77%
FM10, Feeder medium		
Fetal bovine serum, certified	16000-044	10%
GlutaMAX	35050	2 mM
Penicillin–streptomycin	15140-122	100U/mL–0.1 mg/mL
2-Mercaptoethanol	21985-023	0.1 mM
DMEM, high glucose	11995-065	88%
Dispase medium		
Dispase	17105-041	1 mg/mL
Growth medium		to volume
0.1% Gelatin		
Gelatin (Sigma)	G-1890	0.1%
Distilled water, tissue culture grade	15230	to volume
Phosphate-buffered saline	14190-144	1×
Matrigel (BD Biosciences)	354234	Coat with 0.33 mg/mL

Note: all media components are from Gibco unless indicated.

Laboratory) and inactivated using Mitomycin-C or gamma irradiation (See **Chapter 4** for feeder preparation and mitotic inactivation protocols) or commercially available Mitomycin-C inactivated MEFS of the CF-1 strain (Chemicon, #PMEF-P3). Primary MEFs are used between 1 and 5 passages. The feeders are frozen at 5×10^6 cells per vial and one vial is used for one 10 cm plate.

Cell lifter	(Costar #3008)
DNAse I, RNAse-free	(Roche #10776785001)
Cryovials	(Nunc, #377267)

C. Manipulators, optics, cooling devices and injection chambers, injection apparatus, and pipettes for blastocyst injections

Inverted microscope with Hoffman modulation optics and condensers.
Micromanipulation and injection apparatus suitable for microinjection of mouse ES cells into mouse embryos (such as Transfer-Man NK2 micromanipulation and microinjection systems from Eppendorf).
Injection pipettes, 25 micron ID (CustomTip Type IV, Eppendorf #930 00 122-8 or similar). Note: The correct inner diameter of the injection pipette is critical and is larger than that typically used for microinjection of mouse ES cells.
Holding pipettes (VacuTips, Eppendorf #930 00 101-5 or similar)
Media for blastocyst injections is listed in **Table 2**.

D. Reagents for aggregations

60 mm non-tissue culture treated dishes for embryo culture and manipulations
96-well cell culture plates with a conical bottom (Corning Costar #3894)
Media for aggregations is listed in **Table 2**.

Protocols

Mouse strains; generation of embryos and fosters mothers

If mouse–human hybrid embryos are to be transferred to foster mothers for further *in vivo* development, pseudo-pregnant foster mothers should be produced by mating to

Table 2   Aggregation reagents

Material/reagent	Vendor	Catalogue no.	Final concentration
Embryo culture medium			
G2 version 3	Vitrolife	10092	to volume
HSA-solution	Vitrolife	10064	6 mg/mL
Acid Tyrodes solution	Specialty Media	MR-004-D	
Oil for embryo culture	Irvine Scientific	9305	
XVIVO-10 medium			
MEM non-essential amino acids	Gibco	11140-050	0.1 mM
2-Mercaptoethanol	Gibco	21985-023	0.110 mM
XVIVO-10	BioWhittaker	04-380Q	to volume
Outgrowth medium:			
Fetal bovine serum, certified	Gibco	16000-044	15%
Penicillin–streptomycin	Gibco	15140-122	100U/mL–0.1 mg/mL
MEM non-essential amino acids	Gibco	11140-050	0.1 mM
DMEM, high glucose	Gibco	11995-065	83%

vasectomized males. This mating should occur one day after the mating to produce the embryo donor. Blastocysts at E3.5, whether produced by blastocyst injection or aggregation at E2.5 followed by *in vitro* culture to E3.5, are transferred to foster mothers at the equivalent of E2.5. A diagram of the necessary timeline is shown in **Figure 1**.

Timed matings to produce host embryos and pseudo-pregnant foster mothers

1. Day 1: Embryo donors are intraperitoneally (IP) injected with 5 IU of PMSG using a 1 mL syringe with a 26G needle. PMSG is diluted in 0.9% NaCl. In one experiment, ten embryo donors and ten stud males are used.

2. Day 3: 42–48 hr after administration of PMSG, the same females are IP injected with 5 IU of hCG and mated to singly housed males of proven fertility.

3. Day 4:

   a. Copulation plugs are checked the in the morning. The copulation plug is a coagulation of seminal fluid and protein and is visible in the vagina of the female. Place all of the females into a cage and mark the day the animals were plugged. This day is considered 0.5 days post-coitum (E0.5).

   b. Females are checked for natural oestrus and are mated to singly housed vasectomized males. In order to ensure 10 pseudo-pregnant females, it will be necessary to examine approximately 100 animals for oestrus. An oestrus female may be identified by the degree of swelling, pink color and moistness in the vaginal area. Vasectomized males can be conveniently purchased or generated by surgical procedures described elsewhere (Bradley, 1987).

4. Day 5:

   a. Check for the presence of a copulation plug in the pairings with vasectomized males.

   b. Place the females who have successfully mated, i.e. a copulation plug is clearly visible in a separate cage. Indicate on the cage card the date of the plug.

B. Collection and culture of blastocysts and pre-compacting morulae

Collection of E2.5 eight-cell stage host embryos

1. Set up microdrop cultures as follows:

   a. Microdrop cultures are set up by arranging several 20 µL drops of G2v3 culture medium in 60 mm dishes and quickly but gently overlaying with embryo culture grade mineral oil.

b.   The dishes are equilibrated in a tissue culture incubator at 5% CO_2 for at least 30 min before use.

2.   Humanely sacrifice a superovulated embryo donor mouse in the morning of E2 and lay it on its back. Rinse with 70% ethanol. Pinch the skin with micro-dissecting curved forceps and make a lateral incision in the midline area of the abdomen using fine scissors.

3.   Holding firmly above and below the incision, pull firmly in opposite directions until the abdominal area is fully exposed.

4.   To expose the reproductive tract, cut open the peritoneum to reveal the contents of the abdominal cavity. Push the intestines to one side to reveal the U-shaped uterine horns with the ovary and oviduct at the top of each horn.

5.   Cut away as a single unit the fat pad (above the ovary and attached to the kidney), ovary, and oviduct and just below the horn of the uterus and place into M2 medium.

6.   In 3 mL of M2 and 2 mL of PBST (0.05% Tween-20 in PBS) cut away the fat pad, ovary and most of the oviduct using micro-dissecting scissors and micro-dissecting forceps.

7.   Using a pair of micro-dissecting forceps, hold the uterus with one of the pair and extrude the content therein in the direction of the oviduct with a second of the pair.

8.   Collect and move the pre-compacted eight-cell stage embryos into microdrops using aspirator tube assembly and a pulled glass capillary pipette. Pulled Pasteur pipettes are used to move embryos between microdrops and are made by heating and pulling a calibrated glass capillary pipette over a flame of a micro-burner (see protocol for preparing glass tools).

9.   The embryos can be used at this point for preparing aggregations with human ES cells or cultured overnight for use in blastocyst injections.

Procedure for collection of E3.5 early blastocyst host embryos

1.   Set up microdrop cultures of G2v3 medium as in the procedure for collection of E2.5 embryos.

2.   To expose the reproductive tract, follow steps 2 through 4 in the procedure for collection of E2.5 embryos above.

3.   Cut the horn of the uterus just above the cervix and above the fat pad above the ovary. Release the uterus by cutting through the mesenteries and place on a piece of absorbent paper towel. Grasp the uterus and cut away the mesenteries and blood vessels.

4. Move to a dish with M2 medium.

5. Insert a 25G needle and syringe loaded with 3 mL of M2 and 2 mL of PBST into the oviduct end of the uterus and hold the uterus on the needle with the curved micro-dissecting forceps.

6. Flush the blastocysts through the uterus towards the cervix end with about 1 mL of media.

7. Collect and move the early blastocyst stage embryos into microdrops of G2v3 medium using aspirator tube assembly and a pulled glass capillary pipette.

8. The embryos can be incubated at 5% CO_2 until used for microinjection of human ES cells.

Preparation of human ES cells for blastocyst injection

Cells that have been maintained on MEF feeders or on Matrigel in MEF conditioned medium have been used for these protocols. We have included protocols for the passaging and harvesting methods that we used to produce cells for blastocyst injections and morula aggregations. Generally, cells grown on MEF feeders have been used for blastocyst injections and cells grown on Matrigel have been used for the morula aggregations. However, we have had success generating chimeras with both passaging schemes in the protocols. We have successfully used these protocols with human ES lines derived in house i.e. RUES1 and RUES2.

1. Prepare microdrop cultures as for embryos except use 50 µL drops HUESM supplemented with 20U/mL of DNAse I.

2. Prepare MEF feeder plates if passaging human ES cells. Prior to passaging the cells by microdissection, medium is changed in the well to be passaged and on a fresh feeder plate. Protocols for generating and mitotically inactivating feeders are described in **Chapter 4**. We have also used a commercial source of Mitomycin-C inactivated MEFS.

3. Prepare media: We have used both H1 medium and HESM successfully (see **Table 1**). To prepare complete growth medium, bFGF is added just before feeding. To prepare complete growth medium, bFGF is added just before feeding (see table below). Stocks of bFGF are made to 100 µg/mL in sterile 10 mM Tris-HCl pH7.6 with 0.1% BSA. Aliquots of a convenient volume are frozen at −20°C. For maintenance on MEF feeder layers, bFGF is supplemented to 40 ng/mL before feeding the human ES cells.

4. Prepare glass tools for microdissection: For glass tools, Pasteur pipettes are pulled into hair-thin hooks. The hooked end of the glass tool should be thin enough for the microdissection of the human ES cell colonies (such dissection is also described at length in **Chapter 5**) but thick enough to withstand some pressure during the

dissection. Fine glass needles with hooked ends are forged in two steps over a microburner that can produce a very small flame.

a. Establish a small candle-size flame with a micro-burner made from a Bunsen burner fitted with an 18G needle.

b. For glass hooks, hold a Pasteur pipette at both ends and melt the glass approximately one inch below the taper until the lumen is fused and the glass glows orange. (For pulled pipettes used in embryo manipulations, melt the glass, but do not fuse the lumen.)

c. Quickly remove the glass from the flame while simultaneously pulling the ends away from each other. The glass should be drawn out to a filament without breaking. (For pulled pipettes used in embryo manipulations, stop here and break the filament with an inner diameter slightly larger than the embryos.)

d. Again, while holding a Pasteur pipette at both ends several inches above the flame, slowly lower the drawn filament approximately one inch from the new taper towards the flame while gently pulling at each end. Before reaching the flame, the filament should melt in two while being drawn into a very fine filament.

e. While holding the large end at approximately a 90° angle to and several inches above the flame, lower the filament tip towards the flame. The rising heat should curl the filament tip up forming a "hook".

f. Using a no. 5 watchmakers forceps, trim the end of the filament to finish the end into a clean "hook".

5. Preparing human ES cells for injection or passaging: Ideal colonies or parts of colonies are micro-dissected into clumps of cells using the glass hooks. The hook is used to gently pull apart pieces of the colony. This can also be accomplished by cutting a grid into the colony with the back of the hook and pulling the pieces away from the colony one at a time. For passaging, a chunk size of 100–200 cells is optimal. For injection, move the clumps to a fresh microdrop of media. The clumps are further dissected into clumps of 10–15 cells using glass needles (see **Figure 2F**). The back edge of the glass hook can be used to cut larger chunks into smaller pieces. Movies demonstrating this technique can be found on the Brivanlou lab web site (http://xenopus.rockefeller.edu).

6. For passaging, transfer clumps to fresh feeders: For routine passaging after microdissection, the cell clumps are swirled into the center of the dish and 20 to 50 clumps are transferred to the new feeder wells using P1000 micropipettes. Pre-coat the micropipette tip with the medium so that the cells do not stick. If possible, leave the dishes untouched on a warmed surface (preferably under O_2/CO_2 blood-gas mix) for 15–30 min to allow the clumps to begin attaching to the dish before moving to an incubator. Excessive handling of the new dish will cause the clumps to migrate to the center of the dish rather than remaining evenly distributed across the dish. Complete growth medium is exchanged on the growing

colonies every day. The lines should culture for no more than 6 days to a week. The timing of passage is dependent upon the appearance of differentiation within the colonies — mainly from the center of the colony.

Enzymatic passaging and harvesting of human ES cells for morula aggregation

Human embryonic stem cells can also be grown in feeder-conditioned medium on a substrate of extracellular matrix (ECM). We routinely use Matrigel as a growth substrate for growing RUES1 human ES cells in preparation for aggregation experiments. Here we describe the methods used to prepare these human ES cells for aggregation and the passaging conditions used to maintain them. This protocol can also be used to prepare human ES cells for blastocyst injections. We caution that alternative enzymatic passaging protocols may not yield successful embryo aggregates.

Before passaging, examine the colonies under the microscope and look for any colonies that are differentiated. Spontaneously differentiating areas of the culture can be removed with a glass tool as described in the manual dissection protocol or aspirated using a pipette attached to a vacuum. Several types of differentiation can be morphologically identified in spontaneously differentiating cultures. Look for the center of colonies that show a depression or "crater" appearance. Areas of colonies that have begun forming cystic structures in the center of the colony should also be removed. Also avoid the edges of colonies that do not have a tight border between the feeder layer and the colony. Areas where the human ES cells have started to flatten and polarize can also be removed. However, some differentiation on the borders of the colonies can be tolerated, as these cells will detach during the washing steps. Ideal colonies comprise small, round cells with a high nuclear to cytoplasmic ratio that are randomly organized and have not begun forming structures within the colony.

1. Prepare Matrigel coated plates for passaging and maintenance of cell lines. Matrigel-coated plates can be prepared in a slightly different manner to that described in **Chapter 5**:

   Matrigel is diluted to 0.333 mg/mL in cold XVIVO-10 medium (**Table 2**). All plastics that come into contact with the undiluted Matrigel should be kept as cold as possible. We pre-cool six-well plates, dishes and P1000 filter-tips at — 20°C for about 20 min. Keep the plates on the ice cold platform at all times. Place the entire ice bucket with plates into the refrigerator to coat overnight. Before plating cells, check the coating on the microscope for a meshwork-like single layer matrix. When ready to plate human ES cells, aspirate the Matrigel from the wells using a Pasteur pipette in the corner of the well. Get as much Matrigel off of the dish as possible leaving a thin coating on the surface of the dish. Do not scrape the bottom of the dish. Rinsing the Matrigel-coated plate is not necessary.

2. Prepare conditioned medium for passaging and maintenance of human ES cells: MEFs for feeder plates can be prepared and inactivated as described in **Chapter 4**.

We have also successfully used a commercial source of Mitomycin-C inactivated MEFS. Approximately 5 million cells from frozen vials or 3 million cells from freshly irradiated MEFs are plated on gelatin coated 10 cm plates and allowed to attach overnight in FM10 media (see **Table 1**). HESM growth medium is exchanged and conditioned for 24 hr before use. For maintenance on Matrigel, bFGF is supplemented to 20 ng/mL before conditioning on MEFs. The feeder conditioned medium (CM) can be used immediately, stored at 4°C for a week or frozen at −80°C. The conditioned medium is further supplemented with 40 ng/mL before feeding human ES on Matrigel. Passaged human ES cells are plated in 2 mL of CM per well of a six-well plate. Feeder plates can be used for up to 14 days to generate conditioned medium.

3. Passaging human ES cells:

    a. Replace the growth medium with Dispase or Collagenase (type V) at 1 mg/mL dissolved in growth medium and sterile filtered.

    b. Incubate for about 4–5 min in a tissue culture incubator. Check the progress of the matrix digestion, beginning at about 4 min. The colony borders will begin to peel away from the plate, while the center will remain attached (see **Figure 2A, B**).

    c. Ideally, gently wash the dispase (or collagenase) off of the plate with growth medium twice. The colonies should remain attached to the plate. If they have detached after the dispase incubation transfer all of the colonies and Dispase solution to a conical tube and centrifuge at 150 g (~1,000 rpm) for 4 min and wash the colonies with growth medium. They should get two to three washes total — either on the plate or with centrifugation.

    d. If the colonies remained attached after washing, harvest the colonies with a cell lifter (see **Figure 2C, D**).

    e. Transfer all of the colonies and growth medium to a conical tube and spin to pellet the colonies at 150 g/~500 rpm for 4 min.

    f. Using the CM, resuspend the colonies using a P1000 pipette tip in about 500–700 μL of medium.

    g. Triturate the colonies to clumps with an average size of about 100 cells using the P1000 tip (see **Figure 2E**).

    h. Plate a portion of the clumps at a 1:10 split ratio. However, the ratio will need to be optimized for the confluence of the starting population. If possible, leave the dishes untouched on a warmed surface (preferably under O_2/CO_2 blood–gas mix) for 10 min to allow the clumps to begin attaching to the dish before moving to an incubator. Excessive handling of the new dish will cause the clumps to migrate to the center of the dish rather than remaining evenly distributed across the dish. Good spacing between the colonies will allow proper growth of the colonies.

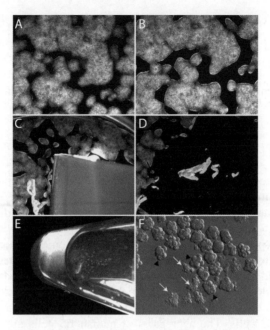

Figure 2 Enzymatic harvesting of human ES cells for passage and aggregation or injection. (**A**) The colonies before treatment. (**B**) The colonies during enzymatic treatment to lift from the matrix. Note the curled edges of the colonies. (**C**) Detaching the colonies with a cell lifter. (**D**) Detached colonies. (**E**) Dissociated colonies in a 15 mL conical tube ready for replating. (**F**) Dissociated colonies of an appropriate size for aggregation next to pre-compaction stage mouse embryos. Black arrowheads point to the mouse embryos. White arrows point to the human ES cell clumps.

4. Maintenance: Complete conditioned growth medium is exchanged on the growing colonies everyday from the MEF plates. Feeding can begin on the second day after passaging. When the colonies get bigger, increase the CM. Cultures in CM on Matrigel can usually grow for 4–5 days before they need passaging. The timing of passage is dependent upon the appearance of differentiation within the colonies — mainly from the center and edges of the colony.

5. Prepare cells for aggregation.

   a. Instead of disaggregating human ES cell colonies into ~100 cell clumps as for passaging, we further dissociate these to form clusters of ~10–15 cells by more vigorous triturating with a P1000 pipette. Alternatively, colonies can be harvested before the dissociation step and microdissected with glass needles as described above for preparing cells for blastocyst injection.

b. Pellet the clumps by centrifugation (150 g/~500 rpm for 4 min).

c. Resuspend clumps in XVIVO-10 medium (see **Table 2**). The clumps should be at a density of about 100 clumps per microliter.

d. Proceed immediately to setting up aggregation plates as described below.

Blastocyst injection

The techniques used for microinjection of human ES cells into mouse blastocysts are based on common protocols described for injecting mouse ES cells into mouse blastocysts (for example, see Bradley, 1987), with some notable differences. Current protocols for injection of murine ES cells into blastocysts begin with murine ES cells in a single cell suspension. Most human ES cell lines, on the other hand, do not tolerate dissociation to single cells. They must instead be manipulated as aggregates of about ten cells and micromanipulation strategies must accommodate this limitation. The technique we describe here uses the simplified micromanipulation set-up currently used for mouse ES cell injection, but uses larger diameter injection needles than those used for mouse ES cells. The aggregates of human ES cells are gently drawn into the injection pipette. An embryo is aligned on a holding pipette with the embryonic pole oriented towards the bottom of the field. The sharp tip of the pipette is used to quickly, but precisely, puncture through the zona pellucida and trophoblast layers at the level of the equator and into the blastocoel cavity. The human ES cells are then slowly expelled into the cavity and the tip withdrawn. The larger internal diameter of the injection pipette reduces, but does not eliminate, physical stress on the human ES cells within the aggregates, while remaining small enough to inject the expanded mouse blastocyst without severe damage. Using these methods, we have seen an average incorporation rate of 14% when outgrowths are analyzed after 6 days of outgrowth (James *et al.*, 2006). Though our experience with *in vivo* development of the hybrid embryos is limited, in one *in vivo* experiment, three of 28 implanted embryos contained human ES cells at E8.5. However, only one of these embryos demonstrated normal development. At this time, the efficiency of incorporation is low and much work remains to optimize these protocols. It may also be possible to use other manipulation methods designed for the transfer of cell aggregates (Gardner, 1971; Gardner and Johnson, 1975).

Setting up the injection plates and pipettes

1. Prepare plates for injection with embryos in a microdrop under oil in injection medium. The microdrop plate should have two drops in the center of the dish. The top drop should have HUES medium to hold human ES cell clumps and the bottom drop should have G2v3 medium (see **Table 2**) to hold the embryos. Also prepare additional microdrop dishes with G2v3 medium.

2. Place the plate on the microscope stage.

3.   Prepare micro-injection pipettes, being sure to purge air bubbles.

4.   Position the pipettes in the injection plate with the coarse adjustment controls.

5.   Focus on a group of embryos in a microdrop.

6.   Without adjusting the focus, bring the ends of the pipettes into the focal plane with the coarse adjustment controls.

7.   Adjust the oil–medium interface in the injection and holding pipettes to a point near the end of the pipettes.

Injecting the blastocysts with human ES cells

1.   Pipette the human ES cell aggregates into the reserved microdrop in the injection plate.

2.   Move the field to the human ES cell microdrop.

3.   Aspirate a clump by carefully drawing the clump into the tip of the pipette.

4.   Move the field to the microdrop containing embryos.

5.   Position a blastocyst on the holding pipette buy expelling medium from the holding pipette to rotate the embryo into position.

6.   Position the equator of the embryo in the focal plane.

7.   Move the injection pipette with the human ES cell clump at the tip into the focal plane and with the sharp tip in line with the equator of the embryo.

8.   Penetrate the zona pellucida and trophectoderm with the injection pipette in a single quick stabbing motion towards the holding pipette. This movement takes practice in order to prevent penetrating too far into the opposite trophectoderm and zona pellucida layers nearest the holding pipette. However, too slow of a motion will collapse the blastocoel and prevent full penetration into the cavity.

9.   Once the injection pipette tip is in the blastocoel cavity, gently begin to expel the cell clump. Just as the clump leaves the pipette tip, pull the pipette out of the cavity with a quick motion opposite that performed in the previous step. With practice, this can be done without pulling the clump out of the cavity with the tip. As the sharp point of the injection pipette dulls or becomes clogged with human ES cell debris, the tip will begin to hold onto the human ES cell clumps. The injection pipette must be replaced.

10. Release the embryo from the holding pipette, segregating the injected blastocysts from the uninjected or failed injected embryos.

11. Continue procedure on remaining blastocysts, stopping every 30 min to remove successfully injected blastocysts for incubation in a microdrop of G2v3 medium in a separate plate in a CO_2 incubator.

12. Embryos are transferred to day 2.5 pseudo-pregnant foster mothers within 4 hr of injection as detailed in the surgical procedures.

H. Morula aggregation

The protocol for aggregation of human ES cells with mouse pre-compaction stage embryos is based on those described for aggregation of the embryos with mouse ES cells (for example, see Bradley, 1987; for example, see Nagy and Rossant, 1999). Mouse embryos prior to compaction in the early afternoon at E2.5 will aggregate and incorporate mouse ES cells with high efficiency if the zona pellucida is removed and the ES cells placed in close contact with the compacting embryo. Generally, mouse ES cells are dissociated to clumps of two to four cells and gently maneuvered into small depressions with compacting mouse embryos. These are incubated *in vitro* until the blastocyst stage before transfer to suitable fosters. This provides the advantage of pre-screening the embryos for incorporation if the ES cells are fluorescently labeled. Human ES cells will also aggregate with mouse embryos if healthy cells can make contact with the murine blastomeres. Due to the low survival of human ES cells after trypsinization, human ES cells are disaggregated by pipetting or microdissection. However, the damaged cells on the edges of the resulting cell clumps can prevent the aggregation by limiting access of the healthy cells to the compacting blastomeres of the embryo. To overcome this limitation, we use centrifugation in conical bottom 96-well plates to bring the cells into contact with the embryos. Incorporation rates using this protocol average 38% and range from 5% to 50%. Mouse ES cells aggregated in parallel show incorporation rates of 95–100%.

Removal of the zona pellucida from eight-cell mouse embryos

1. Set up 60 mm plates with microdrops of Acid Tyrode's solution and G2v3 medium (see **Table 2**).

2. Select eight-cell mouse embryos and place them in one of the microdrops of G2v3 medium.

3. Using a mouth pipette filled with Acid Tyrodes solution transfer 10–20 embryos into a drop of Acid Tyrodes under oil. Rinse by gently pipetting the embryos in and out of the microdrop.

4. Move the embryos to a fresh drop and watch for the zona pellucida to dissolve and disappear.

5. While the embryos incubate, quickly dispel the Acid Tyrodes from the pipette and refill with G2v3 medium.

6. Just as the zona pellucida disappears, move the embryos to a drop of G2v3 medium.

7. Rinse the embryos and transfer to a new drop.

8. Continue processing the remaining embryos.

9. Let the embryos recover for 1 hr in the incubator in G2v3 medium.

Setting up the aggregation plates for morula aggregations

1. While the embryos are recovering from the Acid Tyrodes treatment, harvest the human ES cells (see above) for aggregation.

2. Human ES cells dissaggregated to small clumps of 5–10 cells each are suspended in gassed XVIVO-10 medium (see **Table 2**) at a density of about 100 clumps per microliter.

3. 5 µL of human ES cells are dispensed into the bottom of a 96-well conical bottom tissue culture plate. 12 wells are processed at a time to prevent evaporation.

4. Overlay the 5 µL drop with 100 µL of light mineral oil.

5. Incubate the assembled plate in a tissue culture incubator at 5% CO_2 for 20 min to equilibrate the pH of the medium.

6. Retrieve the embryos from the incubator and transfer a single embryo into each well of the 96-well plate. The embryos are transferred in as small a volume as possible as the reduced volume improves the development of the embryo.

7. Briefly centrifuge the plate at 300 g/∼1,000 rpm for 1 min to bring the embryos in contact with the human ES cells.

8. Gently examine the plate under a stereomicroscope to ensure that the embryos have pelleted with the human ES cells.

9. Return the plate to the incubator and leave undisturbed overnight. The embryos will have progressed to early blastocyst stage by mid-morning of the following day.

10. Embryos can be briefly examined for incorporation of human ES cell if GFP labeled human ES cell lines are used in the aggregation. Gently wash the unaggregated human ES cells and debris from the embryos by pipetting and releasing them a few times in the well with a pulled Pasteur pipette that has an inner diameter slightly larger than the blastocyst. Transfer the embryos to fresh microdrop of G2v3 medium under oil and examine under epifluoresence. Minimize exposure to UV to 30 s per embryo.

11. The manipulated embryos can be further cultured as outgrowths (see outgrowth protocol below) or transferred to foster mothers (see transfer protocol below) in the afternoon of E3.5 for further development *in vivo*.

I. *In vitro* culture and outgrowth of hybrid embryos

Mouse embryo–human ES cell hybrids can be further cultured *in vitro* to assay for early germ layer formation and differentiation capability of the human ES cells (Sherman, 1975). This can be done by plating the embryos from either the blastocyst injection procedure or aggregation procedure on adhesive substrates in serum containing medium that allows for attachment and outgrowth. Though further germ layer differentiation of the mouse embryos is disorganized, this assay may be used to follow early cell fate choices.

In vitro outgrowth culture of hybrid embryos

1. Coat four-well plates with Matrigel as described in the enzymatic passaging protocol above.

2. Aspirate the Matrigel and add 0.5 mL of Outgrowth medium (see **Table 2**) to each well.

3. Gently transfer the embryos individually to wells with a pulled transfer pipette.

4. Culture the embryos for 6 days at 5% CO_2.

5. Check wells for outgrowth of the embryos.

6. Harvest the embryos by gently scraping the outgrowths from the matrix using glass hooks (see preparation of glass tools above).

7. Fix outgrowths in filtered 4% paraformaldehyde in PBS for immunofluoresence staining in suspension. The embryos can be processed using established protocols for immunofluorescent analysis of mouse embryos.

Surgical procedures: anesthesia, transfer of manipulated embryos

For *in vivo* analysis, the manipulated embryos are transferred at the blastocyst stage to pseudo-pregnant foster mothers to promote further organized embryonic development. This is accomplished by surgical procedures that allow access to the uterus, transfer of the manipulated embryos to the uterus, closing the wounds, and recovery of the foster mothers.

Embryo transfer to the foster mice

1. Prepare 2.5 *post coitum* pseudo-pregnant females for surgery. Administer 2.5% Avertin, 125–250 mg/kg of body weight (~0.5 mL/adult female mouse) and wait several minutes for the anesthetic to take effect. Pinch the end of the tail of the

animal to determine if it is fully anesthetized. If there is no response then proceed. Avertin is made as follows:

  a. To make a 100% stock solution, dissolve 10 g of 2,2,2-tribromoethanol in 10 mL of 2-methyl-2-butanol (*tert*-amyl alcohol) at 37°C. Store at 4°C in the dark.

  b. To make a 2.5% stock, add 0.625 mL of 100% Avertin in 10 mL of water drop-wise while vortexing. Add 10 mL of water to this and add another 0.625 mL dropwise while vortexing. Bring volume to 50 mL with water. The solution may need to be warmed to 37°C to dissolve. Store at 4°C in the dark.

2. Retrieve the embryos using the aspirator tube assembly-pulled glass capillary pipette from the micro-well by aspirating columns of media and air in the following order: light mineral oil, air, M2 medium, air, embryos, air and medium. The length from the tip of the pipette until the first air bubble is approximately 3 cm. We transfer 8–10 embryos per uterine horn. It is most convenient to load enough embryos for one transfer per pipette.

3. Surgical procedure to expose the uterus

  a. Lay the animal on the stage of the dissecting microscope.

  b. Wipe with an alcohol prep pad.

  c. Pinch the skin with micro-dissecting curved forceps and make a lateral incision with fine scissors of approximately 1 cm long just below the last rib.

  d. Wipe with an alcohol prep pad.

  e. Locate the ovary and cut the peritoneum.

  f. Retract the fat pad (attached to the kidney) with a clamp and lay it on the body.

4. Transfer of embryos

  a. Gently hold the uterus with curved forceps and using a 25G needle make a hole through the uterine wall into the lumen in the area of the horn of the uterus, making sure to avoid the blood vessels.

  b. Remove the needle and insert the capillary pipette containing the embryos. This should be done without loosing sight of the opening made with the needle.

  c. Blow gently to release the embryos into the lumen and stop before the third air bubble enters the lumen.

5. Closing the wound

  a. Remove the capillary pipette and remove the clamp.

b.   Return the uterus into the peritoneal cavity.

c.   Clip the wound closed with Reflex Autoclip.

6.   Recovery: allow mice to recover from anesthesia in a warm cage or on a heating pad. They should recover within 1 hr.

Husbandry precautions for human–mouse hybrids generated in vivo

We have not attempted term delivery of the hybrids at this point and have not found contribution of the human ES cells to germ cell populations. However, considering the small number of mice we have used for *in vivo* transfers, it remains possible for human ES cells to contribute germ cells in the hybrid mice. To prevent inadvertent mating of the hybrids and possible fertilization with human ES-derived gametes, the United States National Academies of Science guidelines (http://www.nap.edu/catalog/11278.html) recommend strict separation of male and female hybrid mice if embryos are brought to term.

Analysis of chimeras from in vitro and in vivo procedures

Embryos can be processed for immunofluoresence using procedures similar to those used for analysis of normal mouse embryos. Identifying incorporated human cells within the embryos is facilitated by a ubiquitous GFP label in the human ES cells. Immunofluoresence for a human nuclei-specific marker (Chemicon, #MAB1281) can also be used. However, in our experience, this can sometimes result in ambiguous identification of the human cells due to cross-reactivity to antigens in the extra-embryonic tissues of the mouse. An alternative is to use FISH for human chromosomes or chromosomal features. We have followed the manufacturer's protocols with success. Handling the small preimplantation or peri-implantation embryos in these procedures is most easily accomplished by transferring the embryos through solutions in four-well dishes using mouth pipettes. If BSA or another protein source is used in the solutions, sticking of the embryos to the mouth pipette is usually not a problem. Embryos from pre-implantation stages to E8.5 can be examined in whole mount by confocal imaging as live specimens or after fixation and staining by immunofluorescent techniques. Later stages can be fixed in 4% paraformaldehyde and sectioned with a vibratome for imaging. All solutions should be filtered to reduce staining artifacts and improve the quality of the confocal images.

Induction of teratoma in SCID mice

The ability for human ES cells to generate teratomas (Keller, 2005; Spagnoli and Brivanlou, 2006) in immuno-compromised mice is used as a diagnostic criteria for *bona fide* embryonic stem cells. In this *in vivo* assay, human ES cells are engrafted into immuno-compromised adult mice in various tissues to generate teratomas. The resulting tumors are routinely analyzed by histology for the various derivatives of the three primary germ layers. With the exception of the host vasculature within the tumor, teratomas are predominantly derived from the human ES cell graft (Gertow *et al.*, 2004). In the case of the vasculature, it was noted that both human graft-derived cells and host derived mouse cells can contribute to the vessel structures. Frequently, other differentiated and organized tissue can be found in the tumors. This can include, for example, neural tissue and retinal pigmented epithelium, muscle, cartilage, bone, and epithelial cells of the endoderm and ectoderm. However, many of these tissues may be immature and definitive identification of the mature tissue can be difficult. The assistance of a trained pathologist in evaluating the tissues is highly recommended.

Teratoma can be generated at various sites in adult SCID mice by subcutaneous, intraperitoneal or intramuscular injection, implantation under the kidney capsule or beneath the testis capsule (Pera *et al.*, 2003; Przyborski, 2005). As the site of implantation may also influence the growth and differentiation of the teratoma (Przyborski, 2005; Cooke *et al.*, 2006), it is recommended that several sites be tested to access the developmental potential of the human ES cells. Two protocols for implantation of human ES cells into immuno-compromised mice are provided below. The protocols for subcutaneous, intraperitoneal, and intramuscular injection are similar and have the advantage of being technically simple to perform and do not require surgical manipulation of the mice. The final protocol for implantation under the kidney capsule has the advantage of allowing more control over the location of implantation and better retention, vascularization, and tracking of the grafted cells. Alternatively, implantation of cells under the testis capsule is also commonly performed (Pera *et al.*, 2003). The surgical procedure is similar to that used in a vasectomy to locate the testis.

Subcutaneous, intraperitoneal, and intramuscular engraftment of human ES cells in SCID mice

1. Human ES cells are harvested as for passaging as described in the procedures for preparing cells for microinjection or aggregation with approximately 100–200 cells per clump.

2. The human ES cells are suspended in a small volume of media and loaded into syringes fitted with an 18G needle. Load the cells into the syringe before attaching the needle.

3. The suspension is injected at subcutaneous, intraperitoneal or intramuscular sites:

    a. For subcutaneous injection, target the needle beneath the skin on the rear flank.

    b. For intraperitoneal injections, target the abdomen.

    c. For intramuscular injection, penetrate the muscle of a single rear leg to minimize discomfort and alteration of the mobility of the mouse.

4. Monitor the mice and the site of injection weekly for 6–22 weeks. The mice should be weighed weekly and watched for signs of infection during the incubation period.

5. Teratomas can be recovered by dissection with surrounding tissue and usually arise between 6–8 weeks after grafting. They are fixed in formalin and sent for histological examination by a pathology service. Alternatively, they can be embedded for cryosectioning and processed for immunohistochemical detection of germ layer markers.

Implantation of human ES cells beneath the kidney capsule

1. SCID mice of 5–6 weeks of age are prepared for surgery by administration of Avertin as described above for anesthetizing foster mothers for embryo transfer (as described later in this chapter) and are monitored for full anesthetization.

2. While the mice are being anesthetized, human ES cells are harvested as for passaging except that the colonies are not dissociated.

3. 10–20 human ES cell colonies are suspended in a 50–100 µL of medium and loaded into pulled capillary pipettes outfitted with a mouth aspiration unit and similar to those used for transfer of embryos but with a slightly smaller inner diameter. Set the pipette aside attached to the mouth aspiration apparatus but within close reach.

4. When the mice are fully anesthetized, place an animal on its abdomen on sterile towels. Swab the back of the mouse with 70% ethanol from the base of the tail to approximately 4 cm up the spine.

5. Grasp the skin with blunt forceps and make a 2 cm incision with fine surgical scissors beginning approximately 2–3 cm from the base of the tail.

6. Separate the dermis from the body wall by blunt dissection with closed scissors.

7. With the animal placed on its side, the incision can now be used as a window to locate the dark red kidney through the body wall.

8. When the kidney is located, make a small incision in the body wall with fine blunt-tip scissors slightly shorter than the length of the kidney. Avoid major blood vessels or nerves.

9. With the thumb on the abdomen and forefinger on the opposite side of the kidney, apply gentle pressure beneath the kidney to expose and pop it out of the hole in the body cavity. The kidney should remain exteriorized if the incision in the body wall was not made too large.

10. Keep the capsule moist with PBS during the following procedures. It is also helpful to have the mouthpiece of the mouth aspiration unit in you mouth at this

point and all surgical tools in reach. The following steps are also be aided by use of the stereomicroscope. Cover the stage with sterile towels.

11. Use a pair of fine (#5) forceps to grasp the capsule and lift it from the parenchyma of the kidney.

12. While holding the capsule, use a 22G needle to make a small hole in the capsule wall. Be careful not to puncture the kidney.

13. While still holding the capsule with one hand, grasp the pulled pipette with the human ES cell suspension and insert the tip of the pipette into the hole made with the needle. Care is taken not to puncture the kidney to prevent bleeding.

14. Gently expel the human ES cell suspension into the space between the capsule and the parenchyma. Some of the media will leak from the hole. However, if the hole is not too large, the cell clumps will remain in the capsule.

15. Gently return the kidney to the body cavity by pulling the edges of the body wall up over the kidney.

16. One or two stitches are used to suture the body wall.

17. Gather the edges of the skin and clamp with surgical autoclips leaving no gaps in the wound.

18. Allow the mice to recover in a clean cage, preferably on a warm surface. A post-operative analgesic can also be administered.

19. Monitor the mice weekly for 6–22 weeks. The mice should be weighed weekly and watched for signs of infection during the incubation period.

20. Teratomas can be recovered by dissection of the kidney and usually arise between 6–8 weeks after grafting. They are fixed in formalin and sent for histological examination by a pathology service. Alternatively, they can be embedded for cryosectioning and processed for immunohistochemical detection of germ layer markers.

Troubleshooting

Why is XVIVO-10 medium used instead of G2v3 media in the aggregation protocol?

We have found that mouse embryos progress *in vitro* in a medium recently described to maintain human ES cells (Li *et al.*, 2005). We have used this medium as a compromise between support of embryo development and survival of the human ES cells. XVIVO-10 medium is an all-human sourced serum-free medium from BioWhittaker originally designed for human lymphocyte culture. It contains no added growth factors. It can be used for human ES cell culture by adding high concentrations of bFGF (80 ng/mL). Though, bFGF is not included during the aggregation. It does not need to be conditioned by MEFs to maintain the human ES cells. The drawback of this medium is that the formulation is proprietary. Most other human ES growth media tested allowed for incorporation of the human ES cells into the embryos. However, we prefer to resuspend human ES cell clumps in XVIVO-10 culture medium for aggregation as the mouse morulae progress to blastocyst stage with appropriate timing and with better efficiency. The osmolarity of this medium, measured to be approximately 290mOsm/kg, may be better suited to the culture of mouse embryos at this stage.

I see little incorporation of the human ES cells into blastocysts after morula aggregation? What happened?

It is helpful to include a control of murine ES cells aggregated in parallel to monitor efficiency and performance of the other components of the aggregation. However, the likely cause is damage to the human ES cells during dissociation, leaving dead cells attached to the outside of the cell clump. Including a small amount of DNAse I in the culture medium after disaggregation can also improve the aggregation.

My human ES cells are not generating teratomas with subcutaneous or intramuscular injection. How can I remedy this?

There are several problems that will lead to a lack of engraftment of the human ES cells that may have nothing to do with their potential. The major cause of failure is damage to the human ES cell during injection through small gage needles. If too much pressure is applied, the cells will lyse. Load the human ES cells into the syringe without the needle and use as large a needle as possible. The human ES cells can also be mixed one to one with undiluted Matrigel before injection. The strain of SCID mice may also make a difference in the success of engraftment (Przyborski, 2005). NOD-SCID mice are probably the best recipients, followed by the SCID-beige strain.

References

Bradley A. (1987). Production and analysis of chimaeric mice. In: *Teratocarcinomas and Embryonic Stem Cells: a Practical Approach*. EJ Robertson, ed. IRL Press, Oxford, pp 113–151.

Brivanlou AH, FH Gage, R Jaenisch, T Jessell, D Melton and J Rossant. (2003). Stem cells. Setting standards for human embryonic stem cells. *Science* **300**: 913–916.

Cooke MJ, M Stojkovic and SA Przyborski. (2006). Growth of teratomas derived from human pluripotent stem cells is influenced by the graft site. *Stem Cells Dev* **15**: 254–259.

Gardner RL. (1971). Manipulations on the blastocyst. In: *Advances in the Biosciences*, vol. 6, G. Raspe, ed. Pergamon, Oxford, pp 279–301.

Gardner RL and MH Johnson. (1975). Investigation of cellular interaction and deployment in the early mammalian embryo using interspecific chimaeras between the rat and mouse. *Ciba Found Symp* **0**: 183–200.

Gertow K, S Wolbank, B Rozell, R Sugars, M Andang, CL Parish, MP Imreh, M Wendel and L Ahrlund-Richter. (2004). Organized development from human embryonic stem cells after injection into immunodeficient mice. *Stem Cells Dev* **13**: 421–435.

Goldstein RS. (2006). Transplantation of human embryonic stem cells to the chick embryo. *Methods Mol Biol* **331**: 137–151.

Goldstein RS, M Drukker, BE Reubinoff and N Benvenisty. (2002). Integration and differentiation of human embryonic stem cells transplanted to the chick embryo. *Dev Dyn* **225**: 80–86.

James D, SA Noggle, T Swigut and AH Brivanlou. (2006). Contribution of human embryonic stem cells to mouse blastocysts. *Dev Biol* **295**: 90–102.

Keller G. (2005). Embryonic stem cell differentiation: emergence of a new era in biology and medicine. *Genes Dev* **19**: 1129–1155.

Li Y, S Powell, E Brunette, J Lebkowski and R Mandalam. (2005). Expansion of human embryonic stem cells in defined serum-free medium devoid of animal-derived products. *Biotechnol Bioeng* **91**: 688–698.

Muotri AR, K Nakashima, N Toni, VM Sandler and FH Gage. (2005). Development of functional human embryonic stem cell-derived neurons in mouse brain. *Proc Natl Acad Sci USA* **102**: 18644–18648.

Nagy A and J Rossant. (1999). Production and analysis of ES cell aggregation chimeras. In: *Gene Targeting: a Practical Approach.* A.L. Joyner, ed. Oxford University Press, Oxford, pp 177–206.

Pera MF, AA Filipczyk, SM Hawes and AL Laslett. (2003). Isolation, characterization, and differentiation of human embryonic stem cells. *Methods Enzymol* **365**: 429–446.

Przyborski SA (2005). Differentiation of human embryonic stem cells after transplantation in immune-deficient mice. *Stem Cells* **23**: 1242–1250.

Sherman MI. (1975). The culture of cells derived from mouse blastocysts. *Cell* **5**: 343–349.

Spagnoli FM and B AH. (2006). Guiding embryonic stem cells towards differentiation: lessons from molecular embryology. *Curr Opin Genet Different* **16**: 469–475.

Tabar V, G Panagiotakos, ED Greenberg, BK Chan, M Sadelain, PH Gutin and L Studer. (2005). Migration and differentiation of neural precursors derived from human embryonic stem cells in the rat brain. *Nat Biotechnol* **23**: 601–606.

10

In vitro differentiation of human embryonic stem cells

A. HENRY SATHANANTHAN[1] AND ALAN TROUNSON[1,2]

[1]*Monash Immunology and Stem Cell Laboratories, Monash University, Melbourne, Australia*
[2]*Australian Stem Cell Centre, Monash University, Melbourne, Australia*

Introduction

Embryoid body formation

The maintenance of undifferentiated embryonic stem (ES) cells is anchorage dependent and generally requires the presence of appropriate embryonic or adult feeder cells, their conditioned medium or even their lysates (Hoffman and Carpenter, 2005; Klimanskaya *et al.*, 2005; Trounson, 2006). Differentiation occurs spontaneously in human ES cells or dissected human ES cell colonies in suspension cultures forming spheroid bodies (embryoid bodies — EBs) with cells representing those of the primary germ lineages, and has been used as a method of producing differentiating progenitor cells of interest in developmental lineage analysis. The systems used to produce EBs include: culture in non-adherent bacterial-grade Petri dishes; in spinner flasks; in semi-solid methyl-cellulose medium; and as cell suspensions in hanging drops of medium on the lids of cell culture dishes (reviewed by Conley *et al.*, 2004a). These methods were designed for mouse EBs and have been further adapted for the culture of human ES cells. It was first demonstrated by Itskovitz-Eldor *et al.* (2000) that disaggregation of human ES cells using collagenase or gentle trypsinization, and subsequent culture in suspension, resulted in regional specific differentiation in the spheroid structures that resulted, which represent human ectoderm, mesoderm and endoderm lineages as determined by specific immunohistochemical markers. These spontaneous differentiation events recapitulate, to some extent, the developmental events in early embryogenesis. However, there are many disorganized events observed along with the regional specification, that include the random distribution of different cell types, presence of non-embryonic lineages such as extraembryonic endoderm, and a lack of integration of tissue types in the absence of memory for axis that characterizes the normal embryonic developmental program of patterning and differentiation. It is reported that hypoxic conditions

(3% and 5% O_2) favors EB formation in suspension culture (Ezashi *et al.*, 2005). It is also apparent that germ stem cells can also be identified in the EBs. Typically visceral endoderm derivatives form in the peripheral layer of human EBs grown for 7 days in suspension culture (Conley *et al.*, 2004a).

Production of "spinner EBs"

The "stirred vessel cultivation" method for deriving human ES cells has been described by Cameron *et al.* (2006). This method involves the treatment of whole or partial human ES cell colonies with dispase (5 mg/mL; Invitrogen) for 5 min to remove the feeder cells, and the culture of the unattached colonies in low adherent culture dishes (Corning Costar Corp., MA) in DMEM (Invitrogen) supplemented with 15% defined fetal bovine serum (FBS; Hyclone, Logan, UT), 1% (v/v) MEM non-essential amino acids (Invitrogen), 1 mM L-glutamine, and 0.1 mM β-mercaptoethanol (Sigma) for 24 hr.

This brief static culture enables EBs to begin to take shape and they are then removed from their culture dishes and allowed to settle in 50 mL conical tubes and the medium changed to remove debris and to aid detachment of EBs that had begun to agglomerate. EBs are diluted about 35 EBs/mL ($\sim$2–3 $\times$ 10^5 total cells/mL).

The EBs are then placed in 250 mL spinner flasks equipped with a 22 µm filtered gas port and a 2.5 cm sampling port. The spinner flasks contain a magnetic stir bar suspended 1 cm from the bottom of the vessel and maintained at 50 g ($\sim$80 rpm) throughout culture and cultured for up to 21 days, replacing the 100 mL working volume every 3–4 days. This method enables a 15-fold expansion in the number of EBs, compared with a fourfold expansion in static suspension culture and the method has been used to scale up hematopoietic differentiation (Cameron *et al.*, 2006).

Production of EBs using rotator flasks

The use of bioreactor cultivation of human EBs in rotating cell culture systems, with very low fluid shear stress, has been reported by Gerecht-Nir *et al.* (2004a). In this system, rotating bioreactors are used to control EB agglomeration.

Human ES cells are dispersed into small clumps (3–20 cells) using 0.5 mM EDTA supplemented with 1% FBS (HyClone). The medium used was KO-DMEM and 80% FSBd (HyClone), Penicillin/Streptomycin (Sigma), 1 mM L-glutamine, 0.1 mM beta-mercaptoethanol, and 1% non-essential amino acid stock (Gibco Invitrogen).

The human ES cells are seeded into the vessels (Gerecht-Nir *et al.*, 2004b) at initial concentrations of 0.1–1.2 $\times$ 10^6/mL. The bioreactors were set to rotate at a speed in which the suspended cell aggregates remained close to a stationary point within the reactor vessel. They achieved a threefold increase in the generation of human EBs over that obtained with static cultures using the slow turning lateral vessels, preventing agglomeration and favoring the production of small EBs that were able to proliferate extensively. The formation of EBs was not line dependent. We mainly used HUES6 and H9 lines.

Production of EBs within artificial scaffolds

It is also possible to derive human EBs in porous alginate scaffolds (Gerecht-Nir *et al.*, 2004b) with high guluronic acid content (Proanal LF 5/60 or LF120; FMC Biopolymers, Drammen, Norway). Small volume (10–20 μL) applied to the scaffolds and centrifuged (300 *g*/~1,000 rpm for 3 min) and addition of 200 μL of EB medium (Gerecht-Nir *et al.*, 2004a).

After 12–24 hr, scaffolds are placed in 12-well plates with 1 mL of fresh medium and cultured for 1 month with minimal agglomeration and EBs occupy the entire pore volume. The cell proliferation was twofold higher than that of static suspension cultures but was less than that obtained using bioreactors, but they induced vasculargenesis to a greater extent than either static or rotating cultures (Gerecht-Nir *et al.*, 2004b).

An innovative microwell culture system for human ES cells utilizing self-assembly monolayer (SAM) formation has been devised by Mohr *et al.* (2006). The microwells are created by photolithography and plasma etching techniques to form PDMS stamps which mold polymer substrates with protein resistant SAMs to inhibit protein adsorption and cell attachment. However, the surfaces of the wells do absorb extracellular matrix molecules that promote cell adhesion and proliferation. Patterned microwells are formed on glass microscope slides by attaching PDMS stamps and the microwells are coated with gold to allow SAM assembly. SAMs of alkanethiols spontaneously form on gold substrates and contain an 11–18 carbon chain capped with a functional group that regulates adsorption of proteins. The semitransparent microwells allow light microscopy.

These functionalized microwells are coated with Matrigel and transferred to non-cell culture six-well plates. Human ES cell colony fragments are pipetted onto the top of arrays and incubated at 37°C for 30 min to allow human ES cells to settle into the microwells in mouse embryonic fibroblast conditioned medium with bFGF. EBs were formed from human ES cell colonies removed from the microwells and cultured in suspension in T75 flasks for 1 week for differentiation. These human ES cell colonies can be formed of defined size with high viability and repeatability for stable EB forming characteristics.

The forced aggregation method of human EB formation using defined numbers of human ES cells described by Ng *et al.* (2005) is a useful method for enabling high throughput studies for determining the multivariate conditions and addition factors involved in optimizing directed differentiation of human ES cells. This technique is now widely used in our own laboratories for studies on ES cell differentiation for variable lineages and cell type-enhanced production systems. The method is described in detail in the methods section below.

Imaging of human EBs by transmission electron microscopy (TEM)

TEM has been an invaluable tool in the study of the fine structure of cells and tissues in biomedical research, including that of human gametes and embryos in assisted reproductive technology — ART (Sathananthan, 1996a, 2000; Sathananthan *et al.*, 1993). This section deals with basic techniques of TEM applicable to human EBs, as well as human ES cell colonies and neurospheres (NS). It is desirable to define the fine structure of embryonic stem cell lineages to supplement the images obtained by phase-contrast, confocal and fluorescent microscopy, in order to compare with the microstructure of cells and tissues widely documented in the literature (Fawcett, 1981; Sathananthan, 1996b). These images are well beyond the resolution of light microscopy (LM) and fluorescent microscopy (FM) and would help confirm their true identity whilst defining their internal morphology. The ultimate goal is to identify stem cells during their formation and differentiation (Sathananthan, 2003; Sathananthan and Trounson, 2005; Sathananthan *et al.*, 2002, 2003). Combined with FM, TEM should be the gold standard in the assessment of ES cell differentiation.

Although time-consuming and expensive, TEM is invaluable if the facilities and expertise are available in your histology laboratory or institution. The human EBs and NS are more solid, spherical bodies and need internal examination in sectioned material. If TEM is not possible, a great deal of information could still be obtained by examining, epoxy-resin sections (1 μm thick) by advanced digital LM. This requires

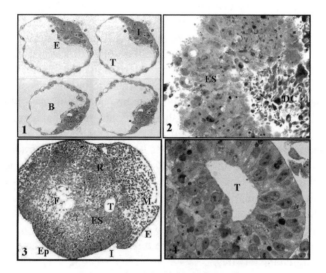

Figures 1–4 Light micrographs of Araldite sections (1 μm) stained with toluidine blue. (1) Hatched human blastocyst (serial sections). B = blastocoele, E = layer of primitive endoderm cells, I = ICM, T = trophoblast (original magnification: × 400) (Reproduced with permission from Sathananthan, 2003). (2) Colony of human ES cells (left) grown on mouse fibroblasts, fixed after 35 passages. Note spontaneously differentiating cells (right) located within the core of the colony (original magnification: × 1,000) (Reproduced with permission from Sathananthan, 2003, Sathananthan and Trounson, 2005). (3) Human EB showing spontaneously differentiating cells within its body. Ep = stratified epithelioid layer, ES = ES cells, F = fat cells, I = invagination of epithelium, M = mesenchyme, R = neural rosette (original magnification: × 100) (Reproduced with permission from Sathananthan, 2003). (4) Neural tube-like structure in EB. Note lumen and stratified epithelium reminiscent of embryonic nerve tube (original magnification: × 1,000) (Reproduced with permission from Sathananthan, 2003).

an ultramicrotome and glass knives. These sections show more structural details than routine paraffin or frozen sections (**Figures 1–4**). Another allied technique that can be used, if available, is scanning electron microscopy (SEM) which explores the surface microstructure of cells at high magnifications, particularly suitable for human ES cell colonies, isolated cells and cell aggregates in culture (Sathananthan and Nottola, 2007).

Our current research is directed toward the induced and spontaneous differentiation of ES cells in extended culture after several passages *in vitro*, particularly in human EBs. The human EB is a more solid structure and shows a more advanced stage of differentiation of cell types than ES cell colonies and appears to correspond to weeks 2 to 5 human embryos *in vivo*, both temporally and structurally. These include some basic cell types such as nerve, muscle, connective, epithelial, digestive, lung and vascular tissues, representing all three primary germ layers — ectoderm, mesoderm and endoderm, which can differentiate spontaneously (Sathananthan and Trounson, 2005). This has been confirmed by sectioning human EBs for TEM. The most common rudiments that appear spontaneously are stratified epithelia, nerve, mesenchyme, cardiac muscle and endoderm. However, there is considerable variation in the differentiation of tissue types in different human EBs, perhaps depending on culture times and the colonies from which they were derived.

Overview of protocols

A rapid method used successfully for human sperm, eggs and embryos is described below that is appropriate for human ES cells and human EBs (Sathananthan, 1996a, 2000; Sathananthan *et al.*, 1993, 2002), since stem cells are a logical progression of embryos produced from assisted reproductive technology (ART). Details of principles and procedures in TEM are given in basic textbooks of electron microscopy (Glauert, 1974; Hayat, 1989), which is beyond the scope of this chapter.

The steps involved in specimen preparation, and imaging are briefly as follows:

(a) Fixation

The main purpose of fixation is to preserve the structure of cells in a more or less life-like condition. Simple chemical fixatives such as buffered glutaraldehyde and osmium tetroxide, used routinely for adult cells, are employed. The cells must be alive and fresh and in the medium they were cultured in. The EB, NS or ES cell colony can be fixed whole at the end of the desired culture sequence or passage in glutaraldehyde, the primary fixative. They must be gently dislodged from the culture dish with a fine needle and fixed in glass or chemical-proof plastic vials. Plastic culture dishes with wells and Falcon bottles are not suitable for fixation. Cells growing in monolayers can be dislodged, gently centrifuged at 160 or 200 g (~600–800 rpm), pelleted in Eppendorf tubes and fixed, as is done for sperm cells (Sathananthan, 1996a, 2000). If the pellet breaks it can be recentrifuged at 600 g (~2,000 rpm) in glutaraldehyde, after fixation. In all cases the specimen must be fixed with minimal medium but never allowed to dry. Water should not be used for washing before fixation. Fixation is the most critical step in specimen preparation that will ensure preservation of the fine structure of cells.

(b) Processing and embedding

The specimen has to be dehydrated and embedded in a hard supporting medium such as an epoxy resin. Araldite or Epon resins give good results. The resin imparts a hard texture to the soft specimen required for thin sectioning. Dehydration involves the step-wise removal of water from the specimen to make it penetrable with the resin. Dehydration should be rapid to preserve fats and glycogen, if present. Ethyl alcohol and acetone are used in our laboratory and the specimen is infiltrated with the resin, then embedded in the resin and polymerized at 60°C to produce a hard block for sectioning.

(c) Sectioning

The block is trimmed and sectioned with an ultramicrotome in a transmission electron microscope or histology laboratory. Glass knives are used for thick sectioning and diamond knives are used for thin sectioning. Glass knives have to be made with a knife maker and diamond knives are available commercially and are very expensive. Beginners should use glass knives for thin sectioning, as well. Thick, survey sections (1 μm) are examined by LM, while thin sections (~70 nm) are examined by TEM. Thick sections are very useful to identify specific cells or tissues for examination by TEM. One of the limitations of TEM is that only a small region of the specimen could be examined in thin sections. Hence serial sectioning is recommended, alternating a series of thick sections with thin sections, until the desired cell or tissue is found.

(d) Staining

Thick sections are stained with toluidine or methylene blue on a hot plate. Sections are mounted on clean, glass slides, dried and stained. Electron stains like uranyl acetate and lead citrate are used routinely for thin sections. These stains contain heavy metals that enhance the electron density of the cells and make them more visible with TEM. The image is formed by scattering of electrons, which produces translucent (light) and dense (dark) areas. The cell membranes, granules and inclusions scatter electrons and appear dense, while the ground cytoplasm allows electrons to pass through and appears translucent producing a black and white image.

(e) Microscopy

Thick sections are examined with a light microscope, while advanced imaging is done with a research microscope with a digital camera, hooked to a computer for image processing and editing. We use Leica QWin, Nikon or Olympus digital microscopes. Alternatively, the sections can be photographed on film and printed or mounted on 35 mm slides. The resolution of photographic images is superior to that of the digital images and they could be scanned on to a computer for editing. Thin sections are examined by TEM. We use Jeol, Hitachi or Philips microscopes but others such as Zeiss are equally good. The transmission electron microscope has to be maintained, and operated by an experienced technician. Lower magnifications ($\times$ 2,000 to $\times$ 5,000) are more useful to image whole cells, while

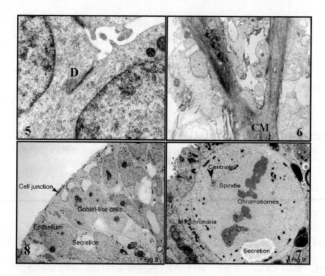

Figures 5–8 Electron micrographs of thin Araldite sections (~70 nm) stained with uranyl acetate and lead citrate (TEM). (5) Two undifferentiated human ES cells in an EB showing the typical structure. The cells have large nuclei with scanty cytoplasm. Note primitive desmosome (D) between the cells (original magnification: × 42,500) (Reproduced with permission from Sathananthan, 2003). (6) Cardiac muscle (CM) in a spontaneously differentiating ES cell colony Note the branching fiber and fine microfilaments within the cell (original magnification: × 12,750) (Reproduced with permission from Sathananthan, 2003). (7) Differentiating ES cell at metaphase of mitosis, showing chromosomes at its equator and a centriole at one spindle pole (original magnification: × 8,750) (Reproduced with permission from Sathananthan *et al.*, 2002). (8) ES cell colony that has spontaneously differentiated showing goblet-like cells with clear secretions. All cells are probably endodermal in origin (original magnification: × 3,500) (Reproduced with permission from Sathananthan *et al.*, 2002).

high magnifications (× 10,000 to × 100,000) help identify minute cell structures like ribosomes, centrosomes, filaments and membranes in cells (**Figures 5–8**). Higher contrast could be obtained with lower voltages of 60 or 80 kV and by using smaller apertures in the TEM. Images are photographed on plate film or 35 mm film or saved on computer.

(f) Developing and printing

This is done in a dark room using standard chemicals and paper. We now use negative digital scanners that can produce computerized images for editing. The images are edited, cropped or colored using the latest versions of Adobe Photoshop or Paint Shop Pro and presented on Microsoft Power Point. The latter is useful for labeling and annotating for presentations. Most images are saved on Tiff format and later converted to JPEG or GIF for online publication or transmission by email or the web (see www.sathembryoart.com for some images).

Materials, preparation of materials, reagents and equipment

Human ES Growth Medium:

1. 400 mL DMEM/F-12 (Gibco Cat # 11320−033)

2. 100 mL (20%) Knockout Serum Replacement (Gibco Cat # 10828−028)

3. 5 mL (0.1 mM, 1%) 10 mM non-essential amino acids (Gibco Cat # 11140−050)

4. 5 mL (1%) L-glutamine (Gibco Cat # 25030−081)

5. 1.1 mL (90 μM) 55 mM 2β-mercaptoethanol (Gibco Cat #21985−023)

Collagenase (4 mg/mL DMEM/F-12)

Pipettes — 10 mL

15 mL centrifuge tubes (conical-bottomed)

Bacteriological Petri dishes

Incubator set at 37°C with a gas phase of 5% CO_2, 5% O_2, 90% N_2

Materials

Equipment

96-well round bottom plates

15 mL sterile tubes

Serological pipettes

Syringe filter units 0.22 μm pore size

70% ethanol

Tissue culture dishes (100 mm)

Iscove's Modified Dulbecco's Medium (IMDM) with Glutamax I (1×)	(Gibco, 31980−030)
Ham's F12 with Glutamax I (1×)	(Gibco, 31762−035)
Chemically defined lipid concentrate (100×)	(Gibco, 11905−031)
Transferrin (10 mg)	(Sigma, T-5391)
α-Monothioglycerol (α -MTG) (12 M)	(Sigma, M-6145)
Insulin−transferring−sodium selenite (ITS-X) (100×)	(Gibco, 51500−056)
Glutamax (100×) (200 mM)	(Gibco, 35050−061)
Protein Free Hybridoma Medium (PFHM-II) (20×)	(Gibco, 12040−077)
Ascorbic acid-2-phosphate	(Sigma, A-8960)
BSA (embryo culture) (100 mg/mL)	(Sigma #A3311, 10%)
Bovine Serum Albumin (BSA)	(Sigma, A-3311)
Trypsin−EDTA	(Gibco, 25200−056)
Glutaraldehyde EM grade 100 mL	(Fluka BioChemika, 79626)
Osmium tetroxide (1 g)	(Fluka Chemika 75630)
Sodium cacodylate (100 g)	(Fluka BioChemika, 20838)
Ethyl alcohol (ethanol) absolute	
Acetone anhydrous	
Araldite Durcupan ACM A, B, C, D (100 mL)	(Sigma-Aldrich Fluka 44611-14)
Epon embedding medium kit	(Sigma-Aldrich Fluka 45359)
Toluidine blue (25 g)	(Sigma-Aldrich 198161)
Methylene blue (25 g)	(Sigma-Aldrich M9140)
Borax 25 g	(Sigma-Aldrich S9640)
Uranyl acetate dihydrate (AR)	(Merck, Germany K 18237473)
Lead nitrate (500 g)	(Sigma-Aldrich L6258)
Sodium citrate 500 g	(Sigma-Aldrich S1804)
NaOH, LiOH, HCl	
Fume cupboard, oven, hotplate, binocular, digital and electron microscopes Rotator	(TAAB Type 1 2 RPM Reading, Berks, UK)
Glass strips for knives	(Proscitech, Australia UM 6425 30 strips)
Glass knife boxes	(Proscitech, Australia U100)
Glass knife boats	(Proscitech, Australia UL072)
Glass knife maker	(Reichert, Leica Microsystems, Australia)
Boats for glass knives	(Proscitech, Australia)
Storage boxes for glass knives	(Proscitech, Australia U100)

Diamond knife (2.4 cm)	(Diatome, SA Bienne, Switzerland)
Ultramicrotome for sectioning	(Reichert Ultracut, Vienna, Austria)
Copper grids for thin sections	(Polaron 100–200 mesh hexagonal)
Grid storage boxes	(LKB-Produktur AB, Sweden 4828 B)
Forceps — stainless steel	(Inox 5 Dumont, Switzerland)
Filters (0.2 μm)	(Acrodisc, Gelman Sciences, USA)

(Consult your local TEM suppliers or search for these products on the internet)

(1) Formation of embryoid bodies by suspension culture

Materials

Healthy karyotypically euploid human ES cell colonies expressing high levels of pluripotential stem cell markers; GCTM-2, Tra-1-60, E-cadherin and Oct-4.

Methods

Suspension culture

1. Aspirate the medium from a stock culture of human ES cells. Add 2 mL collagenase per 25 cm^2 surface area and incubate at 37°C for 20–30 min. Human ES cell colonies should detach, leaving behind most feeders and many spontaneously differentiated cells that may be present.

2. Using a wide bore pipette, transfer detached colonies to a 15 mL centrifuge tube and spin down at 50 g(~100 rpm) for 3 min.

3. Aspirate the supernatant and gently resuspend human ES cell colonies in 10 mL fresh medium. Transfer to a sterile 10 cm bacteriological Petri dish (the surface of bacteriological Petri dishes do not typically facilitate he attachment of EBs). Use the equivalent of 3–5 × 10^6 cells per 10 cm dish (approximately one 25 cm^2 flask of healthy human ES cells.

4. Incubate under 5% CO_2, 5% O_2, 90% N_2 at 37°C, to allow human EBs to form.

5. Feed human EBs every 2–3 days, depending upon density and size. Change the medium by gently transferring the human EBs and the old medium to a clear conical-bottomed tube. If the human EBs stick to the dish, gently wash them off

with a pipette. Allow human EBs to settle for 5–10 min. Aspirate the medium and replace gently, taking care not to break up the EBs. Transfer to a fresh bacteriological Petri dish.

6. Culture human EBs at 37°C under 5% CO_2, 5% O_2, 90% N_2 for the desired number of days.

Hanging-drop suspension culture

1. Harvest and count undifferentiated human ES cells as described.

2. Place 10 mL of PBS in a 100 mm tissue culture dish.

3. Using a multichannel pipettor, make five rows of 30 μL drop (~1,000 cells per drop) on the upturned inner surface of the lid of the 100 mm tissue culture dish.

4. Lift lid and carefully invert it and place the lid on top of the dish containing 10 mL PBS.

5. Carefully place the lift the dish in the incubator at 37°C under 5% CO_2, 5% O_2, 90% N_2 for 7–10 days for embryoid body formation.

(2) Formation of "spin embryoid bodies" (Ng *et al.*, 2005) (Table 1)

Method

1. Ensure all human ES cells are karyotypically normal and express the pluripotential stem cell markers GCTM-2, Tra-1-60, E-cadherin and Oct-4. Expand human ES cells on 75–150 cm^2 tissue culture flasks prior to differentiation trials, cells should reach 60–80% confluency.

2. Passage ES cells using 0.25% trypsin–EDTA/2% chicken serum (Hunter) 1:1 or 1:2 and transfer onto flasks pre-seeded with feeder cells at low density (6.5 × 10^3/cm^2) on the day prior to differentiation.

Table 1 Preparation of chemically defined medium

Ingredient	Final volume 100 mL	Final concentration
IMDM (1×)—without phenol red	42.8 mL	42.8%
Ham's F12	42.8 mL	42.8%
Chemically defined lipid concentrate	1 mL	1%
α—MTG	0.004 mL	5 μM
ITS-X	1 mL	1%
PFHM-II	5 mL	5%
Ascorbic acid	1 mL	0.05 mg.mL
BSA	5 g	5 mg/mL

3. The following day, gently trypsinize the human ES cells into a single cells suspension, wash with phosphate buffered saline

4. Resuspend human ES cells in chemically-defined medium (40,000 cells/mL) supplemented with appropriate growth factors for the desired lineage.

5. Add 100 µL of human ES cell suspension (300–10,000 cells) to each well of 96-well round bottom, low attachment plates (NUNC).

6. Centrifuge plates at 500 *g* (∼1,000 rpm) at 4°C for 5 min to aggregate cells. Set centrifuge to low deceleration to prevent disaggregation.

7. Culture cells in the incubator for 10–12 days for EB formation.

8. Transfer to 96-well flat-bottomed tissue culture plates (NUNC) precoated with gelatin in medium supplemented with appropriate growth factors for enrichment of the desired cell type (see Ng *et al.*, 2005 for hematopoietic differentiation) for the time needed to optimize cell type desired.

(3) Sectioning and imaging for TEM

Note 1:

Glutaradehyde and the buffer are poisonous and hazardous. Cacodylate buffer is a carcinogen and skin and eye irritant. All chemicals including resins must be handled in a fume cupboard, preferably with gloves and goggles. Store fixatives at 4°C in refrigerator or in a cold room.

Note 2:

Osmium is very expensive, poisonous and hazardous. Avoid breathing, skin contact, metal contact and light. Causes eye damage and respiratory irritation. Use extreme caution.

Materials 1. Fixatives

A. Primary fixative: 3% glutaraldehyde in 0.1 M cacodylate buffer (pH 7.3):

    1. 110 mL 0.2 M cacodylate buffer

    2. 110 mL distilled water

    3. 30 mL glutaraldehyde 25%

    Keeps for 3–6 months at 4°C in the dark

    (Phosphate buffer may be substituted for cacodylate buffer.)

B. Post fixative: 1% osmium tetroxide in distilled water:

1. 1 g osmium tetroxide crystals sealed in a glass vial

2. 95 mL distilled water

Stock solutions: glutaraldehyde 25% solution EM grade (store at 4°C in dark):

1. 0.2 M cacodylate buffer

   Score vial with a diamond pen and break inside an amber-colored glass reagent bottle; dissolve 1–2 days by occasional swirling.

Materials 2. Epoxy resin — Araldite (Durcupan) Fluka
Preparation of embedding resin mixture:

1. Use 10 mL plastic measuring cylinder to measure:

   Durcupan ACM — 5 mL

   Durcupan B — 5 mL

   Durcupan C — 0.3 mL

   Durcupan D — 0.3 mL

   Mix in a chemical-proof, plastic vial by vigorous agitation (10–20 min) or rotator.

2. Use immediately or store in a freezer at −10°C (hardens on keeping).

   (Epon resin or Epon/Araldite 1:1 may be used instead for embedding)

Materials 3. Toluidine blue for thick sections

   1. Toluidine blue — 1 g

   2. Borax — 1 g

   3. Distilled water — 100 mL

Dissolve borax in water and then add stain.

Store in 20 mL syringe with a filter.

(Methylene blue can be used instead.)

Materials 4: Electron stains for thin sections

A. Alcoholic uranyl acetate (saturated solution)

   Make the solution fresh just before staining:

1. Dissolve ~600 mg of stain in 5 cm^3 of 70% ethyl alcohol;

2. Shake vigorously in a vial for about 20 min;

3. Filter through 0.2 μm Acrodisc before use

B. Reynold's lead citrate
1. Lead nitrate — 1.33 g

2. Sodium citrate — 1.76 g

3. Carbon dioxide-free — 30 mL

1. Boil 100 mL for 30 min to expel CO_2 and air and cool;

2. Shake lead salts vigorously in 30 mL of in a volumetric flask (30 min);

3. Add 8 mL of 1N NaOH to milky suspension while shaking — clears up;

4. Dilute to 50 mL with distilled water;

5. Store in the dark at 4°C, wrapped in foil (keeps for up to 3 months);

6. Fill 20 mL syringe with stain and filter through 0.2 μm Acrodisc before use.

Note 3:

Both lead and uranium salts are poisonous. Uranium is also radioactive and light sensitive. Avoid contamination with skin. Wash hands well.

Protocols

Fixation

1. Detach ES cells, EBs or NS from Falcon dish with a glass cutter or fine needle.

2. Pick up with forceps or wide-bore Pasteur pipette with minimal medium;

3. Drop into glass vial containing 5 mL Fixative A;

4. Fix for 1 hr and store at 4°C (keeps several weeks to months);

5. Rinse briefly in water (5–10 min);

6. Postfix in Fixative B — 1 mL for 1 hr in dark

7. Process immediately and rapidly remove fixative with a glass pipette and add 5 mL 70% ethyl alcohol.

(Stem cell pellets (2–5 mm) can be fixed and processed in Eppendorf tubes, chemical-proof plastic tubes or glass vials.)

Dehydration and embedding (rapid method)

1. Transfer cells from osmium to 70% alcohol (10 min)

2. Bathe cells in 90% alcohol (10 min)

3. Absolute alcohol — two changes 15 min each (dehydrant)

4. Acetone — two changes 15 min each (dehydrant and resin solvent)

5. Acetone/Araldite mixture (1:1) — 30 min

6. Gently transfer colonies into solid watch glasses (embryological dishes);

7. Process under binocular microscope, if specimen is too small;

8. Remove acetone/araldite mixture with a pipette;

9. Add Araldite mixture (1–2 mL);

10. Infiltrate 3–6 hr, preferably overnight (use rotator, if available);

11. Embed in fresh Araldite in rubber or plastic moulds (use fine needle);

12. Orientate each EB, NS or ES cells at the tip of a well with a fine needle[1].

13. Polymerize in a 60°C oven (48–72 hr).

Note 4:

Swirl often to promote penetration at every step. Never dry the specimen. Some resin components are carcinogenic, use gloves and the fume cupboard. Discard used reagents into a large bottle for proper disposal.

Note 5:

Glass vials may be used for processing if the specimen is large and visible. Plastic vial caps can be used for flat embedding. Beam capsules may be used but orientation is difficult. Monolayers of cells can be grown on scaffolds or collagen membranes, fixed and processed in situ. If the block is not sufficiently hard, repolymerize in the oven for 24 hr.

---

[1] Orientation is critical to section the specimen in the desired plane.

Section cutting and staining

Thick sections:

1. Remove the hardened block from the mould.

2. Trim a cone on all sides with a sharp razor blade under binocular.

3. Shape a rectangle or trapezium with parallel upper and lower edges.

4. Fix the block on to microtome holder.

5. Align, orientate block with upper and lower edges parallel to knife edge.

6. Make fresh glass knives with a knife make and attach a boat or use a histological diamond knife.

7. Cut sections 1 μm thick (manual setting).

8. Mount 5–10 sections on to a clean glass slide with a drop of water.

9. Place slide on a hotplate at 60°C.

10. Dry to attach sections on to slide (1–2 min).

11. Add two or three drops of 1% toluidine blue stain in 1% borax in distilled water.

12. Leave on hotplate 1–2 min till stain steams (do not allow to dry out).

13. Cool and wash by squirting from a wash bottle into a sink.

14. Wipe excess water and allow to dry in air.

15. Examine sections with a light microscope (we do not mount to avoid fading).

16. When the desired cells are obtained cut thin sections.

Thin sections:

Sectioning has to be done by an experienced technician.

1. Replace glass knife with a diamond knife.

2. Fill boat with and focus knife-edge.

3. Re-align block and avoid cutting thick sections.

4. Use automatic cutting mode and cut silver–gold sections (∼70 nm thick).

5. Cut a ribbon of 10–15 sections.

6. Rinse copper grids (200 hexagonal mesh) in 1% HCl acid to etch.

7. With a fine forceps place grid under sections in boat and lift the ribbon upwards. If ribbon breaks group sections with an eyelash mounted on a match.

8. Blot grids of excess water and place on a filter paper in a Petri dish.

9. Dry in 60°C oven for immediate staining (15 min) or leave to dry in air.

10. Wash diamond knife immediately with a jet of distilled water.

Note 6:

Trim the specimen block as small as possible — remove excess resin. Make sure the upper and lower edges are parallel to knife edge to ribbon. Tighten all screws properly on specimen holder and microtome to avoid chatter.

Staining thin sections:

A. Uranyl staining

1. Using a syringe, filter 2 mL of uranyl acetate into a plastic cap.

2. Immerse grids in stain using a fine forceps, sections facing downwards.

3. Stain for 10–15 min. Avoid direct light.

4. Rinse grids in 50% alcohol.

5. Rinse in distilled water.

6. Dry grids on filter paper, sections facing up.

B. Lead staining

1. Using a syringe filter 2 mL of lead stain into a plastic cap.

2. Place cap in a Petri dish with LiOH or few pellets of NaOH (this removes CO_2 that might precipitate the stain).

3. Immerse grids, sections facing downwards.

4. Cover dish and stain for 10–15 min (don't breathe on to dish).

5. Rinse in distilled water (two changes).

6. Dry grids on filter paper, sections facing upwards.

Store grids in grid boxes for storage and examine by TEM.

Note 7:

Hold grids by the edge and bend gently with a forceps to collect thin sections. Hold grids vertically when immersing in electron stains or rinses. Rinse forceps with distilled water between stains and dry with tissue to avoid precipitates. Discard milky lead stain.

TEM examination (briefly)

This should be done by an experienced operator

1. Check vacuum and switch on transmission electron microscope at 60 or 80 kV.

2. Align electron beam each time before use for sharp focusing

3. Insert grid into the column and pump down to restore vacuum.

4. Switch the filament on when high voltage comes on.

5. Remove apertures and examine sections at low magnifications.

6. Select areas of sections to be examined (see survey LM sections).

7. Insert medium or small apertures for better contrast.

8. Switch to higher magnifications and photograph desired cells.

9. Use magnifications of $\times$ 2,000 to $\times$ 10,000 times for routine work.

10. Remove grid, pump down and leave microscope on or shut down.

Note 8:

Use fine stainless forceps for handling specimens. The electron beam needs to be always aligned before imaging. Find sharp focus and good contrast before imaging. Avoid focussing the beam on the same spot for a long time; this darkens the section. Consult your TEM technician and TEM manual for details of use.

References

Cameron CM, WS Hu and DS Kaufman. (2006). Improved development of human embryonic stem cell-derived embryoid bodies by stirred vessel cultivation. *Biotechnol Bioeng* **94**: 938–948.

Conley BJ, AO Trounson and R Mollard. (2004a). Human embryonic stem cells form embryoid bodies containing visceral endoderm-like derivatives. *Fetal Diagn Ther* **19**: 218–923.

Conley BJ, JC Young, AO Trounson and R Mollard. (2004b). Derivation, propagation and differentiation of human embryonic stem cells. *Int J Biochem Cell Biol* **36**: 555–567.

Ezashi T, P Das and RM Roberts. (2005). Low O_2 tensions and the prevention of differentiation of hES cells. *Proc Natl Acad Sci USA* **102**: 4783–4788.

Fawcett D. (1981). *The Cell*. WB Saunders, Philadelphia, PA.

Gerecht-Nir S, S Cohen and J Itskovitz-Eldor. (2004a). Bioreactor cultivation enhances the efficiency of human embryoid body (hEB) formation and differentiation. *Biotechnol Bioeng* **86**: 493–502.

Gerecht-Nir S, S Cohen, A Ziskind and J Itskovitz-Eldor. (2004b). Three-dimensional porous alginate scaffolds provide a conducive environment for generation of well-vascularized embryoid bodies from human embryonic stem cells. *Biotechnol Bioeng* **88**: 313–320.

Glauert A, ed. *Practical Methods in Electron Microscopy*. (1974). North-Holland, Amsterdam

Hayat M, ed. *Principles and Techniques of Electron Microscopy*. (1989). CRC Press, Boca Raton, FL

Hoffman LM and MK Carpenter. (2005). Characterization and culture of human embryonic stem cells. *Nat Biotechnol* **23**: 699–708.

Itskovitz-Eldor J, M Schuldiner, D Karsenti, A Eden, O Yanuka, M Amit, H Soreq and N Benvenisty. (2000). Differentiation of human embryonic stem cells into embryoid bodies compromising the three embryonic germ layers. *Mol Med* **6**: 88–95.

Klimanskaya I, Y Chung, L Meisner, J Johnson, MD West and R Lanza. (2005). Human embryonic stem cells derived without feeder cells. *Lancet* **365**: 1636–1641.

Mohr JC, JJ de Pablo and SP Palecek. (2006). 3-D microwell culture of human embryonic stem cells. *Biomaterials* **27**: 6032–6042.

Ng ES, RP Davis, L Azzola, EG Stanley and AG Elefanty. (2005). Forced aggregation of defined numbers of human embryonic stem cells into embryoid bodies fosters robust, reproducible hematopoietic differentiation. *Blood* **106**: 1601–1603.

Sathananthan AH, ed. *Microscopic Atlas of Sperm Function for ART*. (1996a). National University Hospital and Serono, Singapore.

Sathananthan AH. (1996b). *Atlas of Human Cell Ultrastructure*. CSIRO, Melbourne.

Sathananthan AH (2000) Ultrastructure of human gametes, fertilization, and embryo development. In: *Handbook of In Vitro Fertilization*. A Trounson and D Gardner, eds. 2nd ed. CRC Press, Boca Raton, FL

Sathananthan AH (2003) Origins of human embryonic stem cells and their spontaneous differentiation. *First National Stem Cell Centre Scientific Conference*, Melbourne, Poster 225.

Sathananthan AH, S Gunasheela and J Menezes. (2003). Critical evaluation of human blastocysts for assisted reproduction and embryo stem cell biotechnology. *Reprod Biomed Online* **7**: 219–227.

Sathananthan AH, S Ng, A Bongso, A Trounson and S Ratnam. (1993). *Visual Atlas of Early Human Development for Assisted Reproductive Technology*. National University Singapore and Serono, Singapore.

Sathananthan AH and S Nottola. (2007). Digital imaging of stem cells by electron microscopy. In: *Stem Cell Assays*. M Vermuri, eds. Humana Press, NJ.

Sathananthan AH and A Trounson. (2005). Human embryonic stem cells and their spontaneous differentiation. *Ital J Anat Embryol* **110**: 151–157.

Sathananthan AH, M Pera M and A Trounson. (2002). The fine structure of human embryonic stem cells. *Reprod Biomed Online* **4**: 56–61.

Trounson A. (2006). The production and directed differentiation of human embryonic stem cells. *Endocr Rev* **27**: 208–219.

11

Differentiation of human embryonic stem cells into extraembryonic cell types

CHERYLE A. SÉGUIN[1] AND JONATHAN S. DRAPER[1,2,3]

[1] Department of Developmental and Stem Cell Biology, Hospital for Sick Children, Toronto, Canada
[2] Samuel Lunenfeld Research Institute, Mount Sinai Hospital, Toronto, Canada
[3] Centre for Stem Cell Biology, University of Sheffield, Sheffield, UK

Introduction

Human and murine embryonic stem (ES) cells share a significant range of phenotypic traits but differ in several characteristics, including colony morphology, cell surface antigen expression, and growth factor requirements for the maintenance of self renewal and pluripotency *in vitro*. One further notable difference between ES cells from the two species is the propensity of human ES cells to differentiate into cells representing the extraembryonic lineages in culture; unlike mouse, spontaneous differentiation of both trophectoderm-like cells (Thomson *et al.*, 1998) and extraembryonic endoderm-like cells (Pera *et al.*, 2004) has been reported during routine culture of human ES cells. Although culture conditions of human ES cells differ between labs and cell lines being grown, spontaneous extraembryonic differentiation is generally promoted by media containing fetal bovine serum (FBS) in lieu of serum replacement, especially at low concentrations (**Table 1**). Since the morphological changes associated with extraembryonic differentiation of human cells cannot alone identify a given population, we present a suggested panel of markers and corresponding PCR primer sequences and conditions to help facilitate the characterization of extraembryonic cell types from human ES cells (**Tables 2 and 3**).

Extraembryonic differentiation can alternatively be induced by aggregating human ES cells in suspension to form embryoid bodies (EBs). When murine ES cells are differentiated in EBs they often produce a surface layer of extraembryonic endoderm but do not typically contain cells representative of trophectoderm. In contrast, human ES derived EBs maintained in suspension culture have been shown to secrete elevated levels of the trophectoderm related hormones human CG, progesterone, and

Table 1 Extraembryonic endoderm differentiation medium

Knockout DMEM	Gibco	10829-018
Knockout serum replacement (15%)	Gibco	10828-028
200 mM GlutaMax (1:100)	Gibco	35050-061
Non-essential amino acids (1:100)	Gibco	11140-050
100 μM β-mercaptoethanol	Gibco	21985-023
BMP-4 (20 ng/mL)	R&D	314-BP

Table 2 Primers and PCR conditions for assaying differentiation into extraembryonic cell types

Gene	Primer sequence	Product size	PCR conditions	Reference
Extraembryonic endoderm				
Gata4	F: ACCACAAGATGAACGGCATCAACC R:GGAAGAGGGAAGATTACGCAGTGA	612 bp	94°C 1 min;56°C 1 min;72°C 1 min 30 cycles	(Kuo *et al.*, 1997)
Gata6	F: TCTACAGCAAGATGAACGGCCTCA R:AGGTGGAAGTTGGAGTCATGGGAA	312 bp	94°C 1 min;56°C 1 min;72°C 1 min 30 cycles	(Morrisey *et al.*, 1998)
SOX 17	F: TTTCATGGTGTGGGCTAAGGACGA R: GCTTGCACACGAAGTGCAGATACT	896 bp	94°C 1 min;62°C 1 min;72°C 1 min 32 cycles	(Kurimoto *et al.*, 2006)
HNF4α	F: TGTGTGTGAGTCCATGAAGGAGCA R: AGTCATACTGGCGGTCGTTGATGT	431 bp	94°C 1 min;56°C 1 min;72°C 1 min 28 cycles	(Duncan *et al.*, 1997)
AFP	F: AGAACCTGTCACAAGCTGTG R: GACAGCAAGCTGAGGATGTC	676 bp	94°C 1 min;57°C 1 min;72°C 1 min 30 cycles	(Andrews *et al.*, 1982)
LamininB1	F: ACACTGCATGTGCAGGCATAACAC R: TCACAATGCTGTCCTGTCACCAGA	585 bp	94°C 1 min;56°C 1 min;72°C 1 min 24 cycles	(Dziadek & Timpl 1985)
Collagen IV	F: AGGAATCATGGGCTTTCCTGGACT R: TCGTCGCCTTCTGTACATCTGCAT	585 bp	94°C 1 min;56°C 1 min;72°C 1 min 28 cycles	(Adams & Watt 1993)
Extraembryonic trophectoderm				
Cdx2	F:CCTCCGCTGGGCTTCATTCC R:TGGGGGTTCTGCAGTCTTTGGTC	295 bp	94°C 1 min;60°C 1 min;72°C 1 min 30 cycles	(Beck *et al.*, 1995)
Eomes	F: TCACCCCAACAGAGCGAAGAGG R: AGAGATTTGATGGAAGGGGGTGTC	374 bp	94°C 1 min;58°C 1 min;72°C 1 min 30 cycles	(Ciruna & Rossant 1999)
HCG-B	F: CAGGGGACGCACCAAGGATG R: GTGGGAGGATCGGGGTGTCC	510 bp	94°C 1 min;62°C 1 min;72°C 1 min 30 cycles	(Muyan & Boime 1997)

Table 3   Antibodies for assaying differentiation into extraembryonic cell types

Antibody	Working dilution	Vendor	Catalog number
Extraembryonic endoderm			
Gata 4	1:100	Santa Cruz	sc-9053
AFP	1:100	R&D	MAB1368
laminin	1:80	Chemicon	AB2034
Trophectoderm			
Cdx2	1:400	Biogenex	MU392-UC
hCG	1:400	DakoCytomation	A231

estradiol, with the hCG positive cells localizing to the outer surface of the EBs. The culture of EBs for extended periods of time encapsulated within Matrigel (BD), a mixture of basement membrane factors, further promotes placental endocrine activity and the consistent formation of an outer trophectoderm-like layer (Gerami-Naini *et al.*, 2004). Following 5 to 7 days in suspension culture, human EBs routinely contain an outer epithelial-like layer containing cells that demonstrate characteristics of extraembryonic endoderm, including expression of Gata4 and alphafetoprotein, and the deposition of laminin (**Figure 1**), as detected by confocal immunolocalization. After 1 week in culture, a peripheral cellular layer has been described on cavitating EBs

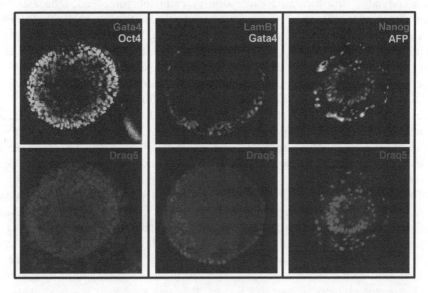

Figure 1   Extraembryonic endoderm differentiation in human embryoid body cultures. Following 7 days in suspension culture, human EBs demonstrate an outer epithelial layer resembling the extraembryonic endoderm. This cell layer can be differentiated from the inner core of Oct4/Nanog positive cells by the immunolocalization of Gata4, alphafetoprotein (AFP), and the deposition of a lamininB1containing basement membrane.

demonstrating cytokeratin-8 and alphafetoprotein expression, suggestive of yolk sac visceral endoderm formation (Conley *et al.*, 2004).

Several studies have demonstrated the induction of extraembryonic differentiation from human ES cells using BMP, a member of the transforming growth factor-β superfamily. BMP-4 treatment of H9 ES cells under serum-free conditions induces trophectoderm differentiation, marked by the expression of early trophoblast markers such as MSX2 and Gata2/3, the secretion of placental hormones hCG, estradiol, and progesterone, and the formation of hCG-positive syncytia (Xu *et al.*, 2002). By contrast, human ES cells treated with BMP-2 (BMP-4 or BMP-7) in the presence of serum and a feeder cell layer appear to undergo an extraembryonic endoderm-like differentiation (**Table 2**). This differentiation is characterized by flattened epithelial cell morphology and the immunolocalization of cytokeratins, laminin and alphafetoprotein. BMP-2 induced differentiation also resulted in the loss of stem cell markers, and the expression of endoderm-associated genes including HNF3α, HNF4, Gata4 and Gata6 (Pera *et al.*, 2004). Together these studies indicate that human ES cell differentiation is likely determined by multiple factors and easily impacted by the *in vitro* culture environment.

Extraembryonic differentiation can also be induced by genetic manipulation of transcription factors associated with the maintenance of pluripotency in undifferentiated ES cells. Down regulation of both Oct4 and Nanog were first shown to cause trophectoderm and extraembryonic endoderm differentiation, respectively, in mouse ES cells (Mitsui *et al.*, 2003; Niwa *et al.*, 2000). Curiously, RNAi-mediated *Oct4* knockdown in human ES cells produces a morphologically heterogeneous cell population which correlates with increased expression of extraembryonic endoderm associated genes, Gata6 and AFP, and also, under serum-free conditions, with the induction of the trophoblast markers Cdx2 and hCG (Hay *et al.*, 2004). Other studies confirm the induction of trophoblast-like gene expression in the absence of Oct4 (Matin *et al.*, 2004). Analysis of human ES cells following RNAi-mediated *Nanog* knockdown also suggested the presence of both extraembryonic endoderm and trophectodermal differentiation (Zaehres *et al.*, 2005). Nanog downregulation in human ES cells not only causes loss of ES-associated cell surface markers and Oct4 expression but also leads to the activation of the extraembryonic endoderm-associated genes Gata4, Gata6, laminin β1, and AFP. Expression of trophectoderm-associated genes Cdx2, Gata2, hCGα and hCGβ were also observed following RNAi-mediated knockdown in human ES and EC cells (Hyslop *et al.*, 2005).

It is interesting to note that several conditions that promote the differentiation of human ES cells towards extraembryonic lineages (i.e. BMP2 treatment, *Oct4/Nanog* knockdown) produce a mixed population of cells containing both trophectoderm and extraembryonic endoderm cells (Pera *et al.*, 2004; Hay *et al.*, 2004; Hyslop *et al.*, 2005). The differentiation of both extraembryonic cell types in these studies may be due to experimental conditions possibly reflecting the heterogeneity of many apparently undifferentiated cultures of human ES cells. It is tempting to speculate that the tendency of human ES cells to produce both extraembryonic lineages in conditions which generate only one extraembryonic lineage in murine ES cells may actually reflect a fundamental difference in the specification of extraembryonic tissues between the two species. Whatever the outcome, these early studies on the differentiation of human extraembryonic cell populations have created a solid platform from which to move forward in the understanding of human extraembryonic stem cells and their role in supporting early human development.

Overview of protocols

We initially discuss how to induce extraembryonic differentiation in human ES cell monolayer cultures, listing different media for the induction of trophectoderm or extraembryonic endoderm. Spontaneous differentiation into extraembryonic cell types is then mentioned, although protocols for generating and fixing embryoid bodies are not included as they have been already outlined in **Chapter 10**. Finally we describe how to assay extraembryonic differentiation, both by PCR analysis and immunostaining.

Materials, reagents and equipment

Culture media for human ES cells

Differentiation of extraembryonic cell types from human ES cells has been reported by several groups following culture of cells in the presence of various growth factors. The following describes the media used to induce extraembryonic differentiation of human ES cells grown on a feeder cell layer (CA1 and CA2 lines derived by A. Nagy, Mount Sinai Hospital, Toronto, Canada).

Protocols

Growth factor based induction of extraembryonic differentiation

The following protocol describes conditions optimized for differentiation of extraembryonic cell types (extraembryonic endoderm and trophectoderm) from the CA1 and CA2 human ES cell lines. Protocols are modified from those previously described (Pera *et al.*, 2004; Xu *et al.*, 2002) and induce heterogeneous differentiation of human ES cells. Based on the culture media used, populations demonstrate a preference towards either trophectoderm (**Table 4**) or extraembryonic endoderm (**Table 1**) differentiation.

Table 4 Trophectoderm differentiation medium

RPMI 1640	Sigma	R0883
Fetal bovine serum (2%)	Hyclone	SH30396.03
200 mM GlutaMax (1:100)	Gibco	35050-061
BMP-4 (20 ng/mL)	R&D	314-BP

Differentiation of human ES cell monolayer cultures into extraembryonic cell types

1. Human ES cell colonies are trypsinized to large clumps (~20 cells or greater) and split 1:6 onto mitotically inactivated mouse feeders, and maintained under standard human ES cell culture conditions for the first 24 hr to permit cell attachment (see the routine protocols section of **Chapter 5** for further details concerning these conditions).

2. The day following passaging, the human ES cell medium is aspirated and cells are rinsed twice with 1xPBS.

3. To induce extraembryonic differentiation, human ES cells are cultured in differentiation media. **Table 4** shows the components of the medium we routinely to differentiate human ES cell cultures into trophectoderm, while the medium described in **Table 1** preferentially promotes differentiation into extraembryonic endoderm.

4. Medium should be changed daily, and the induction of differentiation of human ES cells can be monitored daily via changes in cell morphology as well as the expression of lineage specific markers (see **Table 2**). Trophectoderm differentiation should be detected between 2 to 5 days in culture in trophectoderm differentiation medium. Extraembryonic endoderm differentiation should be detected at the periphery of human ES cell colonies following 24 hr in extraembryonic endoderm differentiation medium, and should become more apparent throughout the culture after 5 days.

It is important to note that although changing culture media can promote differentiation towards one of the two extraembryonic lineages, we find culture of human ES cells in either media outlined above produces a heterogeneous population containing both extraembryonic cell types. The characterization of extraembryonic cell types relies predominantly on alterations in cell morphology and expression of a panel of lineage related markers (trophectoderm: Cdx2, Eomes, hCG; extraembryonic endoderm: Gata4, Gata6, AFP, laminin B1); most individual markers expressed in the extraembryonic lineages are expressed elsewhere in the embryo at later time points (e.g. Cdx2, Gata4, Gata6, AFP and CK7).

Spontaneous extraembryonic differentiation in embryoid bodies

Extraembryonic differentiation can also be spontaneously induced by generating embryoid bodies from human ES cell cultures. Generate embryoid bodies (EBs) in suspension culture as described in **Chapter 10**. Early markers of primitive endoderm lineage can be detected as early as day 4 of culture, through to day 7. Preparation and immunostaining of EBs for differentiation are also described in **Chapter 10**.

Assaying extraembryonic differentiation

PCR analysis

Changes in the gene expression patterns of ES cells following differentiation, or characterization of a given cell population can be examined by PCR-based gene expression analysis. Although no single marker can define a population of cells, PCR enables one to assess the expression of multiple genes from a single sample, thereby enabling one to use a panel of genes to more readily identify the phenotype of a given population of cells.

Table 2 represents human gene-specific primers compiled from the laboratories of Drs J. Rossant and P. Andrews optimized for cDNA reverse transcribed from 0.5–1 µg total RNA. The PCR conditions may have to be optimized for different starting populations of cells and/or RNA extraction and reverse transcription protocols. To facilitate primer optimization and proper analysis of PCR results, the use of appropriate positive and negative controls is required (i.e. human yolk sac carcinoma/choriocarcinoma cells and undifferentiated human ES cells, respectively). References reflect the demonstration of extraembryonic gene expression for each of the markers.

Immunostaining

Preparation and fixation of embryoid bodies has previously been described in **Chapter 10**. We use the following antibodies (in the stated dilutions) to stain EBs and differentiated human ES cell monolayers when assaying for extraembryonic cell types (see **Table 3**).

Troubleshooting

Which media should I use to culture the EBs?

This really depends upon the tissue you are trying to generate. The presence of growth factors and serum can profoundly alter the differentiation of EBs. We often use the human ES medium (without bFGF) that we culture the undifferentiated human ES cells in for routine generation of EBs and find extraembryonic components within the EBs. Use of low serum or the addition of BMPs can also be helpful in generating extraembryonic tissues.

How to feed suspension cultures of EBs?

During maintenance of EB cultures, the medium generally has to be replaced every 2–3 days. This is best accomplished by gently aspirating the culture medium and EBs and transferring into a 15 mL conical bottomed tube. Rinse the culture vessel once with PBS and add to the 15 mL tube to ensure all EBs are retrieved. Pellet the EBs either by gravity (allow to settle for 2–5 min) or centrifugation (30 s at

200 *g* (~600 rpm) is sufficient). Gently resuspend the EBs in fresh culture media, and return to the tissue culture vessel. Dependent upon number and size, EBs can quickly metabolize the medium; we ensure the EBs have plenty of medium, typically 2–3 mL per 10 cm^2 of culture vessel.

Suggestions for working with EBs if too many are being lost during the washing stages of immunostaining.

Manipulating the EBs is most often the most challenging part of immunostaining. 100 µM filter caps designed to filter cells for FACS can be used to prevent EB loss. EBs are placed within these cup shaped mesh filters, and the filters are then moved between wells of PBST for washes.

Alternative ways to image EBs.

Once fixation and staining of EBs is complete, whole EBs can be transferred to moulds, frozen in OCT cyroembedding medium and subsequently sectioned on a cryostat. EBs stained with some primary and secondary antibody combinations may benefit from a brief (<1 min) treatment in 4% PFA to maintain staining integrity. An optional overnight treatment of the EBs in 30% sucrose in PBS prior to OCT embedding can also help retain tissue morphology.

References

Adams JC and FM Watt. (1993). Regulation of development and differentiation by the extra-cellular matrix. *Development* **117**: 1183–1198.

Andrews GK, M Dziadek and T Tamaoki. (1982). Expression and methylation of the mouse alpha-fetoprotein gene in embryonic, adult, and neoplastic tissues. *J Biol Chem* **257**: 5148–5153.

Beck F, T Erler, A Russell and R James. (1995). Expression of Cdx-2 in the mouse embryo and placenta: possible role in patterning of the extra-embryonic membranes. *Dev Dyn* **204**: 219–227.

Ciruna BG and J Rossant. (1999). Expression of the T-box gene Eomesodermin during early mouse development. *Mech Dev* **81**: 199–203.

Conley BJ, AO Trounson and R Mollard. (2004). Human embryonic stem cells form embryoid bodies containing visceral endoderm-like derivatives. *Fetal Diagn Ther* **19**: 218–223.

Duncan SA, A Nagy and W Chan. (1997). Murine gastrulation requires HNF-4 regulated gene expression in the visceral endoderm: tetraploid rescue of Hnf-4(−/−) embryos. *Development* **124**: 279–287.

Dziadek M and R Timpl. (1985). Expression of nidogen and laminin in basement membranes during mouse embryogenesis and in teratocarcinoma cells. *Dev Biol* **111**: 372–382.

Gerami-Naini B, OV Dovzhenko, M Durning, FH Wegner, JA Thomson and TG Golos. (2004). Trophoblast differentiation in embryoid bodies derived from human embryonic stem cells. *Endocrinology* **145**: 1517–1524.

Hay DC, L Sutherland, J Clark and T Burdon. (2004). Oct-4 knockdown induces similar patterns of endoderm and trophoblast differentiation markers in human and mouse embryonic stem cells. *Stem Cells* **22**: 225–235.

Hyslop L, M Stojkovic, L Armstrong, T Walter, P Stojkovic, S Przyborski, M Herbert, A Murdoch, T Strachan and M Lako. (2005). Downregulation of NANOG induces differentiation of human embryonic stem cells to extraembryonic lineages. *Stem Cells* **23**: 1035–1043.

Kuo CT, EE Morrisey, R Anandappa, K Sigrist, MM Lu, MS Parmacek, C Soudais and JM Leiden. (1997). GATA4 transcription factor is required for ventral morphogenesis and heart tube formation. *Genes Dev* **11**: 1048–1060.

Kurimoto K, Y Yabuta, Y Ohinata, Y Ono, KD Uno, RG Yamada, HR Ueda and M Saitou. (2006). An improved single-cell cDNA amplification method for efficient high-density oligonucleotide microarray analysis. *Nucleic Acids Res* **34**: e42.

Matin MM, JR Walsh, PJ Gokhale, JS Draper, AR Bahrami, I Morton, HD Moore and PW Andrews. (2004). Specific knockdown of Oct4 and beta2-microglobulin expression by RNA interference in human embryonic stem cells and embryonic carcinoma cells. *Stem Cells* **22**: 659–668.

Mitsui K, Y Tokuzawa, H Itoh, K Segawa, M Murakami, K Takahashi, M Maruyama, M Maeda and S Yamanaka. (2003). The homeoprotein Nanog is required for maintenance of pluripotency in mouse epiblast and ES cells. *Cell* **113**: 631–642.

Morrisey EE, Z Tang, K Sigrist, MM Lu, F Jiang, HS Ip and MS Parmacek. (1998). GATA6 regulates HNF4 and is required for differentiation of visceral endoderm in the mouse embryo. *Genes Dev* **12**: 3579–3590.

Muyan M and I Boime. (1997). Secretion of chorionic gonadotropin from human trophoblasts. *Placenta* **18**: 237–241.

Niwa H, J Miyazaki and AG Smith. (2000). Quantitative expression of Oct-3/4 defines differentiation, dedifferentiation or self-renewal of ES cells. *Nat Genet* **24**: 372–376.

Pera MF, J Andrade, S Houssami, B Reubinoff, A Trounson, EG Stanley, D Ward-van Oostwaard and C Mummery. (2004). Regulation of human embryonic stem cell differentiation by BMP-2 and its antagonist noggin. *J Cell Sci* **117**: 1269–1280.

Pera MF, MJ Blasco Lafita and J Mills. (1987). Cultured stem-cells from human testicular teratomas: the nature of human embryonal carcinoma, and its comparison with two types of yolk-sac carcinoma. *Int J Cancer* **40**: 334–343.

Thomson JA, J Itskovitz-Eldor, SS Shapiro, MA Waknitz, JJ Swiergiel, VS Marshall and JM Jones. (1998). Embryonic stem cell lines derived from human blastocysts. *Science* **282**: 1145–1147.

Xu RH, X Chen, DS Li, R Li, GC Addicks, C Glennon, TP Zwaka and JA Thomson. (2002). BMP4 initiates human embryonic stem cell differentiation to trophoblast. *Nat Biotechnol* **20**: 1261–1264.

Zaehres H, MW Lensch, L Daheron, SA Stewart, J Itskovitz-Eldor and GQ Daley. (2005). High-efficiency RNA interference in human embryonic stem cells. *Stem Cells* **23**: 299–305.

12, Part A

Directed differentiation of human embryonic stem cells into early endoderm cells

KENJI OSAFUNE[1], ALICE E. CHEN[1] AND DOUGLAS A. MELTON[1,2]

[1]*Department of Molecular and Cellular Biology, Harvard University*
[2]*Howard Hughes Medical Institute, 7 Divinity Avenue, Cambridge, MA 02138, USA*

Introduction

The definitive endoderm, one of three principal germ layers formed during gastrulation in development, gives rise to epithelia of the digestive and respiratory tracts and to the thyroid, thymus, liver and pancreas (Wells and Melton, 1999). Generating cells of definitive endoderm-derived organs such as the pancreas and liver from human ES cells has interested many researchers. However, there has been little success with the efficient generation of pancreatic endocrine cells or hepatocytes from human ES cells that are useful for clinical application (Bonner-Weir and Weir, 2005; Lavon and Benvenisty, 2005). As a first step toward this goal, it would be useful to have an efficient protocol for inducing definitive endoderm.

Many different human ES lines, including ones derived in our laboratory (HUES1-17; Cowan *et al.*, 2004), have been established. Based on our experience (unpublished observations) and reports by others (D'Amour *et al.*, 2005; Hoffman and Carpenter, 2005), human ES cell lines appear to vary in their growth and differentiation potential. The establishment of a reproducible protocol that can be consistently applicable to multiple human ES cell lines is important for generating specific functional cell types including definitive endoderm, as well as for culturing undifferentiated human ES cells.

Conventional differentiation protocols have used multicellular aggregation (Doetschman *et al.*, 1985) based on the idea that *in vitro* formation of the germ layers and their relative spatial organization mimic that of the embryo during development. In addition, conditions have been established for monolayer differentiation of ES cells, reducing the complexity compared to Embryoid body (EB) culture (Nishikawa *et al.*, 1998). Two years ago, a protocol describing the differentiation of human ES cells into definitive endoderm using activin A in a monolayer culture format was reported

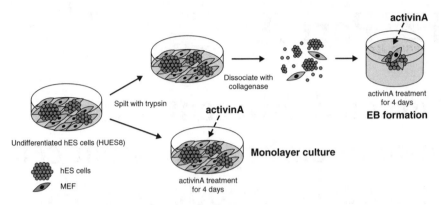

Figure 1 Schematic depiction of two methods for endoderm differentiation described in this chapter.

(D'Amour *et al.*, 2005). Recombinant activin A was used to mimic signaling of the related TGFβ member, Nodal, an essential molecule for endoderm specification during gastrulation in mice (Lowe *et al.*, 2001; Vincent *et al.*, 2003). We have verified that this protocol reproducibly and efficiently generates endoderm cells from HUES cell lines, and that the culture system is more efficient than EB formation. In this chapter, we describe methods for generating endoderm cells both by monolayer culture and by EB formation, combined with activin A treatment. Immunostaining against SOX17, a factor expressed in the developing definitive endoderm that is essential for endoderm differentiation in mice (Kanai-Azuma *et al.*, 2002), is also described.

Overview of protocols

Two protocols for endoderm differentiation are described in this chapter, adherent culture in monolayer format and suspension culture in EB format. Out of four human ES cell lines generated in our laboratory at Harvard (HUES1, 6, 8 and 9), HUES 8 proved to give the best results for endoderm induction. These methods are depicted in **Figure 1**. A quantitative comparison of these protocols indicates that induction of endoderm is more efficient when cells are differentiated via monolayer rather than EB format (**Figure 2**).

Materials, reagents and equipment

Cell culture

Material/reagent	Vendor	Catalog number
DMEM	Invitrogen GIBCO	11965
Knockout-DMEM	Invitrogen GIBCO	10829
RPMI 1620	Invitrogen GIBCO	21870

Material/reagent	Vendor	Catalog number
Phosphate-buffered saline (PBS)	Invitrogen GIBCO	14190
Knockout serum replacement	Invitrogen GIBCO	10828
Plasmanate	Bayer	NDC0026-0613-20
Fetal bovine serum (FBS)	Hyclone	SH30071.03
2-β-Mercaptoethanol	Invitrogen GIBCO	21985-023
Non-essential amino acids	Invitrogen GIBCO	11140050
Glutamax-I	Invitrogen GIBCO	35050-061
Penicillin/streptomycin	Invitrogen GIBCO	15070-063
HEPES buffer	Invitrogen GIBCO	15630-080
0.05% Trypsin/EDTA	Invitrogen GIBCO	25300-054
Collagenase IV	Invitrogen GIBCO	17104-019
Tissue culture plate (24 well)	BD Falcon	35-3047
Ultra low cluster six-well flat bottom plate		
	Corning	3471
Recombinant human basic fibroblast growth factor (bFGF)		
	Invitrogen	13256-029
Recombinant human/mouse/rat Activin A		
	R&D systems	338-AC
Immunocytochemistry		
Superfrost plus microslides	VWR International	48311-703
Paraformaldehyde	Sigma	P6148
Triton-X	Sigma	T8787
Hoechst 33342	Invitrogen Molecular Probes	H3570
Normal donkey serum	Jackson ImmunoResearch	017-000-121
Goat anti-human SOX17 antibody	R&D systems	AF1924
Rhodamine Red-X-conjugated donkey anti-goat IgG (H+L) antibody		
	Jackson ImmunoResearch	705-295-147

Protocols

Cell culture

1. Media for standard culturing of undifferentiated human ES cells and MEFs are described in **Chapter 5** and in Cowan *et al.* (2004).

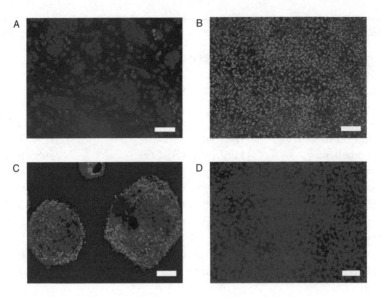

Figure 2 Immunocytochemical analysis of the induction of human ES cells with activin A (100 ng/mL) and low serum. Before treatment, a small number of SOX17-positive cells are seen (**A**). After 4 days of treatment, cells were processed by immunostaining and quantitated (no. of SOX17-positive cells/total no. of cells). More than 70% of the cells are SOX17-positive endoderm in monolayer cultures (**B**), compared to 40% observed in sections of EBs (**C**). Control staining of a monolayer culture without primary antibody (**D**). Red: SOX17; blue, nuclei. Bars: 100 μm. This image is best seen in color; please refer to plate section to see the color image.

2. Medium for the differentiation of human ES cells into endoderm consists of RPMI 1620 supplemented with 2% FBS, 2 mM glutamine, 10 mM HEPES, 500U/mL penicillin/streptomycin, and 100 ng/mL activin A (D'Amour *et al.*, 2005).

Induction of endoderm cells by monolayer culture

1. Aspirate the medium from a 10 cm plate containing human ES cells and wash with 5 mL PBS.

2. Incubate with 1 mL of warm 0.05% Trypsin–EDTA.

3. Add 8 mL of standard human ES culture medium. Using a 10 mL pipet, pipet the cells up and down, five to ten times.

4. Split one third of the cell solution across a 24-well plate seeded with MEFs and maintain cells until they reach to approximately 80% confluence.

5. Once cells have reached ~80% confluency, aspirate the culture medium and wash two times with PBS (1 mL/well).

6. Add 500 μL of differentiation medium containing 100 ng/mL of activin A to each well.

7. Change medium with fresh activin A every 2–3 days.

Induction of endoderm cells by EB formation

1. Aspirate the medium from a 10 cm plate containing human ES cells and wash with 5 mL PBS.

2. Add 7 mL warm DMEM with collagenase IV at a final concentration of 1 mg/mL and incubate at 37°C for 10 min.

3. Aspirate the collagenase solution and add 5 mL standard human ES culture medium.

4. Scrape the cells off the surface with a cell scraper and transfer the cell solution to a 15 mL tube.

5. Wash the dish with 5 mL human ES culture medium and add the wash solution to the same tube.

6. Pipet the cells up and down, five to ten times.

7. Centrifuge at 300 g (~1,000 rpm) for 5 min at room temperature and remove supernatant.

8. Resusped the cell pellet in differentiation medium and distribute across six wells of a low attachment six-well plate.

9. Add differentiation medium to a final volume of 5 mL/well and add activin A at a final concentration of 100 ng/mL.

10. Change medium with fresh activin A every 2–3 days.

Fixation, embedding and sectioning of EBs

1. Harvest EBs at culture day 4 and fix them with 4% paraformaldehyde/PBS at 4°C overnight. Use freshly prepared 4% paraformaldehyde/PBS for best results.

2. Wash EBs with PBS, three times for 15 min at room temperature.

3. Dehydrate EBs in graduated ethanol: 40%, 70%, 80%, and 95% ethanol; each once for 30 min and 100% ethanol; two times for 30 min.

4. Clear EBs in xylene, two times for 30 min.

5. Embed EBs in paraffin after overnight incubation at 60°C.

6.   Section EBs at 10 μm thickness and mount on Superfrost plus microscope slides.

7.   Deparaffinize slides by soaking in xylene; two times for 10 min.

8.   Rehydrate slides in graduated ethanol: 100% ethanol; two times for 5 min and 95%, 80%, 70%, and 40% ethanol; each once for 5 min.

Alternative protocols for fixing, embedding and sectioning EBs are contained in **Chapter 10**.

Immunostaining of monolayer cultures and EB sections for the endoderm marker, SOX17

1.   At day 4 of monolayer culture, fix cells with 4% paraformaldehyde/PBS at 4°C for 20 min after a brief wash, two times with PBS.

2.   Wash fixed monolayer cultures and rehydrated EB slides with PBS, two times for 5 min.

3.   Block monolayer cultures and EB slides by incubating in 5% normal donkey serum/0.1% Triton-X/PBS for 30 min at room temperature.

4.   Incubate with anti-human SOX17 goat antibody diluted with blocking solution at 1:200–1:500 overnight at 4°C or for 4 hr at room temperature.

5.   Wash with 0.1% Triton-X/PBS (PBT), three times for 15 min.

6.   Incubate with Rhodamine Red-X-conjugated donkey anti-goat IgG antibody diluted with blocking solution at 1:200 for 1 hr at room temperature.

7.   Wash with PBT, three times for 15 min.

8.   Counterstain with Hoechst 33342 diluted with PBT at 1:1,000 for 5 min.

9.   Coverslip in anti-fade medium and let samples dry in the dark for 10–15 min.

10.   View with a fluorescent microscope.

Note

Although we use SOX17 as a marker for endoderm in this chapter, *Sox17* is expressed in the extraembryonic endoderm as well as in the definitive endoderm (Kanai-Azuma *et al.*, 2002). Currently, there are no unique markers or functional assays for definitive endoderm. Thus, examination for the absence of extraembryonic markers in combination with the presence of markers for definitive endoderm-derived organs, such as Pdx-1 for pancreas and Albumin for liver, can be used to confirm the formation of definitive endoderm from human ES cells.

Troubleshooting

I cannot get a high induction rate of SOX17-positive cells in monolayer culture: how do I improve the induction rate?

Our experience has shown that the induction rate is somewhat dependent on the density of human ES cells at the beginning of activin A treatment. You should try a titration of low to high human ES cell densities.

Our experiments have also shown that human ES lines exhibit varying proclivities for differentiating into endoderm. Of the lines we have tested, HUES8 consistently gives the highest frequency of endoderm induction.

I get a lot of dead cells or dead EBs after seeding human ES cells onto low attachment plates: what should I do?

If pipetting is too harsh during disaggregation, cell death and lysis are often observed. In that case, you should pipette more gently or decrease the number of pipetting times.

EBs also die or differentiate poorly when human ES cells are infected with mycoplasma. Screening for contamination such as mycoplasma should be regularly undertaken during any human ES cell work (see **Chapters 1, 2 and 5** for more on mycoplasma, consequences of having it and how to detect it).

References

Bonner-Weir S and GC Weir. (2005). New sources of pancreatic beta-cells. *Nat Biotechnol* **23**: 857–861.

Cowan CA, I Klimanskaya, J McMahon, J Atienza, J Witmyer, JP Zucker, S Wang, CC Morton, AP McMahon, D Powers and DA Melton. (2004). Derivation of embryonic stem-cell lines from human blastocysts. *N Engl J Med* **350**: 1353–1356.

D'Amour KA, AD Agulnick, S Eliazer, OG Kelly, E Kroon and EE Baetge. (2005). Efficient differentiation of human embryonic stem cells to definitive endoderm. *Nat Biotechnol* **23**: 1534–1541.

Doetschman TC, H Eistetter, M Katz, W Schmidt and R Kemler. (1985). The *in vitro* development of blastocyst-derived embryonic stem cell lines: Formation of visceral yolk sac, blood islands and myocardium. *J Embryol Exp Morphol* **87**: 27–45.

Hoffman LM and MK Carpenter. (2005). Characterization and culture of human embryonic stem cells. *Nat Biotechnol* **23**: 699–708.

Kanai-Azuma M, Y Kanai, JM Gad, Y Tajima, C Taya, M Kurohmaru, Y Sanai, H Yonekawa, K Yazaki, PPL Tam and Y Hayashi. (2002). Depletion of definitive gut endoderm in Sox17-null mutant mice. *Development* **129**: 2367–2379.

Lavon N and N Benvenisty. (2005). Study of hepatocyte differentiation using embryonic stem cells. *J Cell Biochem* **96**: 1193–1202.

Lowe LA, S Yamada and MR Kuehn. (2001). Genetic dissection of nodal function in patterning the mouse embryo. *Development* **128**: 1831–1843.

Nishikawa SI, S Nishikawa, M Hirashima, N Matsuyoshi and H Kodama. (1998). Progressive lineage analysis by cell sorting and culture identifies FLK1+VE-cadherin+cells at a diverging point of endothelial and hematopoietic lineages. *Development* **125**: 1747–1757.

Vincent SD, NR Dunn, S Hayashi, DP Norris and EJ Robertson. (2003). Cell fate decisions within the mouse organizer are governed by graded Nodal signals. *Genes Dev* **17**: 1646–1662.

Wells JM and DA Melton. (1999). Vertebrate endoderm development. *Annu Rev Cell Dev Biol* **15**: 393–410.

12, Part B

Directed differentiation of human embryonic stem cells into hepatic cells

NETA LAVON

Cedars-Sinai International Stem Cell Institute, Cedars-Sinai Medical Center, 8700 Beverly Blvd., Los Angeles, CA 90048, USA

and

NISSIM BENVENISTY

Department of Genetics, The Hebrew University, Jerusalem 91904, Israel

Introduction

Hepatocytes constitute the primary cells of the liver and perform many crucial functions, including metabolizing dietary molecules, detoxifying compounds, and glycogen storage. Patients with acute hepatic failure or end-stage liver disease must be treated by transplantation of the liver or hepatocytes (Horslen and Fox, 2004). Transplantation of hepatocytes has proven to be efficient, but the availability of human hepatocytes remains limited. For this reason, human ES cells have been proposed as an alternative source of hepatocytes for transplantation.

The ability of human ES cells to proliferate indefinitely in culture allows them to be an unlimited source of cells. Differentiation of human ES cells into hepatocytes may also serve as a model to study processes involved in embryonic development of the liver. Understanding of these developmental processes may help in diagnosis and treatment of liver-associated congenital pathologies. This chapter reviews the various protocols for the differentiation human ES cells into hepatocytes, details our protocol, and emphasizes the complexities in their subsequent characterization.

Human ES cells grown in suspension aggregate to form spheroid clumps of cells called embryoid bodies (EBs) (Itskovitz-Eldor *et al.*, 2000). Cells within the EBs express molecular markers specific for the three embryonic germ layers. The EBs mature by the process of spontaneous differentiation and cavitations, and the cells progressively acquire molecular markers for differentiated cell types. Dissociation of

EBs and plating the differentiated cells as a monolayer has revealed many cell lineages (Itskovitz-Eldor *et al.*, 2004). Addition of various growth factors to a culture of differentiating human ES cells facilitates their differentiation into specific cell types (Schuldiner *et al.*, 2000).

We demonstrated that human ES cells can spontaneously differentiate into hepatic-like cells (Lavon *et al.*, 2004). Human hepatic-like cells were isolated and characterized for their phenotype. Through gene manipulation hepatic cells were labeled, and a homogenous population of differentiated cells was demonstrated. Hepatic-like cells appeared to develop in a niche next to cardiac mesodermal cells, and acidic fibroblast growth factor (aFGF) seemed to play a role in this differentiation. Since many of the genes expressed in the liver are expressed in other tissues as well, it is essential to isolate a homogenous population of cells in order to characterize the hepatic cells properly. Thus, only a subpopulation of cells that expresses several hepatic genes will be characterized as hepatic cells. To characterize a cell as a hepatocyte we should also state its developmental stage, since embryonic, fetal and adult hepatocytes differ in their gene expression and functionality.

Differentiation towards hepatic cells was also demonstrated by using other protocols. Shirahashi *et al.* (2004) added insulin and dexamethasone to EBs cultured on collagen type I and showed that the cells express various endodermal genes. Rambhatla *et al.* (2003) added sodium butyrate to the culture media and induced hepatic differentiation. The resultant cells were morphologically similar to primary hepatocytes and most of the cells expressed liver-associated proteins. Schwartz *et al.* (2005) reported that the addition of fibroblast growth factor 4 and hepatocyte growth factor (HGF) to EBs grown in serum free media and their subsequent attachment to collagen type I plates, produced maximal differentiation into hepatic cells. Baharvand *et al.* (2006) reported differentiation of hES cells into hepatocytes using EBs in 3D collagen type I scaffold with the addition of aFGF, HGF, oncostatin M and dexamethasone. Soto-Gutierrez *et al.* (2006) indicate that following suspension of human ES cells as EBs, growing the resultant cells on poly-amino-urethane-coated, non-woven polytetrafluoroethylene fabric in the presence of bFGF, variant HGF, dimethyl sulfoxide (DMSO) and dexamethasone in a stepwise manner yields hepatocytes. **Table 1** summarizes studies on differentiation of human ES cells into hepatic-like cells, and details their differentiation protocol and the parameters that were used in order to characterize the hepatic cells.

Here we detail our protocol for isolating hepatic cells from human ES cells and the creation of human ES cells lines expressing the enhanced green fluorescent protein (eGFP) reporter gene driven by a hepatic promoter. The protocol details differentiation of human ES cells through EBs and the isolation of eGFP positive hepatic cells using fluorescence activated cell sorting (FACS).

Materials, reagents and materials

Cells

Mitotically inactivated feeders (see **Chapter 4** for how to prepare and mitotically inactivate mouse feeders).

Table 1   Differentiation potential of human ES cells into hepatic cells

Differentiation protocol	RNA markers	Protein markers	Other assays	Reference
Spontaneous (through EBs)	AFP, ALB, APOA4, APOB, APOH, FGA, FGG, FGB,	ALB, AFP	FACS isolation of Alb-eGFP cells	Lavon et al. (2004)
Insulin, dexamethasone, collagen type I (through EBs)	ALB, AAT	ALB	Urea synthesis	Shirahashi et al. (2004)
Na butyrate & DMSO (EBs/monolayer + HCM)	ALB, AAT, AGRP, GATA4, HNF4, TTR, CEBPA, CEBPB	ALB, AAT, CK8, CK18, CK19	ALB synthesis, CYP1A2 activity, PAS	Rambhatla et al. (2003)
FGF4, HGF, collagen type I (through EBs in serum free media)	AFP, ALB, CK18, CK19, GATA4, HNF3B, HNF1, CYP1A1, CYP1A2, CYP2B6, CYP3A4	ALB, HNF1, CK18, HNF3B, ASPGR1	Urea & Alb synthesis, ICG uptake, PROD & CYP2B6	Schwartz et al. (2005)
aFGF, HGF, OSM, Dex. collagen type I 3D scaffold (through EBs)	HNF3B, AFP, TTR, AAT, CK8, CK18, CK19, ALB, CYP7A1, TDO, TAT, G6P	ALB, CK18	Urea & Alb & AFP synthesis PAS, ICG, EM	Baharvand et al. (2006)
bFGF, variant HGF, DMSO, Dex., PAU-coated non-woven PTFE fabric (through EBs)	ALB		EM, ALB & UREA synthesis, lidocaine & ammonia metabolism	Soto-Gutierrez et al. (2006)

Differentiation: EBs — embryoid bodies, DMSO — dimethyl sulfoxide, HCM — hepatocyte culture medium, FGF4 — fibroblast growth factor 4, HGF — hepatocyte growth factor, aFGF — acidic fibroblast growth factor, OSM — oncostatin M, Dex. — dexamethasone, PAU — Poly-amino-urethane, PTFE — polytetrafluoroethylene.

Molecular markers: AFP — alpha fetoprotein, ALB — albumin, APO — apolipoprotein, FG — fibrinogen, AAT — alpha-1-antitrypsin, HNF — hepatocyte nuclear factor, TTR — transthyretin, CEBP — enhancer binding protein, CK — cytokeratin, CYP — cytochrome P450, TDO — tryptophan-2,3-dioxygenase, TAT — tyrosine aminotransferase, G6P — glucose-6-phosphatase, ASPGR1 — asialoglycoprotein receptor.

Functional assay: FACS — fluorescence activated cell sorting, PAS — periodic acid Shiff staining for glycogen, ICG — indocyanine green, PROD — pentoxyresorufin assay, EM — electron microscopy.

Human embryonic stem cells (the protocols detailed below were performed using H9 human ES cell line). Protocols for culturing and passaging human ES cells with trypsin are detailed in **Chapter 5**.

Media

Murine embryonic fibroblast (MEF) medium: as described in **Chapter 4.**
Human ES medium: as described in **Chapter 5.**
EB media: the same as ES media but without the bFGF.

Protocols

1. Creating human ES cell lines carrying the reporter gene eGFP under the albumin promoter

Preparation of DNA for transfection

Plasmid DNA should be linearized by a restriction enzyme that cuts in the vector sequence and not in the construct itself. Complete digestion should be checked and the plasmid DNA is then precipitated, washed and resuspended in sterile double distilled water to a concentration of 1 µg/µL.

Transfection by ExGen 500 (Fermentas #RO511)

1) Human ES cells are plated in a gelatinized six-well dish containing antibiotic resistant MEF (DR4 mice contain resistance genes for neomycin puromycin, hygromycin and 6-thioguanine, Jackson Laboratories #003208). The density should allow that on the following day, there are a large number of small colonies plated uniformly over the well (20–40% confluence).

2) 1 hr prior to transfection, change human ES medium to 1 mL of fresh human ES medium per well.

3) For each well of a six-well tissue culture dish, prepare an Eppendorf tube containing 4 µg of DNA in 100 µL of 150 mM NaCl, vortex briefly, and spin down.

4) Add 13 µL ExGen 500 to each tube (not reverse order) and vortex immediately for 10 s.

5) Allow to stand for 10 min at room temperature.

6) Add 100 µL of ExGen/DNA mixture to each well.

7) To equally distribute the complexes on the cells, gently rock the plate to and fro.

8) Centrifuge the six-well tissue culture dish in a swinging bucket centrifuge for 5 min at 300 g (~1,000 rpm).

9) Incubate at 37°C, 5% CO_2 for 30 min.

10) Wash twice with PBS, add ES medium and return to the incubator.

11) Two days later, selection can be initiated with the appropriate antibiotic.

12) After two more days, massive cell death should be visible. Wash with PBS and replace the ES medium with new antibiotics every 2–3 days. Different human ES cell lines vary in their intrinsic resistance to antibiotic, thus for each cell line the optimal concentration of antibiotic that causes all non-transfected cells on a plate die should be empirically determined.

13) After 5–10 days antibiotic resistant colonies should appear. Using mouth pipette transfer each new colony to separate gelatinized 12-well dishes containing antibiotic resistant MEFs with human ES medium. Maintain the cells in selection media since this ensures that the integrated DNA is not lost during passaging.

14) Expand the cells of each colony and test the cells for the presence of the exogenous DNA segment by PCR. Also test that no eGFP signal appears in the non-differentiated human ES cells (leakiness).

Differentiation of human ES cells by EBs formation

1) Wash a confluent 10 cm plate of human ES cells with PBS and harvest the cells by adding 1 mL of trypsin–EDTA (0.25% trypsin and 0.05% EDTA in Puck's saline A (Gibco BRL 25200-072) for 5 min at room temperature.

2) Add 5 mL of MEF medium, pipette up and down, transfer the cells to a 15 mL conical tube, and centrifuge at 600 g ($\sim$1,600 rpm) for 5 min.

3) The supernatant is aspirated and the pellet is resuspended in EB media by pipetting up and down with a 1 mL pipette.

4) $4-5 \times 10^6$ cells from the cell suspension are then transferred to a sterile (UV-irradiated) non-adherent 90 mm dish containing 15 mL of EB media with antibiotic (Greiner 632 180).

5) The plate is incubated at 37°C at 5% CO_2 for 20 days.

6) It is recommended that for the first 2 days at least, the plate is moved as little as possible. Than, every second day, half of the medium is carefully removed from the plate in such a way as to not take up the EBs, and fresh EB medium is added. This can be accomplished by tilting the plate at an angle. The EBs normally move to the bottom and the medium above is relatively free of EBs. Fresh EB medium containing antibiotic is then added to the plate.

In the lower panel of **Figure 1** (see overleaf; color version in color plate section), we show a heterogeneous culture of differentiated cells derived from an Alb-eGFP human ES cell line. 20 day old EBs were trypsinized and plated on a tissue culture plate in order to analyze the eGFP expression in the differentiated cells.

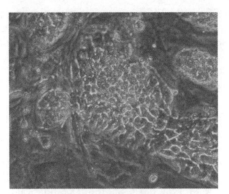

Undifferentiated Alb-eGFPcells

Differentiated Alb-eGFPcells

Figure 1   Undifferentiated and differentiated cells from Alb-eGFP human ES cell
line. Upper panel, colonies of undifferentiated human ES cells on a feeder layer and
no eGFP signal is present. Lower panel,
a heterogeneous culture of differentiated
cells. 20 day old EBs were trypsinized and
plated on a tissue culture plate, some of
the cells are eGFP positive. This image is
best seen in color; please refer to plate
section to see the color image.

Isolation of hepatic-like cells by FACS sorting

1. From a 10 cm plate of 20 day old EBs, transfer the medium with the EBs into
    50 mL tubes and centrifuge the cells for 5 min at 600 g ($\sim$1,200 rpm).

2. Wash once with PBS.

3. Harvest the cells by adding 3 mL of Trypsin-EDTA for 5 min at 37°C at 5% CO_2.
    Twice during the incubation period, take the tube out of the incubator, swirl gently
    and replace inside.

4. Add 10 mL of MEF medium and pipette up and down several times in order to further dissociate the EBs, then centrifuge at 600 g (~1,200 rpm) for 5 min.

5. Carefully aspirate the MEF medium and add 200 µL of PBS.

6. Sort the eGFP positive cells by FACS.

About 5% of the cells are expected to be eGFP positive using this protocol.

The sorted population of cells can be further characterized by microarray analysis or further grown if the sorting is under sterile conditions.

Troubleshooting

The transfection seems to work fine with one of my plasmids but fails when I use another plasmid I have. What should I do?

Sometimes with certain plasmids the transfection does not work. This seems to be related to the purity of the plasmid. Try another cleaning step for the plasmid before or after the linearization and quantify the concentration again.

When I put the cells on a non-adherent plate in order to cause them to create EBs, a lot of cells adhere to the plate although they are not tissue culture treated. What should I do?

We tried many kinds of plates from different brands, and although the cells were not supposed to adhere, some of them did, and this adversely affects EBs formation. Thus, we recommend using the exact catalog numbers detailed above, and do not neglect to UV sterilize them. If problems persist, there are new plates commercially available called Ultra Low Attachment Surface Dishes From Corning Company (catalog number 3262). These work the best, but they are quite expensive.

References

Baharvand H, SM Hashemi, S Kazemi Ashtiani and A Farrokhi. (2006). Differentiation of human embryonic stem cells into hepatocytes in 2D and 3D culture systems in vitro. *Int J Dev Biol* **50**: 645–652.

Horslen SP and IJ Fox. (2004). Hepatocyte transplantation. *Transplantation* **77**: 1481–1486.

Itskovitz-Eldor J, M Schuldiner, D Karsenti, A Eden, O Yanuka, M Amit, H Soreq and N Benvenisty. (2000). Differentiation of human embryonic stem cells into embryoid bodies comprising the three embryonic germ layers. *Mol Med* **6**: 88–95.

Lavon N, O Yanuka and N Benvenisty. (2004). Differentiation and isolation of hepatic-like cells from human embryonic stem cells. *Differentiation* **72**: 230–238.

Rambhatla L, CP Chiu, P Kundu, Y Peng and MK Carpenter. (2003). Generation of hepatocyte-like cells from human embryonic stem cells. *Cell Transplant* **12**: 1–11.

Schuldiner M, O Yanuka, J Itskovitz-Eldor, DA Melton and N Benvenisty. (2000). Effects of eight growth factors on the differentiation of cells derived from human embryonic stem cells. *Proc Natl Acad Sci USA* **97**: 11307–11312.

Schwartz RE, JL Linehan, MS Painschab, WS Hu, CM Verfaillie and DS Kaufman. (2005). Defined conditions for development of functional hepatic cells from human embryonic stem cells. *Stem Cells Dev* **14**: 643–655.

Shirahashi H, J Wu, N Yamamoto, A Catana, H Wege, B Wager, K Okita and MA Zern (2004). Differentiation of human and mouse embryonic stem cells along a hepatocyte lineage. *Cell Transplant* **13**: 197–211.

Soto-Gutierrez A. *et al.* (2006). Differentiation of human embryonic stem cells to hepatocytes using deleted variant of HGF and poly-amino-urethane-coated non-woven polytetrafluoroethylene fabric. *Cell Transplant* **15**: 335–341.

12, Part C

Directed differentiation of human embryonic stem cells into pancreatic cells

HIRAM CHIPPERFIELD

ES Cell International Pte Ltd, 11 Biopolis Way, #05-06 Helios, Singapore 138667

Introduction

The writing of this chapter presents a challenge as there currently is no efficient method for the differentiation of human embryonic stem (ES) cells into insulin-secreting pancreatic β-cells. This lack of an efficient protocol is notable because, in contrast to some other diseases, there exists an effective cell therapy for the treatment of type I diabetes. The transplantation of β-cells contained within pancreatic islets harvested from cadavers is an effective therapy and can result in insulin independence albeit for a period of time (Shapiro *et al.*, 2006). There is, of course, a chronic shortage of donor organs available for transplantation (Balamurugan *et al.*, 2006) so the generation of pancreatic β-cells from human ES cells has the potential to fulfil this therapeutic need (Melton, 2006).

There are many approaches with which one can attempt to generate pancreatic β-cells from human ES cells. Soria and colleagues have successfully used a mouse ES cell line engineered with an antibiotic resistance gene under the control of an insulin promoter to produce β-like cells (Soria *et al.*, 2000); unfortunately this approach is difficult to implement across multiple cell lines and human ES cells are currently less amenable to being genetic engineered. Enrichment for nestin-positive progenitors has also been used as a differentiation strategy for both mouse (Lumelsky *et al.*, 2001) and human ES cells (Baharvand *et al.*, 2006) but the results have been controversial and should be interpreted with extreme caution (Treutelaar *et al.*, 2003; Rajagopal *et al.*, 2003).

Recapitulating the critical events of development *in vitro* is a popular and accepted approach to generating specific cell types from embryonic stem cells. This approach is modeled on the success of Wichterle *et al.* in differentiating mouse ES cells into motor neurons (Wichterle *et al.*, 2002) and follows from the lack of success in finding a useful, single step differentiation protocol for pancreatic β-cells. This section will focus on this approach.

Assuming consistency with normal development, directed differentiation of human ES cells to β-cells can be broken down into several steps: from undifferentiated human ES (Oct4-positive), to definitive endoderm (Sox17-positive), to pancreatic progenitor (Pdx1-positive), to endocrine progenitor (Ngn3-positive) to β-cell (insulin-positive). This scheme, summarized in **Figure 1**, greatly simplifies the developmental biology but it is a useful model for the rational application of differentiation factors. Reviews from the Melton lab are a good starting point for a fuller understanding of pancreatic and β-cell development. (Gu *et al.*, 2003, Grapin-Botton and Melton, 2000)

Due to the lack of a definitive protocol this chapter section provides a framework from which specific protocols can be adapted. It outlines useful reagents and discusses applicable factors at each stage of differentiation. Thus, it constitutes a useful resource for researchers seeking to generate pancreatic β-cells from human ES cells.

Overview of protocol

This protocol outlines a method to differentiate human ES cells into pancreatic β-cells through a step-wise process. **Figure 1** gives a flow chart of the protocol showing the applicable factors, the duration that they should be applied, markers that should be positive and markers that should be negative at each stage of the differentiation. Knowledge of basic human ES cell culture is assumed as it is covered elsewhere in **Chapter 5**.

Materials, reagents and equipment

General equipment

No special equipment is necessary for pancreogenesis; if the laboratory is equipped for human ES cell culture then it should have all the basic items necessary for differentiation (see **Chapters 1 and 5** for general equipment listings). Aside from tissue culture facilities the lab should minimally contain equipment necessary to perform PCR and immunostaining. Ideally, a real-time PCR machine, a good fluorescent microscope for immunofluoresence and a fluorescence activated cell sorting (FACS) machine should also be available.

Antibodies

Useful antibodies are summarized in **Table 1** but a number of them warrant further comment. The author recommends Jackson Immunoreseach Laboratories for all secondary antibodies.

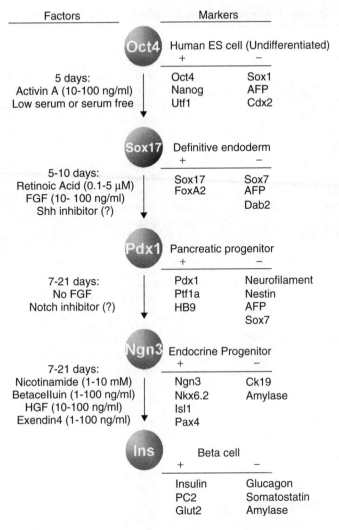

Figure 1 A schematic diagram of the β-cell differentiation protocol and the markers which are expressed at different stages. On the left are the factors and the duration at which they should be applied to direct differentiation from one stage to the next. On the right are markers that should be expressed at each stage (+) and markers that should be absent or minimized (−).

Sox17 and FoxA2: Both the goat anti-Sox17 and goat anti-HNF3β (FoxA2) antibodies from R&D give clear nuclear staining with little background. At a high concentration it is possible to use the Sox17 antibody for FACS though care must be taken as there can be some background when it is used for this technique. FACS analysis for Oct4 and Sox17 is a highly useful way to quantitatively monitor the first stage of differentiation to definitive endoderm.

Table 1   A summary of reagents useful in the differentiation of human ES cells to β-cells

Antibodies	Vendor	Catalog number
Ms anti-Oct4	Santa Cruz	SC-5279
Gt anti-Sox17	R & D	AF1924
Gt anti-FoxA2	R & D	AF2400
Ms anti-AFP (C3)	Sigma	A8452
Gt anti-Pdx1 (A-17)	Santa Cruz	SC-14664
Ms anti-CK19	Chemicon	MAB3238
Ms anti-Ngn3	DHSB	F25A1B3
Ms anti-Isl1	DHSB	39.4D5
Rb anti-C-peptide	Linco	4020-01
Gp anti-Insulin	Linco	4011-01F
Gt anti-Amylase	Santa Cruz	SC-12821
Gt anti-Glucagon	Santa Cruz	SC-7780
Rb anti-somatostatin	DAKO	A0566
Gt anti-Glut2	Santa Cruz	SC-7580

Growth factors	Vendor	Catalog number
Activin A	R&D Systems	338-AC-025
FGF2	Invitrogen	13256-029
FGF10	R&D Systems	345-FG-025
Betacellulin	R&D Systems	261-CE-010
HGF	R&D Systems	294-HG-005
Exendin4	Sigma	E7144
DKK1	R&D Systems	1096-DK-010
DKK3	R&D Systems	1118-DK-050

Chemicals and inhibitors	Vendor	Catalog number
Retinoic acid	Sigma	R2625-50MG
Cyclopamine	Calbiochem	239803
SANT-1	Calbiochem	559303
DAPT (γ-secretase inhibitor IX)	Calbiochem	565770
LY-685,458 (γ-secretase inhibitor X)	Calbiochem	565771
Nicotinamide	Sigma	N0636-100 G
Heparin	Sigma	H4784
Palmitate	Sigma	P9767
Tolbutamide	Sigma	T0891
Forskolin	Sigma	F6886

Tissue culture media and reagents	Vendor	Catalog number
CMRL-Medium 1066	Invitrogen	11530-037
RPMI 1640	Invitrogen	21870-076
Low-glucose DMEM	Invitrogen	10567-014
KnockOut Serum Replacement	Invitrogen	10828-028
N-2 Supplement	Invitrogen	17502-048
B-27	Invitrogen	17504-044
BD Matrigel	BD Biosciences	354234
GFR Matrigel	BD Biosciences	354230
Ultra-low attachment 6-well plates	Corning	3471

Table 1   (*continued*)

Miscellaneous	Vendor	Catalog number
iQ SYBR Green Supermix	BioRad	170-8882
Human insulin ELISA kit	Linco Research	EZHI-14K
Human C-peptide ELISA kit	Linco Research	EZHCP-20K
Human pancreas sections	Spring Bioscience	STS-016
Rat/Mouse Insulin ELISA kit	Linco Research	EZRMI-14K

Abbreviations: Ms (Mouse), Gt (Goat), Rb (Rabbit) and Gp (Guinea pig).

Pdx1: One of the most critical reagents is a good Pdx1 antibody. Unfortunately most published immunostains have used antibodies from Chris Wright. There is a notable absence of good, commercially available Pdx1 antibodies.

Santa Cruz offers a number of Pdx1 antibodies but only the goat anti-Pdx1 (A-17) worked in our hands, giving nice staining on both mouse and human tissue. It is recommended that you test any Pdx1 antibody on sections of human pancreas as a positive control. A number of companies, including Spring Bioscience sell paraffin embedded sections of human pancreas and they are worth the expense as a means of validating antibodies and staining conditions.

Ngn3: A useful endocrine progenitor marker is Ngn3 which also lacks well charac- terized commercial antibodies. Recently the Developmental Studies Hybridoma Bank (DSHB) has started to distribute a monoclonal anti-Ngn3 which has been useful for staining mouse tissue (Zahn *et al.*, 2004). It has not yet been determined if this antibody reacts with the human protein.

Insulin and C-peptide: Generally the claimed species specificities of insulin and C-peptide antibodies correspond with what is found in the laboratory. To ensure that exogenous insulin does not result in false staining it is easiest to use the human specific C-peptide antibody from Linco.

Growth factors

Useful growth factors are summarized in **Table 1.** Ideally, resuspend growth factors according to the manufacturer's recommendations and aliquot them into single use tubes for storage in a freezer. Where this is not practical growth factors should generally be stored at 4°C for no longer than 2 weeks.

All *trans*-retinoic acid (RA)

RA is particularly sensitive to degradation from light and exposure to oxygen. As such it should be stored in single use aliquots dissolved in dimethyl sulfoxide (DMSO) at −80°C, ideally but not essentially under an inert gas. The 50 mg ampoules of RA from Sigma (R2625-50MG) are relatively inexpensive so they can be used to make small batches of aliquots which can be used quickly without significant degradation. At concentrations of greater than 100 mM RA can precipitate if added directly to media so it should be diluted to a working concentration of 10–20 mM in DMSO. Finally, RA is a potent teratogen so due care should be exercised when it is used.

Protocol

Overview

The initial two steps culminating in Pdx1-positive pancreatic progenitors are the best characterized. The maturation of these progenitors is more tentative and requires further investigation. Rather than give a specific recipe that has been optimized for a specific human ES cell line cultured in a specific manner, this protocol aims to provide a toolbox from which reagents and factors can be taken and optimized for a particular setting.

An example application of these tools is shown in **Figure 2**. Collagenase passaged ES3, available from ES Cell International (www.escellinternational.com), were plated onto growth factor reduced (GFR) Matrigel coated plates. GFR Matrigel was used as the growth factors in regular Matrigel can mask the effects of added factors. Endoderm was induced for 5 days by culture with RPMI 1640 plus 10% KnockOut Serum

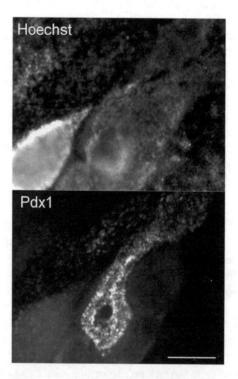

Figure 2   Pdx1 expression in differentiating human ES cells. Differentiated HES3 were stained for Pdx1 (G.p. anti-Pdx1, a generous gift of C. Wright) and counterstained with Hoechst. See text for details of differentiation, scale bar is 100 μM.

Replacement (KOSR) and 50 ng/mL of Activin A. Pdx1 was induced by culturing in RPMI 1640 with 2% B-27, 20 ng/mL FGF10 and 1 μM RA for a further 5 days. At 10 days a small but significant number of cells were positive for Pdx1 as determined by immunostaining and RT-PCR (unpublished observation).

Cell line variation

D'Amour *et al.* demonstrated clear differences between various human ES cell lines in their ability to form definitive endoderm using their protocol (D'Amour *et al.*, 2005a). No comparable analysis has been performed for the induction of Pdx1 or the subsequent differentiation towards β-cells but similar variations are likely to found.

By tailoring differentiation protocols for different lines it may be possible that differentiation can be induced at comparable efficiencies. For example, ES2 grown on human feeders may differentiate more slowly than ES3 grown on MEFs requiring the initial inductive steps to be longer.

General culture conditions

1. Embryoid body or monolayer?

Currently, there is no consensus on what is the best format for differentiation towards the pancreatic lineage. Many protocols incorporate an embryoid body (EB) formation step (Xu *et al.*, 2006; Baharvand *et al.*, 2006) which, due to their three-dimensional structure, may assist in the differentiation of complex structures. Pancreatic induction in monolayer is certainly possible as evidenced by Pdx1 induction by 9 days of Activin A (100 ng/mL) treatment in cells cultured on Matrigel coated plates (Yao *et al.*, 2006).

An advantage of EB formation is that the cells are in a three-dimensional structure that more closely resembles *in vivo* conditions than a monolayer of cells. A major disadvantage is that differentiation factors may not have access to all of the cells within an EB as the inner cells of an EB can be separated from the media by an epithelial layer (Itskovitz-Eldor *et al.*, 2000). A result of this separation is that differentiation may proceed unevenly and at a low efficiency. There are a number of three-dimensional matrices which are designed to combine the advantages of a three-dimensional system with the free exchange of media but it remains to be determined how applicable these will be to pancreatic differentiation (Ouyang *et al.*, 2007, Chen *et al.*, 2006).

If free-floating EB formation is chosen then the best plates are the ultra-low attachment plates from Corning. Bacteriological Petri dishes are a less expensive alternative but after several days EBs will often start to stick to them.

If monolayer differentiation is chosen then suitable substrates include Matrigel, fibronectin, laminin and collagen IV. Unsuitable substrates include tissue culture plastic, gelatine, poly-d-lysine and collagen I.

Feeders

Although feeder-free culture of human ES cells is possible (Yao *et al.*, 2006) most human ES cells cultures are maintained on a layer of mitotically inactivated feeder cells, usually mouse embryonic fibroblasts (MEFs). If they are not depleted these feeder cells can contribute to EBs and persist for extended periods.

Definitive endoderm can be induced in cells cultured on MEFs (D'Amour et al., 2005b) and one study has shown that the addition of extra MEFs during differentiation increases the expression of Pdx1 (Xu et al., 2006) so the presence of feeder cells is unlikely to be deleterious and could be beneficial. However, if feeder cells are not depleted, the rate at which the human ES cells differentiate is likely to be slower as feeder cells are selected to prevent differentiation and the activity of any exogenous factors needs to be interpreted with consideration of possible indirect activity via the feeders.

For collagenase passaged human ES cells, feeder cells can be substantially depleted by passing disrupted human ES cells colonies through a 40 or 70 µM filter; the colony fragments are retained while most of the feeder cells pass through. For trypsin passaged cultures there is no ideal way of removing feeder cells; plating cultures 2×30 min in gelatine coated dishes and retaining the non-adherent cells can deplete feeder cells to some extent.

2. Media

There is presently no strong evidence that any type of medium is better for pancreatic differentiation compared to others. Common media such as DMEM and DMEM:F12 are most widely used in published protocols. CMRL 1066 is considered the best for the culture of human islets making it worthy of consideration (Murdoch et al., 2004).

Fetal calf serum (FCS) is deleterious at the beginning of differentiation towards definitive endoderm and ideally should be avoided in subsequent steps too. FCS is undesirable because it is undefined, there is significant batch-to-batch variation which impedes the development of a standardized protocol and it may mask the effects of any added factors.

"Serum-free" media can be generated by using KnockOut Serum Replacement (KOSR) to replace FCS. KOSR is good at maintaining cells in a healthy condition and is already used in many labs to culture undifferentiated human ES cells so its use during differentiation results in a consistent medium for the cells. The disadvantage of KOSR is that it is a serum-based preparation with defined differentiation factors removed so unknown components are still present that could inhibit some steps of differentiation or interact with added factors. It has been shown that KOSR has a high level of BMP activity (Xu et al., 2005) which may initially inhibit neural differentiation (Xu et al., 2005) but could be deleterious at later stages of differentiation. A good alternative to KOSR are the N2 and B-27 supplements from Invitrogen which together provide a defined serum-free media which nicely supports the growth and differentiation of human ES cells (Yao et al., 2006).

3. Monitoring differentiation

The three main assays that are useful in monitoring differentiations are RT-PCR, immunostaining and FACS analysis.

RT-PCR is a good first step of analysis as it can be quantitative, can be relatively easily performed for any marker, is extremely sensitive and is specific provided care is taken. The main disadvantage of the technique is that, although it can measure the abundance of a transcript in a population of cells, it provides no indication as to the proportion of cells that express a gene. Thus RT-PCR must be combined with immunostaining to avoid the possibility of a very small number of cells giving a positive RT-PCR result.

Table 2 shows useful PCR primer pairs that have successfully been used for SYBR Green real-time PCR using iQ SYBR Green supermix from BioRad with an annealing temperature of 55°C.

To convincingly show that human ES cells have differentiated to β-cells it is necessary to demonstrate co-expression of *Insulin* and *Pdx1* or a similar combination of terminal markers. Therefore good quality immunostaining and the validation of antibodies is essential; this technique can also provide some indication on the proportion of cells that express a particular marker.

FACS allows the quantitative assessment of differentiation efficiency and should therefore be pursued. Unfortunately many antibodies do not work with FACS analysis.

Undifferentiated human ES cells to definitive endoderm

1. **Time:** 4–6 days

2. **General considerations**

The first step towards a pancreatic β-cell is the differentiation from pluripotent human ES cells to definitive endoderm, which in practical terms can be simplified to the induction of Sox17/FoxA2 in undifferentiated cells. This step is covered in detail in **Chapter 12, part A** but it is worthwhile to note that perhaps pancreatic specification can be achieved at this time using a single endoderm specification step (Yao *et al.*, 2006).

Definitive endoderm to pancreatic progenitor

1. **Time:** 5–10 days

2. **General considerations**

The most important and commonly used marker for pancreatic differentiations is Pdx1 (Ipf1). In humans, it is first expressed 26 days post-conception in the nascent pancreatic bud, continues to be expressed in the pancreatic epithelium through development of the pancreas before localizing to the islets and ductal epithelium around 14 weeks post-conception (Piper *et al.*, 2004). The *Pdx1* knockout mouse lacks a pancreas (Jonsson *et al.*, 1994) and this pivotal role in pancreatic specification is likely preserved in humans as a mutation in the gene has been associated with a case of human pancreatic agenesis (Stoffers *et al.*, 1997). Thus the differentiation of definitive endoderm to pancreatic progenitor can be summarized as the differentiation of a Sox17/FoxA2-positive cell to a Pdx1-positive cell.

Table 2   PCR primers for pancreatic lineage genes

Gene	Forward primer 5′	Reverse	Size (bp)
β-actin	CAATGTGGCCGAGGA CTTTG	CATTCTCCTTAGAGA GAAGTGG	126
Amylase	ATTCGCAAGTGGAATG GAGA	CCCATGTCCTCGTTG ATTGT	123
Brachyury (T)	AATTGGTCCAGCCTTG GAAT	CGTTGCTCACAGACC ACAG	112
Chromogranin A	CGGTTTTGAAGATGAA CTCTCAG	GCTCTTCCACCGCCT CTT	77
Glucagon	CCAAGATTTTGTGCAG TGGT	GGTAAAGGTCCCTTC AGCAT	100
Glut2	CATGTCAGTGGGACTT GTGC	CTGGCCCAATTTCAA AGAAG	101
Glucokinase	TCAGAAGCCTACTGGG GAAG	AACTCTGCCAGGATC TGCTC	126
HB9	TGCCTAAGATGCCCGA CTT	AGCTGCTGGCTGGTG AAG	91
Hnf3α (Foxa2)	GGAGCGGTGAAGATG GAA	TACGTGTTCATGCCG TTCAT	122
Insulin	GGGGAACGAGGCTTCT TCTA	CACAATGCCACGCTT CTG	147
Isl1	GCTTTGTTAGGGATGG GAAA	CACGAAGTCGTTCTT GCTGA	104
Kir6.2 (KCNJ11)	GCCTGAGGCTGGTATT AAGAAGT	CTCCCCACCAGACTC TTCCT	66
Ngn3	TTTTGCGCCGGTAGAA AG	GGGCAGGTCACTTCG TCTT	114
Nkx2.2	CGAGGGCCTTCAGTAC TCC	GGGGACTTGGAGCTT GAGT	72
Nkx6.1	CGTTGGGGATGACAG AGAGT	CGAGTCCTGCTTCTT CTTGG	110
Oct4 (Pou5f1)	GGCAACCTGGAGAATT TGTT	GCCGGTTACAGAACC ACACT	118
Pax4	TCTCCTCCATCAACCG AGTC	GAGCCACTATGGGGA GTGAG	115
PC1/3 (Pcsk1)	CTCTGGCTGCTGGCAT CT	CTGCATATCTCGCCA GGTG	68
PC2 (Pcsk2)	GAGAAGACGCAGCCT ACACC	CTGCAAAGCCATCTT TACCC	67
Pdx1 (Ipf1)	CCTTTCCCATGGATGA AGTC	GGAACTCCTTCTCCA GCTCTA	145
Somatostatin	CCCAGACTCCGTCAGT TTCT	ATCATTCTCCGTCT GGTTGG	114
Sox17	CAGAATCCAGACCTGC ACAA	CTCTGCCTCCTCCAC GAA	102
Sur1 (ABCC8)	TGGTCTTAGTGACCCA CAAGC	CCTTCATGGCAATGA TCCA	63
Synaptophysin	CCAATCAGATGT AGTCTGGTCAGT	AGGCCTTCTCCTGAG CTCTT	70

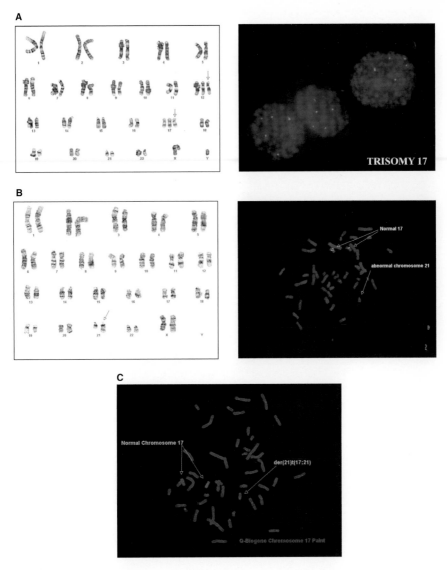

Plate 1 Karyotype analysis of human ES cell line H9 (NIH code WA09) that has been passaged over 20 times with trypsin–EDTA. (**A**) An abnormal female karyotype with partial duplication of the long arm of chromosome 17 resulting in partial trisomy of chromosome 17 and partial deletion of the long arm of chromosome 21. (**B**) FISH analysis on metaphase chromosomes using a cocktail probe that is specific for the HER-2/NEU and TOPO genes on chromosome 17q11.2-q12 together with a chromosome 17 centromere control probe in order to identify additional material present on the long arm of chromosome 21. Both the control cell line and the H9 (p40) human ES cell line showed a normal signal pattern, with no additional signals present on the long arm of the derivative chromosome 21. (**C**) A whole chromosome probe specific for chromosome 17 was used to rule out a break point distal to region 17q12. The chromosome paint probe showed a fluorescent signal on the long arm of the derivative chromosome 21, confirming that the translocation was derived from chromosome 17 with the breakpoint distal to HER-2/NEU locus.

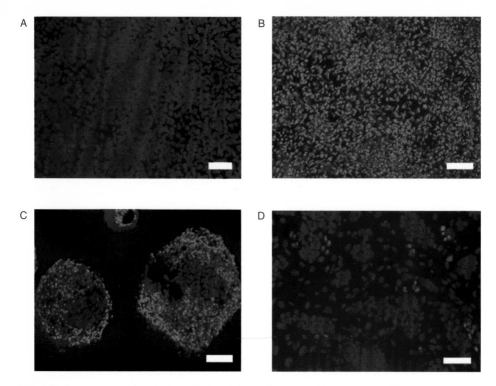

Plate 2 Immunocytochemical analysis of the induction of human ES cells with activin A (100 ng/mL) and low serum. Before treatment, a small number of SOX17-positive cells are seen (**A**). After 4 days of treatment, cells were processed by immunostaining and quantitated (no. of SOX17-positive cells/total no. of cells). More than 70 % of the cells are SOX17-positive endoderm in monolayer cultures (**B**), compared to 40 % observed in sections of EBs (**C**). Control staining of a monolayer culture without primary antibody (**D**). Red: SOX17; blue, nuclei. Bars: 100 μm.

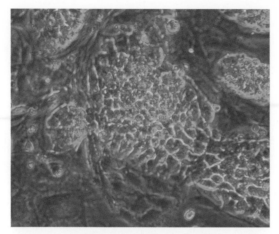

Undifferentiated Alb-eGFPcells

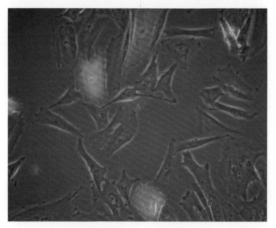

Differentiated Alb-eGFPcells

Plate 3 Undifferentiated and differentiated cells from Alb-eGFP human ES cell line. Upper panel, colonies of undifferentiated human ES cells on a feeder layer and no eGFP signal is present. Lower panel, a heterogeneous culture of differentiated cells. 20 day old EBs were trypsinized and plated on a tissue culture plate, some of the cells are eGFP positive.

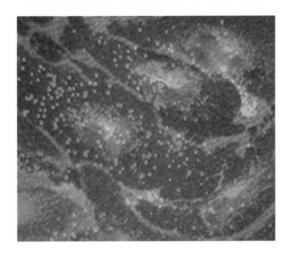

Plate 4 Analysis of human ES cell-derived endothelial cells. Following these three methods for differentiation of human embryonic stem cells, highly enriched populations of endothelial cells will be generated. A proportion of the final cell population should be assayed for endothelial markers. Image shows endothelial cells resulting from Protocol I, following staining with antibodies for PECAM (red), vWF (green) and a nuclear stain DAPI (blue).

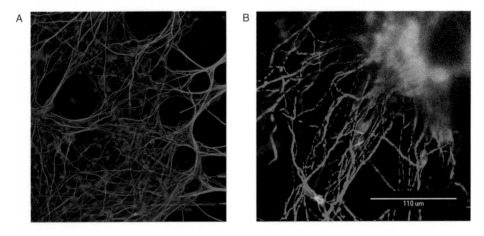

Plate 5 Human ES cells on PA6 for 48 days fixed and stained with (A) Tuj-1 (green) and dapi (blue), and (B) Tuj-1 (green) and HuC/D (red). Scale bars: (A) 205 μm; (B) 110 μm.

In most cases Pdx1-positive cells are pancreatic in nature but there are two possible exceptions that should be considered. There is limited evidence that Pdx1 is expressed in the brain (Schwartz *et al.*, 2000) and the visceral endoderm (McGrath and Palis, 1997). Expression in these regions is not described in the analysis of the Pdx1-LacZ knockin mouse (Offield *et al.*, 1996) but the possibility of non-pancreatic expression cannot be excluded. Ideally Pdx1 should be co-expressed with other pancreatic progenitor markers such as Ptf1a and HB9 in the absence of neural and visceral endoderm markers.

There are a number of factors that induce Pdx1 from definitive endoderm including *all trans*-retinoic acid (RA), FGFs and the absence of sonic hedgehog signaling.

All trans-retinoic acid

All trans-retinoic acid (RA) has many modes of action in the differentiation of human ES cells most notably in neural differentiation. As such care must be taken not to expose the cells to RA too early in the differentiation but only after endoderm has been induced.

A patent application from Cythera offers a useful guide for the application of RA (D'Amour *et al.*, 2005b). After inducing definitive endoderm using their published protocol (D'Amour *et al.*, 2005a), 1 µM RA with 50 ng/mL of FGF10 was added on days 5–10 resulting in a strong induction of Pdx1 as measured by QPCR. 1 µM RA was the highest and most effective concentration of RA that was tested.

In ES3, the Pdx1 induction was highest after 5 days of treatment with 3 µM RA with 50 ng/mL of FGF2 as measured by RT-PCR (unpublished observations).

It should be noted that the standard formulation of B27 contains an undisclosed and presumably low concentration of retinyl acetate, a functional analog of retinoic acid.

FGFs

It is unclear if the FGFs play an inductive or trophic role for Pdx1. FGF10 is the developmentally relevant molecule but FGF2 and possibly other members of this growth factor family can substitute. Whatever their exact mode of action FGFs have a synergistic action with RA in the induction of Pdx1 and should be added for 5–10 days at the relatively the relatively high concentrations of 10–100 ng/mL.

Sonic hedgehog inhibitors

During development there is a notable absence of Sonic hedgehog (Shh) in the nascent pancreatic bud which, along with other evidence (Hebrok *et al.*, 1998, Kim *et al.*, 1997), indicates that the inhibition of Shh signaling may enhance Pdx1 induction. It is possible to inhibit Shh with small molecules such as cyclopamine (0.1–10 µM) and SANT-1. If little or no Shh is present during differentiation the addition of any inhibitors would have little effect.

Pancreatic progenitor to endocrine progenitor

1. **Time:** >7 days

2. **General considerations**

Little is known about the factors that mediate the progression to endocrine progenitors. The inhibition of notch signaling may induce differentiation while β-catenin activity is unlikely to be important for the endocrine pancreas.

Notch inhibition

There is evidence that FGF activity maintains pancreatic progenitors in an undifferentiated state via Notch signaling indicating that perhaps inhibiting FGF or Notch signaling may induce further differentiation. (Murtaugh *et al.*, 2003) (Miralles *et al.*, 2006) FGF inhibition could be achieved through the use of saturating amounts of heparin or inhibitors of downstream kinases. γ-Secretase inhibitors such as DAPT or LY-685,458 inhibit Notch signaling and may also enhance this stage of differentiation.

β-Catenin activity

β-Catenin activity, possibly in response to Wnt signalling, was found to be necessary for the exocrine but not the endocrine compartment of the pancreas (Murtaugh *et al.*, 2005). Although it has not been observed, excessive exocrine differentiation might be modulated using Dkk protein from RnD.

Endocrine progenitor to β-cell

1. **Time:** >7 days

There is no clear guide on the length of time necessary for the maturation of pancreatic progenitors to β-cells. Insulin is first expressed in the fetal pancreas 52 days post-conception (Piper *et al.*, 2004) so it may be necessary to culture the cells for extended periods. In contrast, using an engineered cell line in combination with a selection strategy insulin expression was observed after 28 days of differentiation. (Baharvand *et al.*, 2006) An estimate of expansion/maturation time required is therefore 15–50 days after Pdx1 induction.

General considerations

There are a number of factors described below which are thought to enhance pancreatic β-cell differentiation/maturation. It is unclear if these factors play an inductive role or simply facilitate the β-cell survival and this distinction may not be of practical importance during differentiation.

Nicotinamide

3–5 mM nicotinamide has a protective effect on β-cells (Hoorens and Pipeleers, 1999) and there is evidence that 20 mM nicotinamide assists the maturation of the fetal pancreas. (Sandler *et al.*, 1989) Therefore, despite the lack of strong evidence that it plays a role in the growth or differentiation of human ES derived pancreatic β-cells, it is advisable to add nicotinamide to the maturation media.

Growth factors

Hepatocyte growth factor (HGF) (Balamurugan *et al.*, 2003, Otonkoski *et al.*, 1996), glucagon-like peptide-1 (GLP1)/Exendin4 (Movassat *et al.*, 2002) and betacellulin (Huotari *et al.*, 2002) have been shown to have positive effects on β-cell survival/specification although their effects on differentiating human ES cells have not been conclusively shown.

Evaluating terminal differentiation

Rescue of diabetic animal

The gold standard for human ES cell derived β-cells is the rescue of a diabetic animal. The clearest demonstration of rescue is achieved by transplantation of human ES cell derived β-cells into a streptozotocin induced diabetic animal resulting in normalization of blood glucose levels which is subsequently reversed when the graft is removed. Interpretation of NOD-SCID mouse diabetic rescue can be complicated by possible indirect effects of transplant derived growth factors which could inhibit the degeneration of endogenous islets. Details of this type of surgery are beyond the scope of this chapter but early descriptions of diabetic rescue through islet transplantation may be a useful starting point (Outzen and Leiter, 1981, Mellgren *et al.*, 1986).

Glucose stimulated insulin secretion (GSIS) and C-peptide secretion

A cardinal feature of β-cells is the ability to secret insulin in response to increased glucose concentrations.

1) Culture human ES cell derived β-cells in insulin and serum free low glucose (5.5 mM) media for 1–24 hr to equilibrate the cells.

2) Change media, incubate for 1 hr and take a sample of conditioned media to acquire baseline.

3) Change media to medium or high glucose (10 or 25 mM), incubate for 1 hr and take a sample of conditioned medium.

4) Measure glucose response by assaying for insulin or C-peptide in the conditioned media using a Linco ELISA.

5) B-cells should also respond to the secretagogues: palmitate (500 mM), tolbutamide (100 µM), forskolin (10 µM) and KCl (30 mM)

Troubleshooting

I get a lot of cell death or poor results in serum/feeder-free medium. What is wrong?

An important role of both serum and feeder cells is to provide trophic support to the human ES cells. When either of these supports is removed the cells become more delicate and susceptible to the insults of tissue culture. Where appropriate the addition of supporting or growth promoting factors such as FGF2 or EGF can compensate for this fragility. If feeder-free monolayers are used then the differentiation protocol should be started using undifferentiated cells have been grown to confluence instead of freshly seeded cells.

Do I need to show the Pdx1 is nuclear localized?

Yes. Although there is some evidence that Pdx1 can shuttle between the nucleus and the cytoplasm, Pdx1 is generally expected to be nuclear localized. If you are persistently getting cytoplasmic staining for Pdx1 you need to check specific staining conditions using human pancreas as a positive control.

I get insulin secretion/staining but no C-peptide. What does this mean?

Rajagapol *et al.* (2003) showed clearly that any insulin staining and/or release must be interpreted with extreme care. The simplest way around the problems of insulin uptake from the media and insulin release from dying cells is to assay for C-peptide in place of insulin. If insulin is present in the absence of c-peptide it is likely to artefactual.

Can I use co-culture with embryonic pancreas to mature the pancreatic progenitors?

Certainly, co-culture with murine embryonic pancreas has been used to mature hES derived pancreatic progenitors to C-peptide secreting β-cells (Vaca *et al.*, 2006) providing valuable proof-in-principle that the cells have the potential to become mature cells. However, until the factor(s) responsible for this maturation are identified it is unlikely that this option will suitable for large-scale or clinical applications.

Why are the pancreatic β-like cells not responsive to glucose?

Make sure your GSIS assay is working properly. This is best done using mouse/rat islets and the appropriate ELISA. This is not the perfect positive control as human islets behave somewhat differently from mouse islets but it should give an indication if there is something wrong with your media etc.

If your assay is working properly then it is likely that your cells are simply not glucose responsive. Before β-cells proper are formed, immature insulin expressing cells can be found in the developing pancreas and it may be these cells that are secreting C-peptide in a non-glucose responsive manner. These cells are thought to express multiple endocrine factors so a relatively simple check for them is to co-stain for C-peptide and glucagon.

References

Baharvand H, H Jafary, M Massumi and SK Ashtiani. (2006). Generation of insulin-secreting cells from human embryonic stem cells. *Dev Growth Differ* **48**: 323–332.

Balamurugan AN, R Bottino, N Giannoukakis and C Smetanka. (2006). Prospective and challenges of islet transplantation for the therapy of autoimmune diabetes. *Pancreas* **32**: 231–243.

Balamurugan AN, Y Gu, M Miyamoto, H Hori, K Inoue and Y Tabata. (2003). Effect of hepatocyte growth factor (HGF) on adult islet function in vitro [corrected]. *Pancreas* **26**: 103–104.

Chen SS, RP Revoltella, J Zimmerberg and L Margolis. (2006). Differentiation of rhesus monkey embryonic stem cells in three-dimensional collagen matrix. *Methods Mol Biol* **330**: 431–443.

D'Amour KA, AD Agulnick, S Eliazer, OG Kelly, E Kroon and EE Baetge. (2005a). Efficient differentiation of human embryonic stem cells to definitive endoderm. *Nat Biotechnol* **23**: 1534–1541.

D'Amour KA, AD Algulnick, S Eliazer and EE Baetge. (2005b). PDX1 expressing endoderm Patent Application 11/115.868.

Grapin-Botton A and DA Melton. (2000). Endoderm development: from patterning to organogenesis. *Trends Genet* **16**: 124–130.

Gu G, JR Brown and DA Melton. (2003). Direct lineage tracing reveals the ontogeny of pancreatic cell fates during mouse embryogenesis. *Mech Dev* **120**: 35–43.

Hebrok M, SK Kim and DA Melton. (1998). Notochord repression of endodermal Sonic hedgehog permits pancreas development. *Genes Dev* **12**: 1705–1713.

Hoorens A and D Pipeleers. (1999). Nicotinamide protects human beta cells against chemically-induced necrosis, but not against cytokine-induced apoptosis. *Diabetologia* **42**: 55–59.

Huotari MA, PJ Miettinen, J Palgi, T Koivisto, J Ustinov, D Harari, Y Yarden and T Otonkoski. (2002). ErbB signaling regulates lineage determination of developing pancreatic islet cells in embryonic organ culture. *Endocrinology* **143**: 4437–4446.

Itskovitz-Eldor J, M Schuldiner, D Karsenti, A Eden, O Yanuka, M Amit, H Soreq and N Benvenisty. (2000). Differentiation of human embryonic stem cells into embryoid bodies compromising the three embryonic germ layers. *Mol Med* **6**: 88–95.

Jonsson J, L Carlsson, T Edlund and H Edlund. (1994). Insulin-promoter-factor 1 is required for pancreas development in mice. *Nature* **371**: 606–609.

Kim SK, M Hebrok and DA Melton. (1997) Notochord to endoderm signaling is required for pancreas development. *Development* **124**: 4243–4252.

Lumelsky N, O Blondel, P Laeng, L Velasco, R Ravin and R Mckay. (2001). Differentiation of embryonic stem cells to insulin-secreting structures similar to pancreatic islets. *Science* **292**: 1389–1394.

McGrath KE and J Palis. (1997). Expression of homeobox genes, including an insulin promoting factor, in the murine yolk sac at the time of hematopoietic initiation. *Mol Reprod Dev* **48**: 145–153.

Mellgren A, AH Schnell Landstrom, B Petersson and A Andersson. (1986). The renal subcapsular site offers better growth conditions for transplanted mouse pancreatic islet cells than the liver or spleen. *Diabetologia* **29**: 670–672.

Melton DA. (2006). Reversal of type 1 diabetes in mice. *N Engl J Med* **355**: 89–90.

Miralles F, L Lamotte, D Couton and RL Joshi. (2006). Interplay between FGF10 and Notch signalling is required for the self-renewal of pancreatic progenitors. *Int J Dev Biol* **50**: 17–26.

Movassat J, GM Beattie, AD Lopez and A Hayek. (2002) Exendin 4 up-regulates expression of PDX 1 and hastens differentiation and maturation of human fetal pancreatic cells. *J Clin Endocrinol Metab* **87**: 4775–4781.

Murdoch TB, D McGhee-Wilson, AM Shapiro and JR Lakey. (2004). Methods of human islet culture for transplantation. *Cell Transplant* **13**: 605–617.

Murtaugh LC, AC Law, Y Dor and DA Melton. (2005). Beta-catenin is essential for pancreatic acinar but not islet development. *Development* **132**: 4663–4674.

Murtaugh LC, BZ Stanger, KM Kwan and DA Melton. (2003) Notch signaling controls multiple steps of pancreatic differentiation. *Proc Natl Acad Sci USA* **100**: 14920–14925.

Offield MF, TL Jetton, PA Labosky, M Ray, RW Stein, MA Magnuson, BL Hogan and VC Wright. (1996) PDX-1 is required for pancreatic outgrowth and differentiation of the rostral duodenum. *Development* **122**: 983–995.

Otonkoski T, V Cirulli, M Beattie, ML Mally, G Soto, JS Rubin and A Hayek. (1996). A role for hepatocyte growth factor/scatter factor in fetal mesenchyme-induced pancreatic beta-cell growth. *Endocrinology* **137**: 3131–3139.

Outzen HC and EH Leiter. (1981). Transplantation of pancreatic islets into cleared mammary fat pads. *Transplantation* **32**: 101–105.

Ouyang A, R Ng and ST Yang. (2007). Long-term culturing of undifferentiated embryonic stem cells in conditioned media and three-dimensional fibrous matrices without ECM coating. *Stem Cells* **25**(2): 447–454.

Piper K, S Brickwood, LW Turnpenny, LT Cameron, SG Ball, DL Wilson and NA Hanley. (2004). Beta cell differentiation during early human pancreas development. *J Endocrinol* **181**: 11–23.

Rajagopal J, WJ Anderson, S Kume, OL Martinez and DA Melton. (2003). Insulin staining of ES cell progeny from insulin uptake. *Science* **299**: 363.

Sandler S, A Andersson, O Korsgren, J Tollemar, B Petersson, CG Groth and C Hellerstrom. (1989). Tissue culture of human fetal pancreas. Effects of nicotinamide on insulin production and formation of isletlike cell clusters. *Diabetes* **38** Suppl 1: 168–171.

Schwartz PT, B Perez-Villamil, A Rivera, R Moratalla and M Vallejo. (2000). Pancreatic homeodomain transcription factor IDX1/IPF1 expressed in developing brain regulates somatostatin gene transcription in embryonic neural cells. *J Biol Chem* **275**: 19106–19114.

Shapiro AM, C Ricordi, BJ Hering, H Auchincloss, R Lindblad, RP Robertson, A Secchi, MD Brendel, T Berney, DC Brennan, E Cagliero, R Alejandro, EA Ryan, B Dimercurio, P Morel, KS Polonsky, JA Reems, RG Bretzel, F Bertuzzi, T Froud, R Kandaswamy, DE Sutherland, G Eisenbarth, M Segal, J Preiksaitis, GS Korbutt, FB Barton, L Viviano, V Seyfert-Margolis, J Bluestone and JR Lakey. (2006). International trial of the Edmonton protocol for islet transplantation. *N Engl J Med* **355**: 1318–1330.

Soria B, E Roche, G Berna, T Leon-Quinto, JA Reig and F Martin. (2000). Insulin-secreting cells derived from embryonic stem cells normalize glycemia in streptozotocin-induced diabetic mice. *Diabetes* **49**: 157–162.

Stoffers DA, NT Zinkin, V Stanojevic, WL Clarke and JF Habener. (1997). Pancreatic agenesis attributable to a single nucleotide deletion in the human IPF1 gene coding sequence. *Nat Genet* **15**: 106–110.

Treutelaar MK, JM Skidmore, CL Dias-Leme, M Hara, L Zhang, D Simeone, DM Martin and CF Burant. (2003). Nestin-lineage cells contribute to the microvasculature but not endocrine cells of the islet. *Diabetes* **52**: 2503–2512.

Vaca P, F Martin, JM Vegara-Meseguer, JM Rovira, G Berna and B Soria. (2006). Induction of differentiation of embryonic stem cells into insulin-secreting cells by fetal soluble factors. *Stem Cells* **24**: 258–265.

Wichterle H, L Lieberam, JA Porter and TM Jessell. (2002). Directed differentiation of embryonic stem cells into motor neurons. *Cell* **110**: 385–397.

Xu RH, RM Peck, DS Li, X Feng, T Ludwig and JA Thomson. (2005) Basic FGF and suppression of BMP signaling sustain undifferentiated proliferation of human ES cells. *Nat Methods* **2**: 185–190.

Xu X, B Kahan, A Forgianni, P Jing, L Jacobson, V Browning, N Treff and J Odorico. (2006). Endoderm and pancreatic islet lineage differentiation from human embryonic stem cells. *Cloning Stem Cells* **8**: 96–107.

Yao S, S Chen, J Clark, E Hao, GM Beattie, A Hayek and S Ding. (2006). Long-term self-renewal and directed differentiation of human embryonic stem cells in chemically defined conditions. *Proc Natl Acad Sci USA* **103**: 6907–6912.

Zahn S, J Hecksher-Sorensen, IL Pedersen, P Serup and O Madsen. (2004). Generation of monoclonal antibodies against mouse neurogenin 3: a new immunocytochemical tool to study the pancreatic endocrine progenitor cell. *Hybrid Hybridomics* **23**: 385–388.

13, Part A

Directed differentiation of human embryonic stem cells into cardiomyocytes

CHRISTINE MUMMERY

Hubrecht Laboratory, Interuniversity Cardiology Institute of the Netherlands and the Heart Lung Centre, University of Utrecht Medical School, Uppsalalaan 8, 3584CT Utrecht, The Netherlands

ROBERT PASSIER

Hubrecht Laboratory, Uppsalalaan 8, 3584CT Utrecht, The Netherlands

and

CHRIS DENNING

Wolfson Centre for Stem Cells, Tissue Engineering and Modeling, University of Nottingham, UK

Introduction

Human embryonic stem (ES) cells can differentiate to cardiomyocytes in culture. They are readily identifiable as areas of rhythmically contracting cell clumps in culture which first appear 5–8 days after initiation of the differentiation program. Detailed electrophysiological and immunohistochemical analyses have shown that these clumps contain individual cardiac cells with a spectrum of phenotypes comparable to that found in normal human fetal hearts at 8–10 weeks of gestation (reviewed in Boheler *et al.*, 2002; Passier and Mummery, 2005). Analysis of gene expression patterns during differentiation has also indicated that the program cells undergo as they form cardiomyocytes recapitulates that during embryonic and fetal development (Beqqali *et al.*, 2006; reviewed Wobus and Boheler, 2005). Human ES cells are therefore of interest for studying differentiation of cells during early human heart development as well as identifying genes that may cause abnormal development. In addition, they are useful for studying the physiological and pharmacological properties of human cardiomyocytes, including their responses to cardiac drugs. At some point in the future, they may also represent a source of transplantable cells for cardiac muscle repair by replacement of cardiomyocytes lost during ischemic damage.

The differentiation protocols that are effective in inducing human ES cells to differentiate to cardiomyocytes depend in part on the individual human ES cell line used and in part on the methods used for propagation prior to differentiation. Here several methods for generating and characterizing cardiomyocytes from human ES cells are described, with an indication of which methods have been tried on which cell lines and where possible with which relative efficiency.

Overview of protocols

Many of the methods in current use are based on those developed for murine embryonic stem (ES) cells, namely the formation of aggregates in suspension called embryoid bodies (EBs). Many murine ES cells will start to beat spontaneously between 4 and 10 days of initial aggregation depending on the number of cells in the aggregate or the cell line. It is believed that the formation of an outer layer of (extra) embryonic endoderm on the EBs may be important as a source of differentiation signals since it is known that in normal development of multiple species, endoderm is essential for and signals to the anterior mesoderm during heart formation. The conversion of undifferentiated stem cells to cardiomyocytes is generally a low efficiency process. Much of the literature on improving these efficiencies concerns activating specific developmentally relevant signaling pathways whilst the cells are growing as EBs. The wnt and bone morphogenetic protein (BMP) signaling pathways have proved most potent in this context although addition of ascorbic acid, absence of fetal calf serum, and a variety of modifications to the growth medium or ways of forming cell aggregates have all been reported to enhance efficiency. There are no descriptions to date describing complete conversion of human ES cells into cardiomyocytes; genetic selection methods using cardiac specific promoters in combination with selection markers as reported for murine ES cells are presently under development.

The first report of cardiomyocytes derived from human ES cells (Kehat et al., 2001) appeared almost three years after human ES cells were first derived from blastocyst stage embryos (Thomson et al., 1998). To induce cardiomyocyte differentiation, human ES cells (cell line H9.2) were dispersed using collagenase IV into small clumps (3–20 cells) and grown for 7–10 days in suspension to form EB-like structures, comparable to murine ES cells but apparently without the distinct outer layer of endoderm cells. After plating these EBs onto gelatine-coated culture dishes, beating areas were first observed in the outgrowths four days after plating (i.e. 11–14 days after the start of the differentiation protocol). A maximum in the number of beating areas was observed 20 days after plating (27–30 days of differentiation), with 8.1% of 1884 EBs scored as beating. This spontaneous differentiation to cardiomyocytes in aggregates was also observed by others using different cell lines e.g. H1, H7, H9, H9.1 and H9.2 (Xu et al., 2002). However, in this case approximately 70% of the embryoid bodies displayed beating areas after 20 days of differentiation. On day 8 of the differentiation protocol (growth in suspension followed by plating in culture dishes) 25% of the EBs was beating. A third group also demonstrated spontaneous derivation of cardiomyocytes from human ES cell lines H1, H7, H9 and H14, but in this report 10–25% of the embryoid bodies were beating after 30 days of differentiation (He et al., 2003).

The reasons for these apparent differences in efficiency are not clear. In addition, counting beating EBs may not accurately reflect the conversion of human ES cells to cardiomyocytes since individual EBs may contain significantly different numbers of cardiac cells. Recently, the differentiation of several independent human ES cell lines BG01, BG02 and HUES-7 has been described (Zeng *et al.*, 2004; Denning *et al.*, 2006). The HUES lines 1–17 are of particular interest because they were the first to be derived and maintained by bulk enzymatic passaging using trypsin (Cowan *et al.*, 2004) both on mouse embryonic fibroblast layers or on Matrigel in feeder-free conditions (Denning *et al.*, 2006). Beating cardiomyocytes, immunoreactive for the cardiac marker α-actinin, can be derived by EB formation from these human ES cell lines. EBs are formed either by mass culture of small clumps of human ES cells harvested using collagenase IV or by forced aggregation of defined numbers of human ES cells harvested using trypsin. Both methods are provided as bench protocols here, although compared to EBs formed by the mass culture approach, forced aggregation results in greater EB homogeneity and in up to 13-fold improvement in cardiomyogenesis (Denning *et al.*, unpublished).

An alternative method for the derivation of cardiomyocytes from the hES2 (Reubinoff *et al.*, 2002 and hES3 cell lines has been described by Mummery *et al.* (2002, 2003) and Passier *et al.* (2005). These cell lines are generally passaged by a "cut-and-paste method" (Reubinoff *et al.*, 2000; **Figure 1**) although may also be adapted to enzymatic passage. Beating areas were first observed following co-culture of hES2 cells (Mummery *et al.*, 2002, 2003) with a mouse visceral endoderm-like cell-line (END2). Endoderm plays an important role in the differentiation of cardiogenic precursor cells that are present in the adjacent mesoderm *in vivo*. Earlier co-culture of END2 cells with mouse P19 embryonal carcinoma (EC) cells, a mouse embryonal carcinoma cell line with pluripotent differentiation properties, and with murine ES cells had already shown that beating areas appeared in aggregated cells and that culture medium conditioned by the END2 cells contained cardiomyogenic activity (van den Eijnden-van Raaij *et al.*, 1991; Mummery *et al.*, 2002). For the derivation of cardiomyocytes from human ES cells, mitotically inactivated END2 cells were seeded on a 12-well plate and co-cultured with the human ES cell line HES-2 (**Figure 2**). This resulted in beating areas in approximately 35% of the wells after 12 days in co-culture.

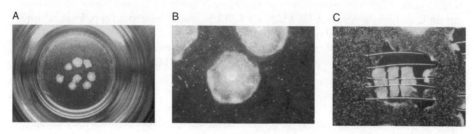

Figure 1 Colonies of hES2 cells "cut" for passage by dispase treatment. Each colony contains ~50,000 cells. Intact colonies are visible as white discs (**A**, enlarged in **B**). Colony cut ready of passage in **C**. See "cut-and-paste" method for HUMAN ES CELLS culture. Photograph courtesy of Dorien Ward-van Oostwaard (Hubrecht Laboratory).

Procedure for differentiation to heart cells:

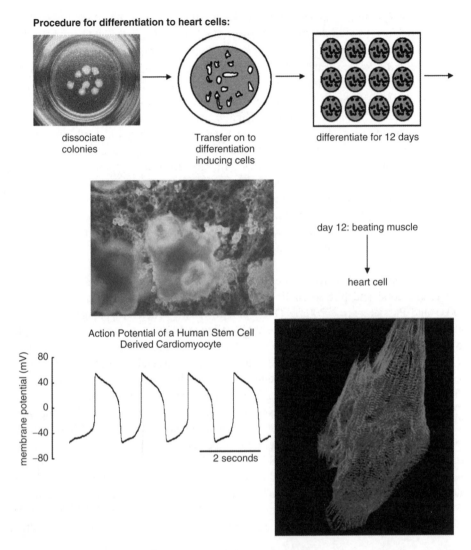

Figure 2 Method for differentiation of hES2 and hES3 cells in co-culture with END2 cells. Confluent monolayers of END2 cells are growth inactivated with Mitomycin-C and small pieces of HUMAN ES CELLS colonies plated on top. First beating muscle is observed 5–7 days later. Clumps contain 20–50% cardiomyocytes. Cells may be dissociated into single cells for immunostaining and/or electrophysiology cultures in monolayer. Electrophysiology courtesy of Leon Tertoolen (Hubrecht Laboratory).

Whilst these methods appear to be effective, all produce cardiomyocytes at low efficiency. Several potential cardiogenic factors have been tested in human ES cells. No significant improvement in cardiomyocyte differentiation has been achieved by adding dimethyl sulfoxide, retinoic acid (Kehat *et al.* 2001; Xu *et al.*, 2002) or BMP-2 (Mummery *et al.*, 2003; Pera *et al.*, 2004). It is not clear whether these factors do not play a role in cardiac differentiation of human ES cells, or whether differentiation protocols were not optimal. One factor that has been described as enhancing cardiomyocyte differentiation of human ES cells is the demethylating agent 5′deoxyazacytidine. Treatment of human ES cell aggregates with 5′deoxyazacytidine induced enhanced cardiomyocyte differentiation and upregulated the expression of cardiac α myosin heavy chain, as determined by real time RT-PCR, up to two-fold (Xu *et al.*, 2002).

The presence of fetal calf serum during differentiation also has important effects on differentiation efficiency. In most reports to date, serum has been present in the culture medium most probably because absence of serum is reported to be detrimental to maintenance of primary cardiomyocytes (Piper *et al.*, 1988) However, serum may contain inhibitory factors and differentiation efficiency has been described as being dependent on serum batch. For example, Sachinidis *et al.* (2003) observed a 4.5-fold upregulation in the percentage of beating mouse EBs after changing to a serum-free differentiation medium. We also recently observed a greater than 20-fold increase in cardiomyocyte yield in hES2–END2 co-cultures in serum-free medium (Passier *et al.*, 2005) which was further enhanced by ascorbic acid. The phenotype of the majority (~90%) of cardiomyocytes derived using this co-culture protocol show greatest similarity to human fetal ventricular cells although atrial and pacemaker like cells are also observed (Mummery *et al.*, 2003). This serum-free protocol is described here for both hES2 and hES3 cells. The method has also proved effective on all of four recently derived human ES cell lines when above passage 10 (see van de Stolpe *et al.*, 2005 for NL-HES1) but was ineffective in BG01 (Denning, unpublished). hES3 cells also differentiate to cardiomyocytes as aggregates in suspension in the presence of serum-free medium conditioned by END2 cells for 4 days.

Materials, preparation of materials, reagents and equipment

Buffers and culture media

MEF medium and human ES cell medium are made up as described in **Chapters 4 and 5** respectively.

Materials

Mitomycin-C treatment of END2 cells for human ES cell coculture.

1. END2 cells: can be obtained from C. Mummery at the Hubrecht Laboratory after completion of a Material Transfer Agreement (www.niob.knaw.nl), email: christin@niob.knaw.nl

2. END2 culture medium: Invitrogen DMEM/F12 (1:1) cat.no. 31331-028 with GlutaMax + penicillin–streptomycin + 7.5% serum (Cambrex) + mam.non-essential amino acids (11100)

3. Mitomycin-C stock

4. PBS (-/-). Without Ca, mg.

5. Trypsin–EDTA

6. 25 cm^2 tissue culture flask coated with 0.1% gelatine. Gelatine 0.1%: dissolve 0.5 g gelatine (Sigma G1890) in 25 mL distilled H$_2$O and autoclave. Add the hot gelatine solution to 475 mL distilled H$_2$O and filter through STERICUP 0.22 μm (Millipore SCGV05RE).

7. 175 cm^2 flask coated with 0.1% gelatine

8. 12-well plates coated with 0.1% gelatine

9. Cover slips treated with 0.1% gelatine, in a 12-well plate

Protocols

Derivation of MEF feeders

MEF feeders are derived as previously described in **Chapter 4**. Feeders should be tested for mycoplasma prior to further use. Feeders do not need to be mitotically inactivated if they are only to be used to generate feeder conditioned medium (i.e. human ES cells will not be seeded directly on them.)

Preparation of feeder-conditioned medium

Feeder-conditioned medium is prepared as previously described in the Extended Protocols section of **Chapter 5**.

Mitomycin-C treatment of END2 cells for human ES cell coculture

1. On day 1 (preferably Monday), seed a 25 cm^2 tissue culture flask coated with 0.1% gelatine with END2 cells in END2 culture medium (see materials). END2 cells should be split 1:8 from a confluent flask.

2. On day 5, seed a 175 cm^2 flask coated with 0.1% gelatine with END2 cells using all the cells from the previous 25 cm^2 flask. Add 40 ml of medium.

3. On day 8, the 175 cm^2 flask should be 100% confluent and is ready for Mitomycin-C treatment, as described for MEFs (see above).

4. Add 5 μL Mitomycin-C per millilitre of medium (discard 15 ml of medium already present and add to remaining 25 ml) from 2 μg/μL stock solution to the culture medium concentration (see Materials). Final concentration = 10 μg/μL in medium.

5. Incubate flasks at 37°C for at least 2.75 hr, maximally 3 hr.

6. Aspirate medium, wash wells once with END2 culture medium, followed by two washes with PBS (-/-).

CAUTION: the waste containing Mitomycin-C is highly toxic.

7. Trypsinize and count the cells.

8. Plate END2 cells at a density of 175,000 cells/mL in 12-well plates 1ml/well coated with 0.1% gelatine in END2 culture medium, with or without 15 mm diameter cover slips, as required for further experimentation.

Human ES cell differentiation to cardiomyocytes

I. Passaging (hES2 and hES3) using the cut-and-paste method

1. The cut-and-paste method has been reported to be the most supportive of karyotypic stability but most human ES cell lines can be adapted to bulk culture methods (trysinization, collagenase) and can be cultured on MEFs or human feeders cells (e.g. foreskin fibroblasts). In our hands, hES2 and hES3 cells are passaged routinely once a week on MEFs. At passage, colonies are usually 0.5–2 mm diameter but will not become confluent. hES2 and hES3 cells were cultured in non-standard human ES cell medium (see **Table 1**), this has mainly been done for historical reasons and these lines can be adapted readily for growth in standard human ES cell medium.

2. Glass needles are sterilized by autoclaving and just prior to passage, are heated over a flame, pulled and broken to give two cutting ends.

3. Cultures are inspected and plates with undifferentiated human ES cell colonies selected.

4. Using the glass needles, colonies are cut (as for a pizza) and undifferentiated "pieces" are selected for passage to new plates (**Figure 1,** see also the Routine Protocols section of **Chapter 5**). Removal of the differentiated areas is essential

Table 1 Culture medium for hES2 and hES3 cells

Material/reagent	Vendor	Catalog number
DMEM (high glucose)	Invitrogen	Cat #11960-044
L-Glutamine—200 mM (1:100)	Invitrogen	Cat #25030-024
Penicillin–streptomycin (1:250)	Invitrogen	Cat #15070-022
Non-essential amino acids (1:100)	Invitrogen	Cat #11140-035
Insulin, transferrin, selenium (ITS) (1:100)	Invitrogen	Cat #41400-045
2-Mercaptoethanol (1.8 μL/mL medium)	Invitrogen	Cat #31350-010
20% fetal calf serum	Hyclone/Perbio	
Once mixed, filter medium through Stericup-GV filter unit (Millipore : SC GVU05RE).		

for long term propagation of undifferentiated cells. 4× magnification on a (stereo) dissecting microscope is the most convenient working enlargement.

5. Medium is removed and 0.5 mL of dispase solution (10 mg/mL in standard human ES cell medium, freshly prepared and filter sterilized; Invitrogen Cat #17105-041) is added. The plate is placed in an incubator for approximately two minutes, or left on a heated microscope stage if available.

6. Two 35 mm diameter dishes with pre-warmed phosphate buffered saline (PBS; Invitrogen Cat #14040-091) are removed from the incubator.

7. Undifferentiated colony pieces are picked up from the dispase-treated plates using a p200 pipette with yellow tip, transferred to a dish of PBS (+/+) and subsequently to the second dish. This results in two washes with PBS (+/+).

8. The colony pieces are evenly distributed over a newly prepared MEF feeder organ culture dish with 1 mL standard human ES cell medium (see **Chapter 5** — Routine Protocols), with nine pieces per plate. Pieces should not be placed too close to the side of the dish or to one another, to allow the colonies sufficient space to grow and remain accessible for next passage.

9. The organ dish is then placed carefully in an incubator at 37°C and 5% CO_2, and refreshed daily with standard human ES cell medium.

Feeder-free culture of human ES cell (BG01 and HUES-7) using trypsin passaging

1. Culture of human ES cells by trypsin passaging on Matrigel in feeder-free conditions provides a route to rapidly expand cultures and reducing the amount of labor required to maintain the lines. In addition, transfection using chemical methods such as lipofection is more efficient in the absence of feeders and maintenance of transgenic human ES cell lines does not necessitate drug resistant feeder cells.

2. Cultures should be maintained at high density, split at a ratio of ∼1:3 to 1:6 and passaged after approximately 3 days depending on growth rates of individual human ES cell lines.

3. Aspirate the medium from a Matrigel coated T25 flask (see Notes) and rinse once with PBS. Matrigel coated flasks should not be allowed to dry out at any time otherwise human ES cells attachment will fail.

4. Seed 1.5 million trypsin passaged human ES cells in MEF-conditioned medium (CM) and incubate at 37°C in a humidified atmosphere containing 5% CO_2.

5. Aspirate medium and refresh with CM every day until human ES cells are confluent.

6. To passage human ES cells, aspirate medium and wash once with PBS. Add 0.5 mL 0.05% trypsin and incubate at 37°C for exactly 1 min. Add 4.5 mL human ES cell medium, transfer to a 15 mL falcon tube, count cells and centrifuge at 100 g/∼300 rpm

for 4 min. Aspirate medium and resuspend pellet in an appropriate volume of CM. Finish by seeding 1.5 million human ES cells to a fresh, Matrigel coated T25 flask in a total volume of 5 mL CM.

hES2 and hES3 cardiomyocyte differentiation by END2 co-culture

1. Refresh Mitomycin-C-treated END2 cells in 12-well plates with human ES cells medium without FCS (see **Table 1**) at least 1.5 hr before plating the human ES cell colony pieces.

2. Prepare dishes for washing undifferentiated human ES cell colony-pieces by filling (a) six 35 mm diameter dishes with PBS and (b) two organ dish with 1 mL human ES cell medium with foetal calf serum (FCS); place in incubator at 37°C and 5% CO_2.

3. Starting material for two 12-well human ES cell-END2 co-culture plates is twelve organ culture dishes each with 9–10 colonies of either hES2 or hES3. Detach human ES cell colonies from MEFs by adding 0.5 mL dispase and placing in the incubator for 7 min.

4. Collect all undifferentiated human ES cell colonies from the 12 organ dishes using a P1000 pipette with a blue tip and distribute them for washing over three dishes of PBS prepared previously.

5. Transfer the colonies to three new PBS dishes (to remove MEFs attached to the colonies).

6. Transfer the colonies to two organ dishes containing 1 mL human ES cell medium.

7. Break colonies into pieces by firmly pipetting up and down (1–3 times depending on the size of the colonies) against the bottom of the dish using a P1000 pipette.

8. Transfer small *clumps* of human ES cells to two 12-well plates containing confluent Mitomycin-C-treated END2 cells (see Materials).

9. Refresh medium on days 5, 9 and 12 (or 5, 8 and 11, depending on the weekend).

10. Score beating areas by microscopic examination 12 days after plating. Beating generally starts on days 5–7 and is maximal on day 12. Immunostaining with antibodies against α-actinin (mouse monoclonal, Sigma; dilution 1:800), Tropomysin (mouse monoclonal; Sigma; dilution 1:400) indicates the number of cardiac cells in the beating clumps. This is usually 20–25% of the cells. Quantification may alternatively be carried out by Western blotting (using e.g. Troponin I antibodies; rabbit polyclonal AB1627, Chemicon International; dilution 1:100).

Note

Murine ES cells also respond to END2 co-culture. Co-culture of murine ES cells with END-2 cells starts with *single cell suspensions*: cells start to aggregate and after about

3 days aggregates attach to END2 cells and grow out and differentiate. Beating areas appear after about 8 days (Mummery *et al.*, 2002). Co-culture with murine ES cells is usually in serum-containing medium, which improves survival.

Differentiation of human ES cells by mass culture of embryoid bodies

1. Prepare 90 mm dishes by seeding with 3.6 million Mitomycin-C inactivated MEFs. Allow to attach overnight.

2. Wash MEFs with PBS and seed approximately 2.5 million human ES cells (see *note 3*) to the MEF feeder layer 10 mL human ES cell medium.

3. Refresh the human ES cell medium daily until human ES cell colonies of at least 0.2 mm in diameter are evident.

4. Aspirate and add human ES cell medium supplemented with 1 mg/mL Collagenase IV for 15 min at 37°C.

5. Transfer the human ES cell/MEF mixture to a 30 mL Universal tube (see *note 4*) and pipette vigorously to break up human ES cell colonies. Centrifuge at 100 *g*/~300 rpm for 5 min

6. Resuspend the pellet in 10 mL CM and transfer to a 90 mm dish with a surface not pre-treated for tissue culture use (i.e. bacteriological dishes from Sterilin) to initiate culture in suspension.

7. Perform a medium change on day 2 of differentiation by transferring the developing EBs to a 30 mL Universal tube. Allow EBs to sediment for 5 min and then carefully aspirate the medium. Resuspend EBs in 10 mL CM and return to the original 90 mm dish.

8. On day 4 of differentiation, sediment EBs as above. Resuspend in 10 mL Differentiation medium (**Table 2**) and transfer to a fresh 90 mm dish with a surface not pre-treated for tissue culture use. Perform a differentiation medium change after 3 days (i.e. on day 7 of differentiation).

9. On day 10 of differentiation, transfer individual EBs to 96-well plates in Differentiation medium.

10. Change medium every 3–4 days and score for beating areas from day 11 of differentiation.

Table 2   Human ES cell differentiation medium-forced aggregation method

DMEM	Invitrogen	21969-035
FCS (20%)	Hyclone Perbio	CH30160.03
GlutaMax (1:100)	Invitrogen	25030-024
Non-essential amino acids (1:100)	Invitrogen	11140-035
2-Mercaptoethanol (100 µM)	Sigma	M7522

Note

1. This mass culture method has been effective in generating EBs from multiple human ES cell lines, including HUES7, BG01, NOTT1 and NOTT2.

2. Human ES cells should be seeded onto MEFs within 1 to 4 days of treatment with Mitomycin-C.

3. The source of human ES cells can either be from cultures of human ES cells maintained on MEFs or from human ES cells on Matrigel in feeder-free conditions. The latter can be used for human ES cells lines that form EBs poorly directly from Matrigel and may be useful for transgenic lines that have been produced in feeder-free conditions.

4. Colonies of human ES cells from some lines will not completely detach from the feeder layer after Collagenase IV treatment and may need to be harvested with a cell scraper.

Differentiation of human ES cells by forced aggregation of embryoid bodies

1. Increased homogeneity is observed in EBs generated by forced aggregation compared with EBs formed by mass culture. Therefore, forced aggregation provides a suitable platform to test the efficacy of different cardiomyogenic factors in parallel conditions.

2. Aspirate the medium from a T25 flask of human ES cells maintained on Matrigel in feeder-free conditions. Wash with PBS, add 0.5 mL of 0.05% trypsin and incubate for 1 min at 37°C.

3. Add 4.5 mL human ES cell medium. Transfer to a 15 mL falcon tube, count cells and centrifuge at 100 g/~300 rpm for 4 min.

4. Resuspend cells at 30,000 cells/mL in CM. Use a multichannel pipette to transfer 100 µL (3,000 human ES cells)/well of a V-bottom 96-well plate (V-96).

5. Centrifuge V-96 plate at 95 g/~280 rpm for 5 min at room temperature and then transfer V-96 plates to the incubator.

6. On day 4 of differentiation, transfer EBs to a 30 mL Universal tube. Allow EBs to sediment for 5 min and then carefully aspirate the medium. Resuspend EBs in 10 mL Differentiation medium Table 2 transfer to a 90 mm dish with a surface not pre-treated for tissue culture use. Perform a Differentiation medium change after 3 days (i.e. on day 7 of differentiation).

7. On day 10 of differentiation, transfer individual EBs to U-bottom 96-well plates in Differentiation medium.

8. Change medium every 3–4 days and score for beating areas from day 11 of differentiation.

Note

1. Forced aggregation has been effective in generating EBs from multiple human ES cell lines all cultured in common feeder-free conditions on Matrigel, including HUES7, BG01, NOTT1, NOTT2 and H1.

2. Differences in the formation and growth of EBs have been observed between human ES cell lines. The example given above recommends using 3,000 cells/well in CM and incubating for 4 days. However, it is recommended that 1,000 and 10,000 cells/well should also be tested. Similarly, incubation for 2 days or 6 days in the V-96 well plates should also be tested. Thus, the most rigorous strategy is to seed plates in triplicate with either 1,000, 3,000 or 10,000 cells/well (i.e. nine plates), with one plate examined for each density on day 2, day 4 and day 6 of formation.

3. If EBs maintained in 90 mm dishes in differentiation medium (Step 6 above) start attaching to the plastic, then it is advisable either to maintain them in suspension by gentle pipetting each day or to transfer them to U-bottom 96-well plates prior to day 10 of differentiation (Step 7 above).

Dissociation of human ES cells-derived cardiomyocytes

1. Isolate beating areas from co-culture plates by cutting them out with scissors and collect excised tissue in standard human ES cell medium with FCS.

2. Three buffers are required for dissociation (**Table 3**). To start dissociation, transfer excised tissue pieces to a dish or 1 well of a 12-well plate with low-Ca (buffer 1) using a P1000 pipette with blue tip; leave for 30 min at room temperature.

3. Transfer cell clumps from low calcium buffer into enzyme buffer (buffer 2) and incubate at 37°C for ~45 min (cover dish with Parafilm before transferring to the CO_2 incubator).

4. Transfer cell clumps from enzyme buffer into KB buffer (buffer 3). Shake gently in KB buffer (buffer 3) at RT for 1 hr at 100 rpm on non-pivoting shaker.

5. Transfer cell clumps from KB buffer into human ES cell culture medium with 20% FCS to promote attachment and survival. Break up the cell clumps by pipetting up and down (2–4 times) against the bottom of the dish using a P1000 pipette. The degree of dissociation required depends on the particular experiments that will be done with the cardiomyocytes:

- For transplantation into animals or immunofluorescent staining, it is sufficient to obtain a mixture of cell clumps and single cells.

- For electrophysiology single cells are required.

Table 3 Dissociation buffers for human ES cell-derived cardiomyocytes

Three buffers are needed: volumes are for making *100 mL of each buffer*

	low Ca(1)	enzyme(2)	KB(3)
NaCl	12 mL (1 M)	12 mL (1 M)	—
CaCl2	—	3 µL (1 M)	—
K2HPO4	—	—	3 mL (1 M)
KCl	0.54 mL (1 M)	0.54 mL (1 M)	8.5 mL (1 M)
Na2ATP	—	—	2 mmol/L
MgSO4	0.50 mL (1 M)	0.50 mL (1 M)	0.50 mL (1 M)
EGTA	—	—	0.1 mL (1 M)
Na pyruvate	0.50 mL (1 M)	0.50 mL (1 M)	0.50 mL (1 M)
Glucose	2 mL (1 M)	2 mL (1 M)	2 mL (1 M)
Creatine	—	—	5 mL (0.1 M)
Taurine	20 mL (0.1 M)	20 mL (0.1 M)	20 mL (0.1 M)
Collagenase A	—	1 mg/mL	—
HEPES	1 mL (1 M)	1 mL (1 M)	—
pH corr.	NaOH	NaOH	
pH	6.9	6.9	7.2

Buffers should be sterilized by filtration and can be stored at −20°C.

Note 1: when making buffer 3 (KB), leave out the glucose, otherwise a precipitate forms at −20°C; add glucose just prior to use.

Note 2: when making buffer 3: add K2HPO4 as the last step, otherwise a precipitate forms.

Note 3: the efficiency of dissociation of the beating clusters into individual cardiomyocytes may depend on the batch of collagenase A used and batch testing may be advisable. We use collagenase from Roche.

6. For electrophysiology and immunofluorescent staining, seed dissociated beating areas in standard human ES cell medium, on gelatine-coated (0.1%) coverslips or culture plates at 37°C, as required.

7. Allow cells on coverslips to recover for at least 2 days and up to 1 week for electrophysiology experiments.

8. For transplantation into mouse heart, dissociated cells are kept in suspension in human ES cell medium at 4°C on ice. 10^5–10^6 cells are injected into the left ventricle in a volume of maximally 15 µL of medium. Larger volumes cause scarring of the cardiac tissue.

Troubleshooting

My human ES cells differentiate on the MEFs during expansion. What am I doing wrong?

Your MEFs may not be good enough and/or appropriate for the human ES cell line you are using. The mouse strain used for MEF isolation critically determines self-renewal of some human ES cell lines. In our hands hES2 and hES3 grow preferentially on 129Sv strain MEFs, HUES1 and 7 on CD1 or 129Sv strain MEFs. Both grow on human foreskin fibroblasts.

We generally use one feeder density (170,000–175,000 per 2.45 cm^2) although lower densities have been described. Feeder density has been suggested to influence colony morphology: on high density feeders, colonies may appear "domed" rather than flat.

Preparation of MEF feeders for human ES cell culture may depend on breeding efficiency and the number of embryos per mouse and initial growth of the isolated primary fibroblast cells. Cell growth rate determines feeder quality (faster growth generally indicates better ability to support self-renewal of human ES cells) and the number of confluent flasks that is converted into MEF cryovials up to passage 4 to 5. For human ES cell culture, these vials are thawed weekly and grown up until confluence prior to Mitomycin-C treatment, and subsequently used for human ES cell passaging.

I would like to have some MEFs "on hand". Can I store them?

Gamma-irradiated feeders (irradiated in suspension with 2500 Rad) can also be used and frozen and stored. For human ES cells lines passaged by cut-and-paste for best karyotypic stability, Mitomycin-C-treated feeders are used fresh. Feeder dishes and plates should be used after 48 hr but not used for transferring human ES cell colonies if older than 4 days. The passage number up to which MEFs can be used differs per lab. In general (low) passage 4 cells are preferable although some labs are able to use feeder cells up to passage 7.

I'm having problems storing Matrigel. What can I do?

Some human ES cells lines will grow on Matrigel with feeder-conditioned medium, e.g. in our hands the HUES and Bresagen lines under DII. To coat T25 flasks with Matrix Growth Factor Reduced Matrigel (VWR; Cat: 41100AB), thaw a 0.5 mL aliquot of from −80°C and immediately dilute at 1:100 to 50 mL cold (direct from fridge) DMEM:F12 base medium. Immediately add 5 mL diluted Matrigel to each T25 flask and allow to polymerize for 45 min at room temperature or overnight at 4°C. Scale volumes of diluted Matrigel according to culture vessel surface area.

Stock aliquots of Matrigel are frozen at low temperatures (−80°C and −20°C), liquid at 0°C and polymerization occurs at above 4°C. Therefore it is critical to dilute Matrigel as soon as it defrosts otherwise polymerization will occur prematurely.

Matrigel coated flasks/plates can be sealed with parafilm and stored at 4°C for up to 1 month. It is advisable to check the storage surface is level with a spirit level; sections of the culture vessel not bathed in medium will become dry.

Brief trypsin treatment for 1 min will necessitate that the flask is tapped in order to achieve complete cell detachment. This ensures that the majority of cells are released in small clumps (rather than as single cells), a factor may reduce the likelihood of karyotypic abnormalities arising.

My human ES cells do not differentiate to cardiomyocytes very well. Any tips?

The efficiency of differentiation towards cardiomyocytes in mouse and human ES cells may be variable. Important factors are the initial number of cells per embryoid body or human ES cell clump and, if the differentiation is not in serum-free medium, the FCS batch. It is advisable to test different batches of FCS for differentiation efficiency and choose the batch, which performs best for all subsequent differentiation assays.

Some human ES cell lines respond best in END2 co-cultures (hES2; NL-hES1–4), other respond better when grown as aggregates in END2 conditioned medium (hES3), as described previously for pluripotent P19 embryonal carcinoma cells (van den Eijnden-van Raaij *et al.*, 1991). Human ES cell medium is conditioned for 4 days by a confluent END2 cell monolayer.

My human ES cell-derived cardiomyocytes do not re-attach and die after dissociation. What can I do about it?

Cardiomyocytes are very fragile. If pipetting is too rigorous, cells will fail to recover either in culture or *in vivo*.

References

Beqqali A, J Kloots, D Ward-van Oostwaard, C Mummery and R Passier. (2006). Genome-wide transcriptional profiling of human embryonic stem cells differentiating into cardiomyocytes. *Stem Cells* **24**: 1956–1967.

Boheler KR and AM Wobus. (2002). Differentiation of pluripotent embryonic stem cells into cardiomyocytes. *Circ Res* **91**: 189–201.

Denning CT, C Allegrucci, H Priddle, MD Barbadillo-Munoz, D Anderson, T Self, NM Smith, CT Parkin and LE Young. (2006). Common culture conditions for maintenance and cardiomyocyte differentiation of the human embryonic stem cell lines, BG01 and HUES-7. *Int J Dev Biol* **50**: 27–37.

He JQ, Y Ma, Y Lee, JA Thomson and TJ Kamp. (2003). Human embryonic stem cells develop into multiple types of cardiac myocytes: action potential characterization. *Circ Res* **93**: 32–39.

Kehat I, D Kenyagin-Karsenti, M Snir, H Segev, M Amit, A Gepstein, E Livne, O Binah, J Itskovitch-Eldor and L Gepstein. (2001). Human embryonic stem cells can differentiate into myocytes with structural and functional properties of cardiomyocytes. *J Clin Invest* **108**: 407–441

Mummery C, D Ward-van Oostwaard, P Doevendans, R Spijker, S van den Brink, R Hassink, M van der Heyden, T Opthof, M Pera, AB de la Riviere, R Passier and L Tertoolen. (2003). Differentiation of human embryonic stem cells to cardiomyocytes: role of coculture with visceral endoderm-like cells. *Circulation* **107**: 2733–2740.

Mummery C, D Ward, CE van den Brink, SD Bird, PA Doevendans, T Opthof, A Brutel de la Riviere L, Tertoolen, M van der Heyden and M Pera. (2002). Cardiomyocyte differentiation of mouse and human embryonic stem cells. *J Anat* **200**: 233–242.

Passier R and CL Mummery. (2005). Cardiomyocyte differentiation from embryonic and adult stem cells. *Curr Opinion Biotechnol* **16**: 498–502.

Passier R, D Ward-van Oostwaard, J Snapper, J Kloots, R Hassink, E Kuijk, B Roelen, A Brutel de la Riviere, C Mummery. (2005). Increased cardiomyocyte differentiation from human embryonic stem cells in serum-free cultures. *Stem Cells* **23**: 772–780.

Pera MF, J Andrade, S Houssami, B Reubinoff, A Trounson, Stanley EG, D Ward-van Oostwaard and C Mummery.(2004). Regulation of human embryonic stem cell differentiation by BMP-2 and its antagonist noggin. *J Cell Sci* **117**: 1269–1280

Piper HM, SL Jacobson, P Schwartz. (1988). Determinants of cardiomyocyte development in long-term primary culture. *J Mol Cell Cardiol* **20**(9): 825–835.

Reubinoff BE, MF Pera, CY Fong, A Trounson and A Bongso. (2002). Embryonic stem cell lines from human blastocysts: somatic differentiation in vitro. *Nat Biotech* **18**: 399–404.

Sachinidis A, C Gissel, D Nierhoff, R Hippler-Altenburg, H Sauer, M Wartenberg and J Hesheler. (2003). Identification of platelet-derived growth factor BB as cardiogenesis-inducing factor in mouse embryonic stem cells under serum free conditions. *Cell Physiol Biochem* **13**: 423–429.

Thomson JA, J Itskovitch-Eldor, SS Shapiro, MA Waknitz, VS Marshall and JM Jones. (1998). Embryonic stem cell lines derived from human blastocysts. *Science* **282**: 1145–1147.

van den Eijnden-van Raaij AJ, TA van Achterberg, CM van der Kruijssen, AH Piersma, D Huylebroeck, SW de Laat and CL Mummery. (1991). Differentiation of aggregated murine P19 embryonal carcinoma cells is induced by a novel visceral-endoderm specific FGF-like factor and inhibited by activin A. *Mech Dev* **33**: 157–165.

Wobus AM and KR Boheler. (2005). Embryonic stem cells: Prospects for developmental biology and cell therapy. *Physiol Rev* **85**: 635–678.

Xu C, S Police, N Rao and MK Carpenter. (2002). Characterization and enrichment of cardiomyocytes derived from human embryonic stem cells. *Circ Res* **91**: 501–508.

Zeng X, T Miura, Y Luo, B Bhattacharya, B Condie, J Chen, I Ginis, I Lyons, J Mejido, RK Puri, MS Rao and WJ Freed. (2004). Properties of pluripotent human embryonic stem cells BG01 and BG02. *Stem Cells* **22**: 292–312.

13, Part B

Directed differentiation of human embryonic stem cells into endothelial cells

CARRIE SOUKUP[1], SHULAMIT LEVENBERG[2] AND ONDINE CLEAVER[1]

[1]Department of Molecular Biology, University of Texas Southwestern Medical Center, 5323 Harry Hines Blvd, Dallas, TX 75390-9148, USA
[2]Faculty of Biomedical Engineering. Technion, Haifa 32000, Israel

Introduction

In vitro derivation of endothelial cells from stem cells

Murine and human ES cells

Embryonic stem (ES) cells are capable of differentiating into endothelial cells, both *in vivo* and *in vitro*. ES cells are remarkable cells derived in the laboratory from the inner cell mass of the mammalian blastocyst (Martin, 1981). When isolated, murine ES cells are immortal and pluripotent. They can contribute to all embryonic lineages, including the germ line, when injected into a blastocyst and allowed to develop (Bradley *et al.*, 1984). They also maintain their pluripotency indefinitely when cultured *in vitro* (Nagy *et al.*, 1993). In recent years, many "directed" differentiation methods have been reported in which murine ES cells can be preferentially driven into specific differentiated cells types, including neural progenitors (Brustle *et al.*, 1997), hematopoietic lineages (Kennedy *et al.*, 1997), cardiomyocytes (Klug *et al.*, 1996) and endothelial cells (Wang *et al.*, 2004; Levenberg *et al.*, 2002; Gerecht-Nir *et al.*, 2003; Itskovitz-Eldor *et al.*, 2000). ES cell lines have also been derived from other species, including humans, which was initially accomplished in 1998 (Thomson *et al.*, 1998) and has steadily gained momentum since then (Cowan *et al.*, 2004). Like murine ES cells, human ES cells have since been shown to possess the same pluripotency and ability to differentiate into all three germ layers, including mesodermal lineage differentiation into endothelial cells (Lavon and Benvenisty, 2003; Levenberg *et al.*, 2002; Gerecht-Nir *et al.*, 2003; Wang *et al.*, 2004).

Spontaneous differentiation of endothelial cells into EBs

Fortuitously, although endothelial differentiation is a complex process in the embryo, it is essentially recapitulated during the spontaneous differentiation of embryoid bodies (EBs). An embryoid body is an aggregation of cells that is formed when ES cells differentiate in suspension culture and that develops derivatives of all three germ layers (ectoderm, mesoderm and endoderm) (Itskovitz-Eldor *et al.*, 2000; Cowan *et al.*, 2004). Human embryoid body (EB) differentiation is similar in many respects to that of mouse EB differentiation. In murine EBs, endothelial differentiation occurs spontaneously upon removal of leukemia inhibitory factor (LIF), which results in the formation of wide variety of tissues, including heart, neurons, blood and vasculature (Bailey and Fleming, 2003; Risau *et al.*, 1988). Work has demonstrated that LIF is not sufficient to keep human ES cells in an undifferentiated state, rather basic fibroblast growth factor (bFGF) is required (Thomson *et al.*, 1998; Draper *et al.*, 2004). Thus, human EB differentiation occurs upon removal of bFGF.

A number of studies have characterized vascular development and gene expression in differentiating human EBs (Gerecht-Nir *et al.*, 2005; Wartenberg *et al.*, 1998; Bautch, 2002). By day 6 in human EB formation, key haemato-endothelial gene expression begins, including PECAM, Flk-1, Scl, GATA1 and GATA2 (Zambidis *et al.*, 2005). By day 10, mature endothelial markers begin to be expressed, coinciding with the physical formation of vascular tubes and networks (Vittet *et al.*, 1996). One important difference between mouse and human EBs is that some common endothelial markers are expressed in undifferentiated human ES cells, such as Flk-1 and Tie-2, emphasizing the need to distinguish the two systems (Kaufman *et al.*, 2001). However, most endothelial markers are largely restricted to the vasculature in both differentiating mouse and human endothelium. Such markers include PECAM, GATA-2, CD34, and VE-cadherin, all of which are expressed in early human EBs (Levenberg *et al.*, 2002; Vittet *et al.*, 1996). These markers increase in expression levels as differentiation proceeds, reaching their highest level in day 13–15 human EBs. At this time, an extensive vascular network can be recognized in human EBs (Levenberg *et al.*, 2002).

"Directed" or "induced" differentiation of endothelial cells from human ES cells

As an alternative to spontaneous endothelial differentiation during EB development, endothelial precursors have also been cultivated using two-dimensional culture techniques. Several groups have shown that culture of human ES cells directly on ECM proteins (including collagen IV) or co-culture with ECM-secreting cells increases hematopoietic (Kaufman *et al.*, 2001; Vodyanik *et al.*, 2005) and endothelial differentiation (Gerecht-Nir *et al.*, 2003). The enrichment of endothelial cells achieved using this directed differentiation method is further increased by the addition of angiogenic factors such as VEGF and bFGF (Gerecht-Nir *et al.*, 2003). Endothelial yield from such two-dimensional techniques is initially lower than that acquired via EB formation (~10% compared to ~80%), but further sorting results in highly enriched endothelial populations (Wang *et al.*, 2004; Levenberg *et al.*, 2002; Gerecht-Nir *et al.*, 2003). Given the ease of culture and the abundance of cells obtained, it is a viable alternative to EB differentiation.

Isolation and selection of endothelial cells

Separation of endothelial cells using cell surface molecules

In all ES cell differentiation protocols covered here and elsewhere, varying proportions of endothelial cells are generated. For instance, Protocol II in this chapter results in 99% endothelial cells, white Protocol III results in 20–80%. However, it is important to note that they never comprise a homogeneous population and must be separated away from other cells that have taken on different fates. A standard set of markers has been used to isolate and select for endothelial cells. In particular, expression of PECAM/CD31, or VE-cadherin, either alone or in combination with the ability to take up Dil-labeled acetylated low-density lipoprotein (Dil-Ac-LDL) have been used as key markers to isolate developing endothelial cells. In addition, other common markers commonly used include: CD34 (which marks both endothelium and hematopoietic progenitor cells), Flk-1, Tie-1, Tie-2, Flt-1 (VEGFR1), von Willebrand factor (vWF) and endothelial nitric oxide synthase (eNOS). In protocols presented here, two of three protocols depend on sorting based (in part) on PECAM selection and enrichment.

Characterization of human ES cell-derived endothelial cells

Endothelial markers

Although there is increasing recognition of the inherent molecular and structural heterogeneity of endothelial cells from one tissue to the next (Gerritsen, 1987; Conway and Carmeliet, 2004), a number of molecules are expressed across many different types of endothelia and can thus be used as common endothelial markers. Furthermore, many surface markers and functional separation techniques used to isolate and select endothelial cells are also reliable markers used to characterize vascular endothelium. Primarily, the cell adhesion molecules PECAM and VE-cadherin, as well as the receptors Flk-1 and Tie-2, are the most widely accepted markers used to distinguish endothelial cells from other cell types. In addition, expression of vWF, eNOS, integrin a_5v_3, and Flt-1 (VEGFR1) are used to characterize functional blood vessels (Levenberg et al., 2002).

In vitro angiogenesis assay

A unique feature of endothelial cells is their cellular ultrastructure and their ability to coalesce into a coherent vascular organ that makes them recognizable as "endothelial". When seeded onto extracellular matrix (ECM) molecules, such as Matrigel or collagen IV, endothelial cells will assemble themselves into a branching network of cells, forming junctions and cord-like structures (Kuzuya and Kinsella, 1994; Nicosia and Ottinetti, 1990). This morphological phenomenon is further accentuated when cells are mixed into a three-dimensional matrix gel of collagen or Matrigel. In this case, cell cords will actually reorganize to form tube-like arrangements, resulting in a lumenized vascular plexus. This inherent ability to mimic in vitro, what they normally do in vivo, makes endothelial cells particular well suited for laboratory study. Exogenous growth factors can be added to two- or three-dimensional culture to evaluate their effects cell growth and behavior. Genetic manipulation of endothelial cells (transgenic, gene targeted, RNAi) can also be carried out in conjunction with functional assays such as wound healing (scratch assay), migration (chemotactic assay), tube formation (culture in Matrigel), and angiogenesis assays (Matrigel plug assay).

In vivo blood vessel differentiation

Many methods have been used to analyze endothelial function and behavior *in vivo*. Vasculogenic potential can be tested by injecting cells into the early embryo (mouse E7.5-E8.5) and looking for incorporation into the native plexus (Hatzopoulos *et al.*, 1998). Endothelial cells have also been injected into damaged and ischemic tissue to evaluate their ability to integrate and participate in angiogenesis and vascular repair (Kalka *et al.*, 2000; Kocher *et al.*, 2001). Most relevant to this handbook, endothelial cells derived from human ES cells have also been successfully cultivated on polymer scaffolds (Nor *et al.*, 2001). These scaffolds can then be implanted subcutaneously into SCID mice, where the implanted *in vitro* generated vessels will anastomose with the host mouse vasculature (Levenberg *et al.*, 2002).

Therapeutic implications of human ES cell-derived endothelial cells

The availability of human ES cells represents an extraordinary opportunity for the development of cell-based therapies that could be applied to a wide range of human diseases. The ability of human ES cells to differentiate *in vitro* into endothelium provides an unlimited source of tissue for both laboratory and clinical studies. Since human ES cells-derived human EBs contain putative hemangioblasts, this offers an excellent system for *in vitro* studies of cell and developmental lineages, allowing for clonal analysis of bi-potential cells. In addition, human EBs develop well-defined vascular beds that grow and adapt during EB growth, providing an extremely accessible platform to study angiogenesis, and anti-angiogenesis agents. Because human ES cell-derived endothelial cells can be generated as large and homogenous populations, this would facilitate high-throughput screening of small molecules that can target growing human EB blood vessels, providing a research model for tumor angiogenesis. Using this system, endothelial cells and candidate anti-angiogenic factors can then be tested *in vitro*, using matrix-based angiogenesis assays, or *in vivo*, using Matrigel plugs, corneal angiogenesis, or other animal models. In addition, cultures of human ES cell-derived endothelium are also particularly amenable to genetic manipulation, including transgenic vectors or targeted gene ablation. This provides an opportunity to directly assess the importance of vascular genes for chosen applications.

Human ES cell-derived endothelial cells can also be used for a variety of medical applications, such as injection to repair ischemic tissues or building of engineered tissues. Improperly vascularized tissues are of particular medical interest, and result from occlusion of blood vessels. Organs and tissues that undergo vascular failure suffer severe cellular damage due to impaired access to oxygen and metabolites. Direct injection of stem cells with and without the addition of VEGF has suggested that human ES cell-derived endothelial cells may facilitate vascular repair. Such repair has shown promise in clinical settings (Min *et al.*, 2002; Wang *et al.*, 2001; Orlic *et al.*, 2001; Tomita *et al.*, 2002). In addition, the growing interest in artificially generated tissues for transplantation underlines the need for ensuring adequate integration of host and transplanted vessels for blood circulation. Although great strides have been made in differentiating tissues and engineering biodegradable scaffolds to shape laboratory generated tissues, for instance bone and cartilage (Sharma and Elisseeff, 2004) or rapid and functional vascularization of these tissues both before and after implantation.

Overview of protocols

Here, we present three protocols for inducing human ES cells to differentiate into endothelial cells (**Figure 1**). These three methods form the basis for recent advances in endothelial differentiation of human ES cells (Wang *et al.*, 2004; Levenberg *et al.*,

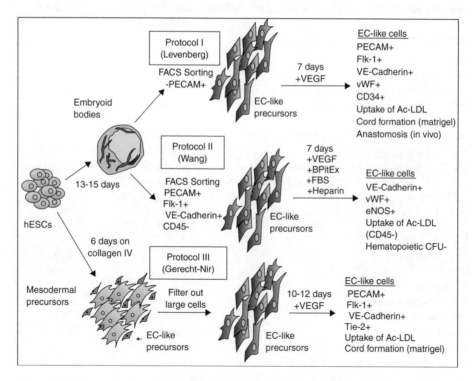

Figure 1 Schematic diagram of three protocols for human ES cell differentiation of endothelial cells. Outline of the key steps in each protocol presented in this chapter. Protocol I (Levenberg) and II (Wang) make use of spontaneous differentiation of embryoid bodies followed by FACS sorting of endothelial precursors. These are then cultured in endothelial differentiation conditions in the presence of VEGF. Protocol III (Gerecht-Nir) makes use of monolayer differentiation of ES cells on collagen IV, followed by size separation of cells and differentiation in the presence of VEGF.

2002; Gerecht-Nir *et al.*, 2003) and make use of defined culture conditions, spontaneous or directed differentiation, endothelial selection and subsequent endothelial amplification. Although the protocols use slightly different techniques, the common result is the generation of highly enriched populations of human stem cell-derived endothelial cells.

Materials, reagents, and equipment

Protocol I — Three dimensional cell sorting model (Levenberg *et al.*, 2002) (Table 1)

Table 1   Protocol I

Material/reagent	Vendor	Catalog number
Cells		
Growth-arrested mouse embryonic fibroblasts (MEFs)	Cell Essentials (Boston)	PMEF-CF
Human ES cells (H1 or H9 lines have been used)		
Culture media		
MEF medium		
90% Dulbecco's Modified Eagle Medium (DMEM)	Invitrogen	12430-104
10% Fetal Bovine Serum (FBS)	HyClone	SV3001403
ES cell medium		
78.3% Knockout DMEM (KO-DMEM)	Invitrogen	10829-018
20% KO serum replacement	Invitrogen	10828-028
1% Non-essential amino acids (NEAA)	Invitrogen	11140-050
0.5% L-glutamine (200 mM in 0.85% NaCl stock)	Invitrogen	25030-081
0.2% beta-mercaptoethanol (55 mM in DPBS stock)	Sigma-Aldrich	G1890
10^3 U/mL Leukemia inhibitory factor (LIF)	Sigma-Aldrich	L5283
5 ng/mL Basic fibroblast growth factor (bFGF)	BD Biosciences	350060
Knockout medium		
80% KO-DMEM	Invitrogen	10829-018
20% FBS	HyClone	SV3001403
1% NEAA	Invitrogen	11140-050
1 mM L-glutamine	Invitrogen	25030-081
0.1 mM BME	Sigma-Aldrich	M7522
5 mM EDTA in DPBS	Sigma	E6758
Endothelial medium		
Endothelial growth medium (EGM-2)	Clonetics	CC-3156
*All media is sterile and filtered through a 0.22 µM membrane.		
Plasticware		
Sterile tissue culture plates	Corning	430167
Low-adherence Petri dishes	Corning	3262

Table 1 *(continued)*

Material/reagent	Vendor	Catalog number
Other		
Mitomycin C	Sigma	M4287
Gelatin	Sigma	G1890
Collagenase, type IV	GIBCO/BRL	17104-019
Trypsin/EDTA (0.25%/1 mM)	Gibco	25200-056
Trypsin inhibitor	Gibco	17075-029
1× Phosphate buffered saline (PBS)	Gibco	14190-144
Fluorescent-labeled PECAM1 monoclonal antibody	PharMingen	30884X
Flow cytometer & associated analysis software		

Protocol II — Three dimensional cell sorting model (Wang *et al.*, 2004) (Table 2)

Table 2 Protocol II

Cells		
Growth-arrested mouse embryonic fibroblasts (MEFs)	Specialty Media	PMEF-CFL
Human ES cells (H1 or H9 clone)		
Culture Media		
MEF medium		
80% KO-DMEM	Invitrogen	10829-018
20% KO serum replacement	Invitrogen	10828-028
1% NEAA	Invitrogen	11140-050
1 mM L-glutamine	Invitrogen	25030-081
0.1 mM BME	Sigma-Aldrich	M7522
4 ng/mL human recombinant bFGF	Invitrogen	13256-029
Differentiation medium		
80% KO-DMEM	Invitrogen	10829-018
20% non-heat inactivated FBS	HyClone	SV3001403
1% NEAA	Invitrogen	11140-050
1 mM L-glutamine	Invitrogen	25030-081
0.1 mM BME	Sigma-Aldrich	M7522
Endothelial growth medium		
80% Medium-199	Sigma-Aldrich	M 4530
20% FBS	HyClone	SV3001403
50 µg/mL bovine pituitary extract	Invitrogen	13028-014
10 IU/mL heparin	Leo Pharma Inc.	
5 ng/mL hVEGF	R&D Systems	293-VE-050
*All media is sterile and filtered through a 0.22 µM filter membrane (Millipore).		
Plasticware		
Six-well sterile tissue culture plates	Corning	3516
Six-well low attachment tissue culture plates	Corning	3471

(continued overleaf)

Table 2   *(continued)*

Other		
40 μm cell strainer	BD Biosciences	352340
70 μm cell strainer	BD Biosciences	352350
0.22 μm (pore size) sterile membranes	Corning	431212
Matrigel	BD Biosciences	354234
Collagenase B	Roche	11088807001
Collagenase, type IV	Invitrogen	17104-019
Fluorochrome conjugated PECAM1 monoclonal antibody	BD Biosciences	558094, 558068
Fluorochrome conjugated VE-cadherin monoclonal antibody	Alexis Biochemicals	ALX-210-232
Fluorochrome conjugated Flk-1	Abcam	AB9530
7-Amino actinomycin-D (7-AAD)	Immunotech	183422
Fluorescence activated cell sorting (FACS) Calibur flow cytometer		
Cell Quest software	BD Biosciences	
Fibronectin	Sigma	F1141
Cell dissociation buffer	Invitrogen	13151-014

Protocol III — Two dimensional model of directed differentiation (Gerecht-Nir *et al.*, 2003) (Table 3)

Table 3   Protocol III

Cells		
Growth-arrested mouse embryonic fibroblasts (MEFs)	Chemicon	PMEF-CF
Human ES cell lines (H9.2, H13, I6, or I9 clone)		
Media[1]		
Components of MEF and standard human ES media are listed in **Chapter 5 — Routine Protocols**.		
Differentiation medium		
90% Alpha MEM medium	Gibco	32561037
10% defined FBS	HyClone	SV3001403
0.1 mM BME	Sigma-Aldrich	M7522
Human recombinant VEGF165	R&D systems	293-VE
*All media are sterile and filtered through 0.22 μM filter membrane (Millipore) membrane.		
Plasticware		
10 cm^2 sterile tissue culture plates	Corning	430167
Other		
0.1% gelatin	Sigma	G1890
40 μm mesh strainer	Falcon	352340
EDTA	Promega	H5031
Type IV collagen coated Petri dishes	BD Biosciences	354453
Type IV collagenase	Invitrogen	13151-014

[1]Cells were generated in Itskovitz-Eldor lab, where Protocol III was developed.

Protocols

A. — Embryoid body formation and spontaneous differentiation

A.1. Protocol I — Three dimensional cell sorting model (Levenberg et al., 2002)

Overview

Human ES cells are grown on MEFs, differentiated into human EBs, and then dissociated and sorted for endothelial cells. Briefly, human ES cells are expanded on growth arrested MEFs (refer to **Chapter 5**) in an undifferentiated state. To induce differentiation, human ES cells are removed from the feeders, seeded onto non-adherent plates and grown in the absence of LIF. These cells will begin to form embryoid bodies (EBs); spontaneously differentiating into many cell lineages, including endothelial cells. The endothelial population is purified from 13–15 day old EBs with FACS sorting, using an antibody to the surface molecule PECAM. Endothelial cells isolated using this method are up to ~80% pure as assayed by endothelial gene expression and other endothelial characteristics such as correct organization of endothelial junctions, stress fiber formation, uptake of acetylated LDL, formation of cord-like structures and lumen formation in Matrigel, as well as microvascular formation in SCID mouse implants. These cells can be successfully cultivated for up to seven passages in EGM-2 media. Of interest, these cells have been seeded on highly porous PLLA/PLGA biodegradable polymer scaffolds and these "sponges" develop organized vessels when implanted into SCID mice.

Levenberg (2002)

Undifferentiated hES cells on MEF feeder (KO-medium)

↓

Dissociate w/collagenase

↓

Transfer to non-adherent Petri dish (-LIF and bFGF)

↓

13-15 day EB culture

↓

FACS sorting for PECAM⁺ cells

↓

2%of total EBvolume yields ~80% EC purity

Protocol I

Preparation of MEFs

1. Coat sterile tissue culture plates with 0.1% gelatin (autoclaved).

2. Plate growth-arrested feeder MEFs onto 10 cm^2 gelatin-coated plates at 100% confluence in MEF media.

3. Allow MEFs at least 2 hr to attach. Do not use after 1 week.

Human ES cell propagation and passaging

1. Plate human ES cells on MEFs in knockout medium. Cells are not counted, since they are not dissociated into single cells, but rather kept as small clumps. However, passaging should be done at 1:4. (See **Figure 2A**).

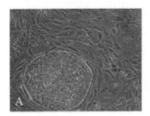

Figure 2 Human embryonic stem (ES) cell colony appearance on MEFs. When plated on MEF feeders, hESCs should remain undifferentiated. Colonies should maintain smooth, translucent edges. Colonies should also not be allowed to grow to confluence, since this will promote differentiation. (**A**) Example of colony used in Protocol I. (**B**) Example of colony used in Protocol II.

2. Grow cultures in 5% CO_2 and passage every five days. Change media daily.

Human ES cell differentiation

1. Disaggregate cells in 1 mg/mL collagenase or trypsin/EDTA. Collagenase is preferred.

2. Transfer ES cells to 10 cm^2 low-adherence Petri dishes to allow their aggregation. Split one ES cell plate to five Petri dishes. Do not dissociate cells completely to single cells. Keep ES cells as small aggregates.

3. Culture in knockout media without LIF or bFGF for 13–15 days. Change media every 2 or 3 days. These cells will spontaneously differentiate and form embryoid bodies (EBs) (see **Figure 3A**).

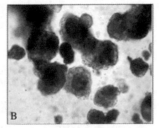

Figure 3 Human embryoid body (hEB) appearance in suspension culture. When human ES cells are dissociated to small groups of cells and plated on low adherence plates, they will spontaneously form aggregates or embryoid bodies. These hEBs should appear as small, rounded and increasingly cystic structures. (**A**) Example of an hEB used in Protocol I. (**B**) Example of hEB used in Protocol II.

Selection of endothelial cells

1. Dissociate day 13–15 human EBs with 0.25% trypsin/EDTA. To do this, collect EBs into a 15 mL tube and allow them to settle. Remove medium. Wash with 1× PBS. Add 5 mL trypsin/EDTA. Place tube on shaker in 37°C incubator for approximately 5 min. Aid the dissociation by pipetting rather roughly with a 5 mL sterile pipette. Once EBs start to come apart, add 5 mL of trypsin blocker.

2. Wash with 10 mL PBS containing 5% FBS. Follow with two washes in PBS, prior to antibody staining.

3. Incubate cells with fluorescent-labeled PECAM antibody diluted in PBS/FBS on ice. Use 10–20 μL of PECAM antibody for approximately 1 million cells.

4. Isolate PECAM(+) cells with a flow cytometer. The vast majority of cells will be alive at this point.

Culture endothelial cells

1. Plate about 30,000–50,000 sorted cells on a 24-well 0.1% gelatin-coated tissue culture plate. Since the numbers of live cells will be lower due to FACS sorting, collect cells in either a 12- or 24-well plate.

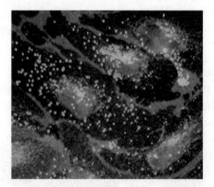

Figure 4 Analysis of human ES cells derived endothelial cells. Following these three methods for differentiation of human embryonic stem cells, highly enriched populations of endothelial cells will be generated. A proportion of the final cell population should be assayed for endothelial markers. Image shows endothelial cells resulting from Protocol I, following staining with antibodies for PECAM (red), vWF (green) and a nuclear stain DAPI (blue). This image is best seen in color; please refer to plate section to see the color image.

2. Culture sorted cells in endothelial growth medium, EGM-2.

3. To propagate these cells, passage before confluency (at approximately 70–90% plate coverage). Split 1:4.

4. Change media every 2–3 days.

5. Cells can now be characterized for endothelial markers. Most cells will express the endothelial differentiation markers PECAM, VE-cadherin and vWF (see **Figure 4**).

A.2. Protocol II — Three-dimensional cell sorting model (Wang et al., 2004)

Overview

Human ES cells are grown in feeder-free cultures, differentiated into human EBs, which are then dissociated and sorted for endothelial cell-like precursors. Briefly, MEF-conditioned media (MEF-CM) is collected and supplemented with bFGF. hES cells are then transferred to collagen IV-coated plates in supplemented MEF-CM. On the day of passage, undifferentiated hES cells at confluence are dissociated and transferred to low attachment plates to induce differentiation and EB formation. EBs are allowed to differentiate for 13–15 days and then dissociated with collagenase B in cell dissociation buffer, followed by physical disruption through a cell strainer. Cells are labeled with monoclonal antibodies to CD45 and PECAM, and sorted using a FACSCalibur flow cytometer and Cell Quest software. Cells isolated sorting by this method have been demonstrated to also be positive for VE-cadherin, CD34 and Flk-1 and have the functional ability to take up Dil-Ac-LDL (Wang *et al.*, 2004). These cells have therefore been called "CD45^{neg}PFV" cells, due to their characterization as CD45-/PECAM+/Flk-1+/VE-cadherin+. This profile can be assumed during the procedure, but it should also be tested following sorting. Finally, the isolated CD45^{neg}PFV subset is then seeded on fibronectin-coated plates and cultured in endothelial growth media. From this procedure, an almost pure population of endothelial cells can be obtained.

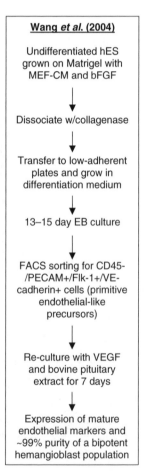

Wang *et al.* (2004)

Undifferentiated hES grown on Matrigel with MEF-CM and bFGF

↓

Dissociate w/collagenase

↓

Transfer to low-adherent plates and grow in differentiation medium

↓

13–15 day EB culture

↓

FACS sorting for CD45-/PECAM+/Flk-1+/VE-cadherin+ cells (primitive endothelial-like precursors)

↓

Re-culture with VEGF and bovine pituitary extract for 7 days

↓

Expression of mature endothelial markers and ~99% purity of a bipotent hemangioblast population

Protocol II

Preparation of MEFs

1. Culture MEFs to confluency in MEF medium. MEFs are prepared from 12 day old embryos (CF1 mouse, Charles River Canada. Preparation protocols are from the websites of WiCell or Geron. Also see **Chapter 4**).

2. Radiate confluent MEFs with 40Gy irradiation to arrest growth. Either X-ray irradiator or isotope irradiator can be used. Follow the irradiation instructions. The actual irradiation dose will depend on the distance between the cells and the head of the irradiator and the depth of the medium added.

3. Each day for 7 days, collect MEF conditioned media (MEF-CM).

4. Filter the MEF-CM through a 0.22 μm sterile membrane and store at −20°C for no longer than 1 month.

Human ES cell propagation and passaging

1. Coat sterile tissue culture plates (six-well plate with approximately 10 cm^2/well) with 1 mL of 1:30 diluted Matrigel. Put the plate at 4°C overnight. The dilution of Matrigel should be performed on ice to avoid gel formation.

2. Change MEF-CM daily, and passage the human ES cells weekly at 75–95% confluence by dissociation for 5–10 min with 200U/mL collagenase IV to maintain undifferentiated growth.

3. Plate 9×10^4–1.7×10^5 human ES cells/cm^2 onto the Matrigel coated plates in MEF-CM supplemented with 8 ng/mL bFGF (approximately 1:3 to 1:6 passage, e.g. evenly dividing one well of human ES cells at 80–90% confluency into 3–6 wells). Colonies of undifferentiated ES cells should have smooth and translucent edges (see **Figure 2B**).

Human ES cell differentiation

1. To form EBs, treat human ES cells at approximately 75–90% confluency with 200U/mL collagenase IV for 5–10 min to dissociate cells.

2. Manually assist dissociation of colonies by gentle scraping with 10 mL pipette. This should generate small cell clusters (avoid dissociation to single cells).

3. Transfer one to two wells of human ES cell clusters to one well of ultra low attachment plate in 4 mL of differentiation media.

4. Culture human EBs for 10 days with media changes every 3–5 days. The cystic human EBs can be seen from 3 to 5 days of differentiation (see **Figure 3B**).

Selection of endothelial cells

1. Collect human EBs in a 15 mL centrifuge tube at day 13 to 15 and centrifuge at 130 g ($\sim$300 rpm) for 3 min. Aspirate human EB medium. Add 2–4 mL of 0.4U/mL pre-warmed collagenase B, transfer the cell suspension to the original human EB culture wells, and dissociate for 2 hr in a 37°C incubator. Do not nutate or mix.

2. Transfer each well to a centrifuge tube and centrifuge at 450 g (1,700 rpm) for 3 min. Aspirate collagenase B and wash once with PBS, followed by treatment with 2 mL of cell dissociation buffer for 10 min in a 37°C water bath.

3. Dissociate human EBs by gentle pipetting (triturate 40–50 times) with 1,000 μL pipettor (setting at 500–700 μL to avoid generating bubbles) and pass through a 70 μm cell strainer.

4. To isolate endothelial cells from other human EB cell types, stain the single-cell suspension with fluorescently conjugated antibodies and sort using a flow cytometer and associated data analysis software. Use approximately 2–5×10^5 cells/mL and stain with fluorochrome-conjugated PECAM-1 (CD31), CD45-APC, and 7-AAD. 7-AAD exclusion is used to identify live cells, and PECAM+/CD45− staining is used to isolate a population of endothelial-like precursor cells, that have the ability to differentiate into mature endothelium (Boldicke *et al.*, 2001)

5. The purity of CD45^{neg}PFV cells should be examined using the same gate setting immediately after cell sorting to confirm purification.

Culture of endothelial cells

1. Seed isolated CD45^{neg}PFV cells (5×10^4 cells/cm^2) on fibronectin-coated plates.

2. Culture in endothelial growth media (the media must be prepared prior to use and can be stored at 4°C only up to 12 hr). Change media every 2 days.

3. The endothelial-like cells will appear starting approximately at day 3. At day 7, the number of committed endothelial cells can be determined by FACS, based on the expression of endothelial surface markers and the formation of endothelial network.

Directed differentiation using two dimensional or co-culture techniques

Protocol III — Two dimensional model of directed differentiation (Gerecht-Nir et al., 2003)

Overview

Human ES cells are grown on MEFs, differentiated as monolayers in flat culture, then dissociated and mechanically sorted for endothelial cell-like precursors, and finally cultured to induce endothelial differentiation. Briefly, human ES colonies are separated into individual cells by treating with EDTA, then passing them through a 40 μm mesh strainer. Human ES cells are then seeded on collagen IV-coated plates in differentiation medium containing bFGF. After 6 days in culture, the resulting differentiated cells are of two distinguishable sizes: larger and smaller. The smaller cells are then selected for using a 40 μm mesh strainer, which excludes the larger cells. These smaller cells represent endothelial-like precursors, which are re-cultured for differentiation. To select for endothelial cells, cells are then plated on type-IV collagen-coated plates in a differentiation medium containing human VEGF and cultured for up to 10–12 additional days.

Protocol III

Preparation of MEFs

1. Coat sterile tissue culture plates with 0.1% w/v gelatin.

2. Plate growth arrested feeder MEFs on the plate at 100% confluency in MEF media.

3. Allow MEFs at least 2 hr to attach. Change media to remove unattached MEFs. Do not use after 1 week.

Human ES cell propagation and passaging

1. Plate 10^5 human ES cells on MEFs in knockout medium.

Gerecht-Nir et al. (2003)

Undifferentiated hES cells on MEF feeders (KO-medium + 20% KO-SR, bFGF, L-glut, NEAA, BME)

↓

Dissociate w/collagenase

↓

Seed at 5–7 × 10⁴ cells/cm² on collagen IV-coated plates for 6 days

↓

Filter through 40 m mesh strainer (mesodermal cells)

↓

Reseed at 2.5 × 10⁴ cells/cm² on collagen IV-coated dishes

↓

Add VEGF165 to induce endothelial differentiation for 10-12 days

↓

~20% selected cells are PECAM+, DilAcLDL+
~90% are SMA+
~80% show endothelial character in Matrigel

---

2. Grow cultures in 5% CO_2 and passage every five days, change media daily. Split once the colonies reach approximately a 70% confluence. Colonies should have smooth, translucent edges. Uneven borders can indicate differentiation.

3. Passage human ES cells by disaggregating colonies in 1 mg/mL collagenase. For undifferentiated cell maintenance, split human ES cells 1:3 to 1:4 and reseed onto fresh feeder cells.

Human ES cell differentiation

1. To differentiate, rinse cells with PBS, and treat human ES cells with 2 mL of 5 mM EDTA in PBS, supplemented with 1% (v/v) FBS for 20 min. To detach cells, add 5 mL ESC media and scrub gently with 5 mL pipette tip.

2. Collect the cells by centrifuging in 15 mL tube for 3 min at 300 g (~1,200 rpm). Aspirate the medium and gently resuspend the cell pellet in 1 mL differentiation media.

3. Separate into single cells using a 40 μm mesh strainer. Count the cells.

4. Seed undifferentiated human ES cells ($5 \times 10^4 - 7 \times 10^4$ cells/cm^2) in "differentiation medium" on collagen-coated six-well dishes.

Selection of endothelial cells

1. After six days of culture, two differentiated cell types can be observed: smaller flat cells, with large nuclei, which have angiogenic differentiation potential, and large flat cells that organize into a fiber-like arrangement. To select for endothelial precursor cells, digest cells with EDTA as mentioned above, and filter them through a 40 μm mesh strainer. The large cells will be largely excluded by this filtering.

2. Re-plate smaller, filtered cells (2.5×10^4 cells/cm^2) on collagen-coated dishes in "differentiation medium" supplemented with human VEGF165 (50 ng/mL) for up to 10–12 additional days. The vast majority of these cells will be small, endothelial-like, differentiated cells. Replace differentiation media every other day. Proceed to analysis of endothelial differentiation using methods that include FACS, immunostaining, RT-PCR, or angiogenesis Matrigel assay.

Culture endothelial cells

1. The resultant human ES cell-derived endothelial cells can be cultured and passaged on type IV collagen-coated dishes for several weeks before they become senescent or transdifferentiate.

2. Growth media should be continually supplemented with VEGF165.

Troubleshooting

Protocols I and II

How many ES cells do I plate on the feeder MEFs?

A single aliquot of frozen human ES cells can be plated on MEFs. In general, colonies should be far enough apart that they do not touch; however, there should be approximately many hundreds of colonies per plate (up to 1,000).

When do I split the ES cell colonies? What will they look like?

Colonies at time of passaging should look approximately as in **Figure 2**, or smaller. In general, when propagating and passaging, colonies should not touch one another or be too packed. If colonies touch, grow too large, or grow into each other, they will start to differentiate in the middle. A defined, smooth border to the colonies is important. Rough borders indicate differentiation

Why are my EBs of varying sizes, falling apart, and/or dying?

Poor human EB formation is often the result of using poor human ES cells. Monolayer human ES cells may not aggregate well during human EB formation. Sometimes, the resulting human EBs will be small and ultimately dissociate and undergo apoptosis. Well defined, healthy human ES cell colonies are required for proper cellular aggregation and human EB formation.

Why are my EBs not completely dissociating?

Inefficiently dissociated human EBs after differentiation is another challenge, which remains a common and unsolved hurdle in human EB experiments. Combined use other enzymes may provide help. Increase of trituration is not advised, since it results in decreased cell viability.

Why are my sorted cells dying?

Once sorted by FACS, cells will exhibit a certain amount of cell death. This is a direct result of the FACS sorting and cannot generally be avoided. After re-plating sorted endothelial cells, the cells take about a week to recover. Once they recover, they should proliferate nicely and appear like small endothelial cells.

Protocol III

Why do different ES clones produce varying proportions of endothelial cells using the same protocol?

Differentiation efficiency may vary using different human ES cell lines. For example, H9.2 was more efficient to derive smooth muscle cells (SMCs) than endothelial cells.

My ES cell colonies appear to be differentiating on the collagen. What could be causing this?

Maintenance of human ES cells in an undifferentiated condition remains a challenge, since it greatly affects subsequent human EB formation. The components of the MEF-CM (e.g. serum replacement, quality of MEF, bFGF activity, etc) may play an important role in human ES cell maintenance. If human ES cells show differentiation properties, as characterized by poor proliferation rate or monolayer colonies, a new batch of MEF cells should be generated for preparation of the MEF-CM.

Why is the yield of PECAM positive endothelial cells so low?

Studies of various medium compositions are currently being preformed to enhance endothelial differentiation towards a purified endothelial cell population. It is likely that different conditions will enhance further the enrichment of endothelial cells acquired with this protocol.

Why do my selected endothelial cells change over time?

Continuous culturing of the differentiated-filtrated human ES cell-derived endothelial cells may result in overtake of SMCs phenotype. Thus, it is recommended endothelial assays or experiments be done in early passage.

Acknowledgments

We are deeply indebted to Lisheng Wang and Sharon Gerecht-Nir for giving detailed advice on their protocols, providing images and continued discussions on the subject of this chapter. We are also grateful to Alethia Villasenor for careful reading of the manuscript. O.C. is supported by the March of Dimes Basil O'Connor Award and the American Cancer Society IRG.

References

Bailey AS and WH Fleming. (2003). Converging roads: evidence for an adult hemangioblast. *Exp Hematol* **31**(11): 987–993.

Bautch VL. (2002). Embryonic stem cell differentiation and the vascular lineage. *Methods Mol Biol* **185**: 117–125.

Boldicke T, Tesar M, Griesel C, Rohde M, Grone HJ, Waltenberger J, Kollet O, Lapidot T, Yayon A, Weich H. (2001). Anti-VEGFR-2 scFvs for cell isolation. Single-chain antibodies recognizing the human vascular endothelial growth factor receptor-2 (VEGFR-2/flk-1) on the surface of primary endothelial cells and preselected CD34+ cells from cord blood. *Stem Cells* **19**(1): 24–36.

Bradley A *et al.* (1984). Formation of germ-line chimaeras from embryo-derived teratocarcinoma cell lines. *Nature* **309**(5965): 255–256.

Brooks PC, Clark RA, Cheresh DA. (1994). Requirement of vascular integrin alpha v beta 3 for angiogenesis. *Science* **264**(5158): 569–571.

Brustle O *et al.* (1997). In vitro-generated neural precursors participate in mammalian brain development. *Proc Natl Acad Sci USA* **94**(26): 14809–14814.

Clark EL and ER Clark. Microscopic observations on the growth of blood capillaries in the living mammal. *Am. J. Anat.*, 1939. **64**: 251–301.

Conway EM and P Carmeliet. (2004). The diversity of endothelial cells: a challenge for therapeutic angiogenesis. *Genome Biol* **5**(2): 207.

Cowan CA, Klimanskaya I, McMahon J, Atienza J, Witmyer J, Zucker JP, Wang S, Morton CC, McMahon AP, Powers D, Melton DA. (2004). Derivation of embryonic stem-cell lines from human blastocysts. *New Engl J Med* **350**(13): 1353–1356.

Draper JS, Smith K, Gokhale P, Moore HD, Maltby E, Johnson J, Meisner L, Zwaka TP, Thomson JA, Andrews PW. (2004). Recurrent gain of chromosomes 17q and 12 in cultured human embryonic stem cells. *Nat Biotechnol* **22**(1): 53–54.

Gerecht-Nir S, Dazard JE, Golan-Mashiach M, Osenberg S, Botvinnik A, Amariglio N, Domany E, Rechavi G, Givol D, Itskovitz-Eldor J. (2005). Vascular gene expression and phenotypic correlation during differentiation of human embryonic stem cells. *Dev Dyn* **232**(2): 487–497.

Gerecht-Nir S, Ziskind A, Cohen S, Itkovitz-Eldor J. (2003). Human embryonic stem cells as an in vitro model for human vascular development and the induction of vascular differentiation. *Lab Invest* **83**(12): 1811–1820.

Gerritsen ME. (1987). Functional heterogeneity of vascular endothelial cells. *Biochem Pharmacol* **36**(17): 2701–2711.

Hatzopoulos AK, Folkman J, Vasile E, Eiselen GK, Rosenberg RD. (1998). Isolation and characterization of endothelial progenitor cells from mouse embryos. *Development* **125**(8): 1457–1468.

Huang PL, Huang Z, Mashimo H, Bloch KD, Moskowitz MA, Bevan JA, Fishman, MC. (1995). Hypertension in mice lacking the gene for endothelial nitric oxide synthase. *Nature* **377**(6546): 239–242.

Itskovitz-Eldor J, Schuldiner M, Karsenti D, Eden A, Yanuka O, Amit M, Soreq H, Benvenisty N. (2000). Differentiation of human embryonic stem cells into embryoid bodies compromising the three embryonic germ layers. *Mol Med* **6**(2): 88–95.

Kalka C, Masuda H, Takahashi T, Kalka-Moll WM, Silver M, Kearney M, Li T, Isner JM, Asahara T. (2000). Transplantation of ex vivo expanded endothelial progenitor cells for therapeutic neovascularization. *Proc Natl Acad Sci USA* **97**(7): 3422–3427.

Kaufman DS, Hanson ET, Lewis RL, Auerbach R, Thomson JA. (2001). Hematopoietic colony-forming cells derived from human embryonic stem cells. *Proc Natl Acad Sci USA* **98**(19): 10716–10721.

Kennedy M *et al.* (1997). A common precursor for primitive erythropoiesis and definitive haematopoiesis. *Nature* **386**(6624): 488–493.

Klug MG, Soonpaa MH, Koh GY, Field LJ. (1996). Genetically selected cardiomyocytes from differentiating embronic stem cells form stable intracardiac grafts. *J Clin Invest* **98**(1): 216–224.

Kocher AA, Schuster MD, Szabolcs MJ, Takuma S, Burkhoff D, Wang J, Homma S, Edwards NM, Itescu S. (2001). Neovascularization of ischemic myocardium by human bone-marrow-derived angioblasts prevents cardiomyocyte apoptosis, reduces remodeling and improves cardiac function. *Nat Med* **7**(4): 430–436.

Kuzuya M and JL Kinsella. (1994). Reorganization of endothelial cord-like structures on basement membrane complex (Matrigel): involvement of transforming growth factor beta 1. *J Cell Physiol* **161**(2): 267–276.

Lavon N and N Benvenisty. (2003). Differentiation and genetic manipulation of human embryonic stem cells and the analysis of the cardiovascular system. *Trends Cardiovasc Med* **13**(2): 47–52.

Levenberg S *et al.* (2002). Endothelial cells derived from human embryonic stem cells. *Proc Natl Acad Sci USA* **99**(7): 4391–4396.

Martin GR, Evans M, Kaufman MH, Robertson E. (1981). Isolation of a pluripotent cell line from early mouse embryos cultured in medium conditioned by teratocarcinoma stem cells. *Proc Natl Acad Sci USA* **78**(12): 7634–7638.

Min JY, Yang Y, Converso KL, Liu L, Huang Q, Morgan JP, Xiao YF. (2002). Transplantation of embryonic stem cells improves cardiac function in postinfarcted rats. *J Appl Physiol* **92**(1): 288–296.

Nagy A, Rossant J, Nagy R, Abramow-Newerly W, Roder JC. (1993). Derivation of completely cell culture-derived mice from early-passage embryonic stem cells. *Proc Natl Acad Sci USA* **90**(18): 8424–8428.

Nicosia RF and A Ottinetti. (1990). Modulation of microvascular growth and morphogenesis by reconstituted basement membrane gel in three-dimensional cultures of rat aorta: a comparative study of angiogenesis in matrigel, collagen, fibrin, and plasma clot. *In Vitro Cell Dev Biol* **26**(2): 119–128.

Nor JE, Peters MC, Christensen JB, Sutorik MM, Linn S, Khan MK, Addison CL, Mooney DJ, Polverini PJ. (2001). Engineering and characterization of functional human microvessels in immunodeficient mice. *Lab Invest* **81**(4): 453–463.

Orlic D, Kajstura J, Chimenti S, Bodine DM, Leri A, Anversa P. (2001). Transplanted adult bone marrow cells repair myocardial infarcts in mice. *Ann NY Acad Sci* **938**: 221–229; discussion 229–230.

Risau W Sariola H, Zerwes HG, Sasse J, Ekblom P, Kemler R, Doetschman T. (1988). Vasculogenesis and angiogenesis in embryonic-stem-cell-derived embryoid bodies. *Development* **102**(3): 471–478.

Sadler JE. (1998). Biochemistry and genetics of von Willebrand factor. *Annu Rev Biochem* **67**: 395–424.

Sharma B and JH Elisseeff. (2004). Engineering structurally organized cartilage and bone tissues. *Ann Biomed Eng* **32**(1): 148–159.

Shih SC, Robinson GS, Perruzzi CA, Calvo A, Desai K, Green JE, Ali IU, Smith LE, Senger DR. (2002). Molecular profiling of angiogenesis markers. *Am J Pathol* **161**(1): 35–41.

Thomson JA, Itskovitz-Eldor J, Shapiro SS, Waknitz MA, Swiergiel JJ, Marshall VS, Jones JM. (1998). Embryonic stem cell lines derived from human blastocysts. *Science* **282**(5391): 1145–1147.

Tomita S, Mickle DA, Weisel RD, Jia ZQ, Tumiati LC, Allidina Y, Liu P, Li RK. (2002). Improved heart function with myogenesis and angiogenesis after autologous porcine bone marrow stromal cell transplantation. *J Thorac Cardiovasc Surg* **123**(6): 1132–1140.

Vittet D *et al.* (1996). Embryonic stem cells differentiate in vitro to endothelial cells through successive maturation steps. *Blood* **88**(9): 3424–3431.

Vodyanik MA, Bork JA, Thomson JA, Slukvin II. (2005). Human embryonic stem cell-derived CD34+ cells: efficient production in the coculture with OP9 stromal cells and analysis of lymphohematopoietic potential. *Blood* **105**(2): 617–626.

Voyta JC, Via DP, Butterfield CP, Zetter BR. (1984). Identification and isolation of endothelial cells based on their increased uptake of acetylated-low density lipoprotein. *J Cell Biol* **99**(6): 2034–2040.

Wang JS, Shum-Tim D, Chedrawy E, Chiu RC. (2001). The coronary delivery of marrow stromal cells for myocardial regeneration: pathophysiologic and therapeutic implications. *J Thorac Cardiovasc Surg* **122**(4): 699–705.

Wang L, Li L, Shojaei F, Levac K, Cerdan C, Menendez P, Martin T, Rouleau A, Bhatia M. (2004). Endothelial and hematopoietic cell fate of human embryonic stem cells originates from primitive endothelium with hemangioblastic properties. *Immunity* **21**(1): 31–41.

Wartenberg M, Gunther J, Hescheler J, Sauer H. (1998). The embryoid body as a novel in vitro assay system for antiangiogenic agents. *Lab Invest* **78**(10): 1301–1314.

Zambidis ET, Peault B, Park TS, Bunz F, Civin CI. (2005). Hematopoietic differentiation of human embryonic stem cells progresses through sequential hematoendothelial, primitive, and definitive stages resembling human yolk sac development. *Blood* **106**(3): 860–870.

13, Part C

Directed differentiation of human embryonic stem cells into osteogenic cells

JEFFREY M. KARP[1], ALBORZ MAHDAVI[1], LINO S. FERREIRA[1,2], ALI KHADEMHOSSEINI[3,4] AND ROBERT LANGER[1,3]

[1]Department of Chemical Engineering, Massachusetts Institute of Technology, Cambridge, MA 02139, USA
[2]Center of Neurosciences and Cell Biology, University of Coimbra, 3004-517 Coimbra and Biocant-Biotechnology Innovation Center, 3060-197 Cantanhede, Portugal
[3]Harvard-MIT Division of Health Sciences and Technology, Massachusetts Institute of Technology, Cambridge, MA 02139, USA
[4]Center for Biomedical Engineering, Brigham and Women's Hospital, Harvard Medical School, Boston, MA 02139, USA

Introduction

Human embryonic stem (ES) cells can be used as an *in vitro* system for studying bone formation and provide a cell source for production of bone (Sottile *et al.*, 2003; Bielby *et al.*, 2004; Cao *et al.*, 2005; Karp *et al.*, 2006; Ahn *et al.*, 2006; Barberi *et al.*, 2005; Olivier *et al.*, 2006). Given the high incidences of bone trauma, cancer, and congenital and acquired disease, which are associated with over 500,000 bone graft procedures each year in North America, it is not surprising that bone is one of the most common transplanted tissue, second only to blood (Langer and Vacanti, 1993). Thus, one of the great challenges facing bone surgeons is to increase bone stock, or the amount of bone available for grafting. Current therapies for regeneration are fraught with many shortcomings including donor site morbidity, lack of suitable graft material, and osteogenic cell sources. The emerging field of bone engineering attempts to augment or replace the current approaches by utilizing the combination of liquid, gel, or solid carriers with a source of osteogenic cells. The scaffold or cell carrier, although temporary, can be engineered to support migration, proliferation and differentiation of osteoprogenitor cells and to aid in the organization of these cells in three dimensions. In one application, progenitor cell numbers are expanded *in vitro*, placed onto biodegradable matrices in combination with factors that stimulate osteogenic differentiation, followed by implantation into a bone defect site (Langer and

Vacanti, 1993; Alsberg *et al.*, 2001). The ideal construct would eventually be replaced with host bone tissue. In addition to therapeutic applications, osteogenic cells and tissue engineered bone constructs derived from human ES cells can be used for screening drug candidates, for fabricating physiological systems to study bone biology, and for developing *in vitro* toxicity assays.

In choosing an appropriate cell source for bone engineering strategies, one must consider the capacity of the chosen cells to regenerate bone tissue and the ability to obtain a therapeutically relevant number of cells. The ratio of mesenchymal stem cells to total bone marrow cells within human bone marrow decreases from ~1:10,000 in newborns to ~1:1,200,000 at the age of 80 (Caplan, 1991). Therefore approaches which make use of bone marrow cells have to account for the scarcity of mesenchymal stem cells within the bone marrow, as well as the need to perform marrow biopsy to obtain the cells. Embryonic stem (ES) cells offer benefits including ease of isolation, ability to rapidly propagate without differentiation, and a greater capacity to give rise to different cell types (Bhatia, 2005; Thomson *et al.*, 1998; Perlingeiro *et al.*, 2001).

Although only few studies have reported osteogenic differentiation of human ES cells (Cao *et al.*, 2005; Ahn *et al.*, 2006), numerous culture strategies have been employed. In general, after expanding the human ES cells in an undifferentiated state, the cells may be (A) differentiated through an EB stage, (B) differentiated by plating them as embryoid bodies (EBs), or (C) differentiated by plating single cells into a 2D cell culture dish, as described in **Figure 1**. Plating of EBs may be useful to achieve cell adhesion (as human ES cell-derived aggregates tend to adhere more efficiently compared to single cells) or to isolate certain populations of cells based on their migration from the aggregate.

Application for human ES cell based bone engineering strategies face numerous challenges, for example directed differentiation towards the osteogenic lineage. Currently, human ES cell-based bone research is centered on elucidating soluble and immobilized cues and respective signaling mechanisms that direct osteogenic differentiation, on characterization and isolation of differentiated progeny cells, and on establishing protocols to improve the expansion and homogeneity of osteogenic cells (Sottile *et al.*, 2003; Bielby *et al.*, 2004; Cao *et al.*, 2005; Karp *et al.*, 2006; Ahn *et al.*, 2006). This chapter focuses on fundamental concepts in osteogenic differentiation and key protocols and techniques for inducing the differentiation of human ES cells into functional osteoblasts. It must be emphasized that osteogenic differentiation of human ES cells is a relatively young field and much work remains to be done. For example, the process of human ES cell differentiation into osteoprogenitors is relatively inefficient and no more than 1:250–1,000 adherent human ES cells gives rise to an osteoprogenitor cell (Karp *et al.*, 2006). Given the assumption that clonal populations of human ES cells have the capacity to differentiate into multiple cells types, presumably individual clones could be forced to differentiate into a pure population of a desired cell type under specific media conditions. Therefore, in addition to the methods described herein, increasing the efficiency of osteogenic cells from human ES cells will likely require exquisite control over the cell microenvironment. Increasing the frequency of osteogenic cells derived from stem cells requires the presence of specific cues supplied by physical stimulation, soluble factors such as bone morphogenetic proteins, contact with other cell types, or

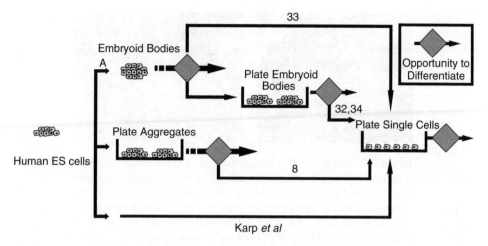

Figure 1 Flowchart of various methods for stimulating osteogenic differentiation of human ES cells. **(A)** Human ES cells are placed into suspension cultures as EBs. After potential addition of differentiated media, the cells can either be plated as a single cell suspension, or plated directly onto a culture dish which may be followed by a single cell step. **(B)** Human ES cells may also be plated directly as cell aggregates. This may be useful to increase the number of adherent cells prior to separation as a single suspension. **(C)** Human ES cells may also be directly plated as a single cell suspension. Opportunities to differentiate may include addition of soluble cues and/or use of immobilized cues. Conditions which may be used to enhance osteogenic differentiation that are not included in this chart include: genetic manipulation, co-cultures, and placement of cells within bioreactors. There is also opportunity at various stages in the chart to purify populations of cells using flow cytometry or other means.

by the substrate chemistry or morphology as recently reviewed (Heng *et al.*, 2004). In addition, emerging microscale technologies for controlling the cellular microenvironment may also be useful for enhancing stem cell differentiation (Maniatopoulos *et al.*, 1988)

Conventional wisdom holds that cells capable of forming bone are more useful for engineering bone tissue than cells that express osteogenic markers yet do not produce bone. Therefore, in addition to using classical stains to identify osteogenic cells, we believe that it is imperative to examine the matrix produced by the cells (Karp *et al.*, 2006). Recently we demonstrated that human ES cells, regardless of being cultured with or without the EB step, can produce many of the hallmarks of *de novo* bone formation including an elaborate cement line matrix and overlying mineralized collagen (Karp *et al.*, 2006) (**Figure 2**). In comparison to frequencies of osteoprogenitors derived form adult mesenchymal stem cells described above, by using standard osteogenic media with the inclusion of an EB step it maybe possible to achieve approximately one osteoprogenitor cell per ∼1,000–2,000 adhered human ES cells. In comparison, when the EB step is omitted, we have demonstrated that the frequency can be increased to one osteoprogenitor cell per ∼250 adhered cells.

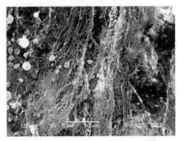

Figure 2 Morphological characterization of bone produced from differentiated human ES cells. Scanning electron micrographs show (left) the deposition of mineralized globular accretions, which are reminiscent of the cement line formed by differentiating osteogenic cells. Collagen fibers can be see anchoring to the underling globular accretions. (right) Mineralized collagen is observed above the cement line matrix as verified by FTIR and EDX analysis. (Figure adapted from Karp *et al.*, 2006).

Overview of protocols

The methods presented here for differentiation of human ES cells to osteogenic cells, with or without the EB step, are complemented with a list of characterization techniques that allow for determination of success of differentiation and functional capacity of the resulting tissue.

Materials, reagents and equipment

OSTEOGENIC CELL CULTURE

Material/reagent	Vendor	Catalog number
α-MEM	Invitrogen	12571-089
Fetal Bovine Serum (FBS)	Invitrogen	10437-028
Dexamethasone	Sigma	D8893-1 MG
Ascorbic acid 2-phosphate	Sigma	A8960
Glycerol 2-phosphate disodium salt hydrate	Sigma	G9891
Penicillin G	Sigma	P3032
Gentamicin	Sigma	G1397-10 ML
Amphotericin B (Fungizone)	Sigma	A2942
Phosphate buffer saline (PBS)	VWR	72060-034
Trypan blue	Sigma	93595
Trypsin–EDTA solution	Sigma	T4049

Material/reagent	Vendor	Catalog number
Non-enzymatic cell dissociation solution	Sigma	C5914
70% ethanol solution	Sigma	E7148

ALKALINE PHOSPHATASE/VON KOSSA QUALITATIVE STAINING

Formalin/formaldehyde	Sigma	11−0705
Disodium hydrogen phosphate (Na_2HPO_4)	Sigma	7907
Sodium phosphate monobasic monohydrate ($NaH_2PO_4H_2O$)	Sigma	S9638
N,N-Dimethylformamide (DMF)	Sigma	D4551
Silver nitrate (AgNO3)	Sigma	S8157
Sodium carbonate anhydrous (Na_2CO_3)	Sigma	S7795
Naphthol AS-MX phosphate disodium salt	Sigma	N5000
Tris-HCl (MW=157.6; pH 8.3; 0.2 M)	Sigma	88438
Red violet LB salt	Sigma	F1625

ALIZARIN RED STAINING

Formalin/formaldehyde	Sigma	11−0705
Alizarin Red S	Sigma	A5533

TETRACYCLINE LABELING

Tetracycline HCl ($C_{22}H_{24}N_2O_8 \cdot HCl$)	Sigma	T7660

QUANTITATIVE DETERMINATION OF ALKALINE PHOSPHATASE AND OSTEOCALCIN

Osteocalcin immunostain 96-well kit	DSL	10−7600
BCA protein kit	Sigma	QPBCA-1KT
Alkaline phosphatase detection kit	Sigma	APF
96-well plates (UV transparent)	Corning	3370

IMMUNOHISTOCHEMISTRY

Osteocalcin monoclonal antibody (human)	R&D	MAB1419
OCT-4 (human)	BD	611202
Alkaline phosphatase monoclonal antibody (human)	Sigma	A2064
Anti-Mouse IgG (whole molecule)-FITC antibody	Sigma	F9137-1 ML
DAPI nuclear stain	Mol. Probes	D21490

SCANNING ELECTRON MICROSCOPY PREPARATION

Sodium cacodylate trihydrate	Sigma	C9722-50 MG
Calcium chloride	Sigma	222313
0.2N hydrochloric acid	Sigma	H9892
Paraformaldehyde	Sigma	P6148

Material/reagent	Vendor	Catalog number
1N sodium hydroxide	Sigma	484024
Glutaraldehyde	Sigma	G5882
Ethanol	Sigma	E7148
Hexamethyldisilazane (HMDS)	Sigma	440191
FOURIER TRANSFORM INFRARED SPECTROSCOPY (FTIR)		
Potassium bromide (KBr)	Sigma	60090

Protocols

Osteogenic cell culture

For simplicity, we focus herein on the differentiation conditions for osteogenesis and not on the culture of human ES cells required prior to this stage; this was discussed in **Chapters 4, 5 and 6**. Typically, for differentiation experiments involving ES cells it is customary as a first attempt to apply culture conditions used for adult mesenchymal systems. Below we describe a widely accepted protocol that we have applied towards human ES cells as described in a recent publication (Karp *et al.*, 2006). The differentiation medium is based on a protocol by Maniatopoulus *et al.* (1988) who derived progenitors from rodent bone marrow. In addition to describing the concentrations of various components, we provide details for preparation of stock solutions.

Standard medium for inducing osteogenic differentiation

Base:	85–90% (v/v) α-MEM
Serum:	10–15% (v/v) fetal bovine serum (FBS)
Dexamethasone:	10^{-7}–10^{-9}M
Ascorbic acid 2-phosphate:	50 μg/mL
Glycerol 2-phosphate disodium salt hydrate:	5 mM (Karp *et al.*, 2003a)
Antibiotics including fungicide:	167 (U/mL) penicillin G
	50 (μg/mL) gentamicin
	0.3 (μg/mL) amphotericin B

Although we provide ranges for some of the components listed above, it is sometimes preferred to use 85% α-MEM, 15% FBS, and 10^{-8} M Dex. Typically we have found that increasing the FBS concentration from 10% to 15% can increase the number of bone nodules in our cultures. Although based on the lot of serum used (as discussed in the section on trouble shooting), it is likely best to test both 10% and 15%, as given the current price of serum, 10% would be more cost efficient.

Preparation of stock solutions

Dexamethasone

Is a synthetic member of the glucocorticoid class of hormones. Aside from being used clinically to treat many inflammatory and autoimmune conditions, it has been demonstrated to interact with specific glucocorticoid receptors leading to stimulation of osteogenic differentiation *in vitro* for progenitor cells derived from multiple tissues and animal sources (Karp *et al.*, 2006; Aubin, 1999; Sarugaser *et al.*, 2005; Qu *et al.*, 1998; Lecoeur and Ouhayoun, 1997; Cornet *et al.* 2002; Karp *et al*, 2003b). However, *in vivo* dexamethasone administration results in decreased osteogenesis (Baron *et al.*, 1992).

1. To make a 10^{-4} M ($10,000\times$) stock solution, add 25.5 mL of absolute ethanol to 1 mg of dexamethasone, mix and store at $-20°C$.

2. To prepare a 10^{-6} M supplement stock solution ($100\times$), aseptically add 1 mL 10^{-4} M dexamethasone solution into 99 mL of culture media (containing α-MEM, FBS and antibiotics) and mix well.

3. Aliquot 2 mL per sterile tube and store at $-20°C$. This can be used at 1% [v/v] when making fresh media.

4. With all supplements, it is best not to refreeze thawed aliquots.

β-glycerophosphate (β-GP)

Is a source of organic phosphate added to the culture medium (Davies, 1996). Inorganic phosphate within α-MEM is in the soluble form typically in the physiological range and is readily available and quickly absorbed by cells. Organic phosphate refers to phosphate that is bound to organic matter such as proteins or glycerol. Organic phosphate must be broken down in order to become soluble phosphate. Osteogenic cells can facilitate this process (cell mediated) using alkaline phosphatase. To avoid non-specific precipitation of mineral, also referred to as ectopic or distrophic mineralization, generally the β-GP concentration should be 3.5–5.0 mM (Maniatopoulos *et al.*, 1988; Aubin, 1999). Furthermore, in some cases, it may be necessary to add the β-GP after significant multi-layering has occurred as previously described (Baksh *et al.*, 2003).

1. To make $100\times$ supplement stock solution (500 mM), dissolve 10.8 g of glycerol 2-phosphate disodium salt hydrate in double distilled water at room temperature and make up to 100 mL.

2. Filter through 0.1 μm filter to sterilize.

3. Aliquot 2–3 mL per tube and store at −20°C. This can be used at 1% [v/v] when making fresh media.

Ascorbic acid (AA)

Ascorbic acid is an important cofactor for formation of hydroxyl praline which plays a key role in the stabilization of the collagen triple helix and thus collagen assembly. Given that collagen is the main organic component of bone, AA is included as a main supplement for osteogenic cultures (Maniatopolous *et al.*, 1988). Although most α-MEM formulations contain AA, the activity is completely lost after about 10 days at 4°C and much more quickly at 37°C (Feng *et al.*, 1977). Ideally, AA deficient α-MEM should be employed to ensure a known concentration. In addition, fresh AA (from frozen aliquots) should be added to media at each reefed, preferably every other day. If cultures are to be fed less frequently (i.e. within bioreactors), a long acting version of AA is preferred. L-Ascorbic acid 2-phosphate is a phosphate derivative of L-ascorbic acid and has more prolonged vitamin C activity in solution than does L-ascorbic acid.

1. To make 100 × supplement stock solution (5 mg/mL) of short acting ascorbic acid (L-ascorbic acid, MW 176.12), add 0.5 g in 0.1 M PBS at room temperature and make up to 100 mL.

2. Filter through 0.1 μm filter to sterilize.

3. Aliquot 2–3 mL per tube and store at −20°C. These aliquots must be freshly thawed prior to use.

Preparation of differentiation medium

1. Add 375 mL of α-MEM to a sterile container.

2. Add 50 mL of fresh 10× antibiotics.

3. Add 75 mL of FBS (for 15% v/v).

4. Add 1% β-glycerophosphate stock (5 mM final concentration).

5. Add 1% dexamethasone stock (10^{-8}M final concentration).

6. Store at 4°C (for up to 2 weeks).

7. Upon each refeed, add 1% ascorbic acid freshly thawed stock (50 μg/mL final concentration).

8. Mix thoroughly.

9. Label with date of prepared antibiotic solution.

Alkaline phosphatase/von kossa qualitative staining

Reagents

10% Neutral formalin buffer (NFB) (store at room temperature)

Formalin/formaldehyde:	100 mL
Na_2HPO_4:	16 g
$NaH_2PO_4H_2O$:	4 g
Distilled water:	to 1L

2.5% Silver nitrate solution (store at room temp in the dark)

$AgNO_3$:	2.5 g
Distilled Water:	to 100 mL

Sodium carbonate formaldehyde (store at $4°C$):

Formalin/formaldehyde:	25 mL
Na_2CO_3:	5 g
Distilled water :	to 100 mL

Other reagents

Naphthol AS MX-PO_4	0.005 g
N,N-Dimethylformamide (DMF)	200 µL
0.2 M Tris-HCL pH 8.3	25 mL
Red Violet LB salt	0.03 g
Distilled water	25 mL

Protocol

1. Remove media from dishes, rinse once in cold PBS.

2. Fix in 10% cold NFB for 30 min in chemical hood.

3. Remove buffer and rinse dishes in distilled water 3×.

4. Leave in distilled water for 15 min.

5. While waiting, prepare APase reagent using following protocol:

In an Eppendorf tube, dissolve 0.005 g of naphthol in 200 µl of DMF. Add to graduated cylinder with 25 mL of Tris-HCL and 25 mL Distilled water. Add 0.03 g of Red Violet salt to solution and filter with Whatman's No.1 filter paper immediately prior to adding to dishes to be stained.

6. Remove distilled water from cells and add APase reagent and incubate for 45 min at room temperature.

7. Rinse in distilled water 3–4 times.

8. Remove distilled water and stain with 2.5% silver nitrate for 30 min in the dark.

9. Remove silver nitrate and rinse with distilled water four times.

10. Prior to examination or drying dish, the color of the mineralized nodules can be deepened by adding sodium carbonate formaldehyde to the dish for 30 s to 2 min. One must observe closely as the black color may become too intense thus preventing analysis.

11. Remove sodium carbonate formaldehyde and rinse with slowly running tap water for 1 hr (do not let water fall directly onto dish, instead sink dishes within a large plastic tub).

12. Image/count nodules. Positive alkaline phosphatase staining appears bright red and positive von Kossa appears dark brown to black. The dish may be dried for indefinite storage.

13. A good negative control for the Von Kossa staining is to treat a test dish or well with 10% formic acid for 10 min prior to step 8. The formic acid should dissolve the mineral component of the matrix (i.e. the calcium phosphate) and thus the test dish should show a negative reaction.

Alizarin red staining

1. Fix cells in 10% NFB for 20–30 min.

2. Rinse cells 3× with distilled water.

3. Add 2% (w/v) solution of alizarin red in distilled water for 30 s to 5 min.

4. Rinse thoroughly with distilled water.

Calcium deposits should appear bright orange–red.

Tetracycline staining

Tetracycline staining

1. Prepare 900 µg/mL Tetracycline HCL solution in PBS and pass through a sterile filter (can store for 2 weeks at 4°C).

2. 24–48 hr prior to medium change or termination of the culture, add 1% of stock tetracycline solution to medium.

3. For analysis, rinse samples with PBS 3×

4. Fix in 100% cold ethanol (−20°C) for 2 hr.

5. Air dry samples and immediately image for green fluorescence.

Quantitative determination of alkaline phosphatase and osteocalcin

Given the availability of numerous kits available for quantifying alkaline phosphatase (a non specific early marker of osteogenesis and osteocalcin (a later marker of osteogenesis) expression, we have chosen to omit specific protocols here. Details for the protocol we used in a recent study with human ES cells can be found elsewhere (Karp *et al.*, 2006). For osteocalcin analysis, it is possible to analyze its release into the media, or osteocalcin that becomes entrapped within the produced matrix through a homogenization step. Similarly, alkaline phosphatase can be measured either within the media or within the cell membranes through a homogenization step.

Immunohistochemistry

1. Transfer cell suspension to the wells of a chamber slide or slides of your choice. The choice of slide design is often dictated by the experiment. Some slides have four wells, some have eight, some are glass, and some are plastic. Glass is recommended because the slide becomes more versatile, and reduces the photo bleaching during the fluorescence or confocal microscopy analyses.

2. Allow cells to grow to confluence with the addition of fresh media.

3. Rinse cells 2× with PBS buffer.

4. Fix cells in freshly made 4% (v/v, in PBS) paraformaldehyde for 30 min.

5. Rinse cells 2× with PBS buffer. Do not let the cells dry at any step.

6. Incubate the cell preparations with 0.25–0.5% Triton X-100 in PBS for 10 min to permeabilize the membranes.

7. Rinse the cells 2× with PBS buffer.

8. Block the cells with 2–5% normal serum in PBS for 30 min (normal serum should be the same species as the secondary antibody is raised). This step is required to block non-specific binding of immunoglobulins. Alternatively, blocking with 1% (w/v, in PBS) BSA solution may be used.

9. Rinse cells 2× with PBS buffer.

10. Incubate the fixed cells with anti-human monoclonal antibodies, for 1 hr (*Note*: antibody concentration should be determined by titration of the stock solution and testing on a known positive specimen. Usually, working concentrations are in the range of 10–20 μg/mL. However, depending on the source of antibodies, this concentration could vary significantly). It is advisable to run the appropriate negative controls. Negative controls establish background fluorescence and non-specific staining of the primary antibodies. It should be isotype-matched, not specific for cells of the species being studied and of the same concentration as the test antibody. The degree of autofluorescence or negative control reagent fluorescence will vary with the type of cells under study and the sensitivity of the instrument used.

11. Rinse cells 2× with PBS buffer.

12. Add fluorescence conjugated secondary antibody in the appropriate concentration, and recommended for the monoclonal antibody used previously, for 30 min.

13. Rinse cells 2× with PBS buffer.

14. Add DAPI nuclear staining (2 μg/mL, in PBS) for 10 min.

15. Remove the DAPI solution, add a drop of fluorescence mounting media and cover the slide with a coverslip.

16. Examine the cells under the microscope.

Scanning electron microscopy (SEM)

Karnofsky's Fixative (helps to preserve cell morphology for electron microscopy)

For 10 mL

2.5 mL 8% paraformaldehyde (final concentration of 2.0% (v/v) paraformaldehyde).

1 mL 25% glutaraldehyde (final concentration of 2.5% (v/v) glutaraldehyde).

5 mL 0.2 M cacodylate buffer.

Make up to 10 mL with distilled water, adjust pH to 7.2–7.4 with 1N NaOH or 1N HCl.

8% Paraformaldehyde (make up fresh on the day for better staining)

a. Dissolve 2 g of paraformaldehyde in 25 mL of water that is already at 60°C (do not let temperature go above 60°C).

b. Stir for 10 min.

c. Add 1–2 drops of 1N NaOH and wait 10 s (the solution should go from cloudy to clear — add more NaOH if required).

d. Let cool until the solution reaches room temperature.

e. Filter.

0.2 M *Sodium cacodylate buffer*

*All work with sodium cacodylate should be performed in a fume hood.

Sodium cacodylate tryhydrate	8.56 g
Calcium chloride	25.0 g
0.2N hydrochloric acid	2.5 mL

Dilute to 200 mL with distilled water, pH 7.4

SEM preparation

1. Prior to fixation, culture substrates should be washed 2–3 × with α-MEM and then with 0.1 M cacodylate buffer 2–3 × (dilute 0.2 M cacodylate buffer with distilled water).

2. Fixation is best carried out for a minimum of 2 hr in Karnovsky's fixative at 4°C.

3. Rinse with 0.2 M cacodylate buffer three times,

4. Dehydrate in graded alcohols (50%, 70%, 80% 90%, 95% and 100%) for 10 min each. The final step should include at least 2–3 wash steps in 100% anhydrous ethanol.

5. Replace ethanol with 100% HMDS (in a fume hood) and let stand for 30 min.

6. Repeat Step No. 5.

7. Remove HMDS and air dry in fume hood or desiccator.

8. Cut samples to appropriate size and shape for SEM.

9. Mount samples on aluminum stubs using carbon tape.

10. If observation of the matrix/culture surface is desired (i.e. cement line matrix), the overlying cell layers and the collagenous matrix can be partially removed by applying small blasts of compressed air.

11. To reduce charging, apply a small amount of colloidal silver or carbon paint and leave to dry. This paint is useful to bridge the culture substrate to the metal stub — this any charging that occurs can be quickly transferred away from the sample to the metal stub.

12.   Sputter coat with gold, platinum/palladium, or carbon. Carbon is useful if energy dispersive X-ray analysis is to be performed to reduce chance of interfering peaks.

13.   Store samples in a desiccator.

Energy dispersive X-ray analysis (EDX)

Typically instrumentation for this is attached to a scanning electron microscope or a transmission electron microscope and can be quite useful for elemental analysis and mapping. For example, the presence of calcium and phosphorous can be determined. Furthermore, semi-quantitative calcium to phosphate ratios (Ca:P) can be obtained by integrating the area under the Ca and P peaks. When performing such analysis, it is important to have a positive control consisting of crystalline hydroxyapatite which should have a ratio of 1.67:1.

Fourier transform infrared spectroscopy (FTIR)

*It is imperative for the fixing cells prior to FTIR that PBS is not used. The phosphate in the buffer can interfere with the phosphate signal from the sample.

1.   Remove part of the cell culture substrate (approximately 0.5–1.0 mg), using a spatula, and place the sample into an agate mortar containing approximately 100 mg of KBr.

2.   The sample and KBr must be ground to a fine powder until it sticks to the mortar.

3.   Take the powder sample and place it into a FTIR die-set. Press the powder, using a hand or hydraulic presser, into a pellet with a thickness of about 1 mm. A good KBr pellet is transparent. Opaque pellets give poor spectra, because little of the infrared beam passes through them. White spots in a pellet indicate that the powder is not ground well enough, or is not dispersed properly in the pellets.

4.   Before, placing the pellet in a FTIR sample holder, a reference background should be taken. In general, 128 scans between 4,000 and 450 cm^{-1}, with a resolution of 2 cm^{-1}, are acquired.

5.   Place the pellet in a sample holder and acquire the spectra.

Analysis

Positive identification of bone nodules

To ensure positive identification of bone nodules formed in culture, we recommend using a variety of assays (**Figure 3 I**). Prior to such studies, it is important to understand the hallmarks of *de novo* bone formation as illustrated in **Figure 3 II and Figure 3 III**)

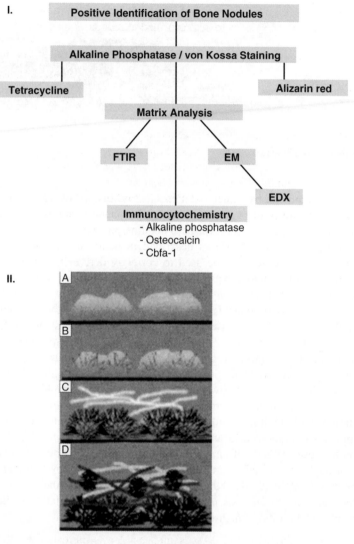

Figure 3 I. Experimental flow chart showing steps involved in the positive identification of bone nodules. After first confirming the presence of bone nodules with an alkaline phosphatase/von Kossa stain, the matrix should be examined with FTIR and/or with electron microscopy. Immunocytochemical analysis using a variety of osteogenic markets can be used to justify the results, but on their own are not sufficient to conclude the presence of bone nodules. II. Cascade of *de novo* bone formation on a solid surface. (**A**) Differentiating osteogenic cells initially secrete an organic matrix that is rich is non-collagenous proteins which mediates (**B**) the nucleation and formation of calcium phosphate crystals. (**C**) Collagen fibers assemble and anchor to this cement line matrix. (**D**) The overlying collagenous matrix is mineralized (Adapted from Davies, 1996). III. A cross-section through a bone nodule illustrates the various phases of *de novo* bone formation and the associated cellular phenotypes (adapted from).

III.

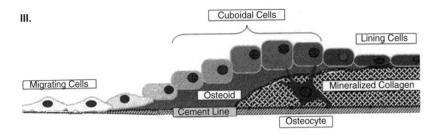

Figure 3   *(continued)*

and what information can be gained from each assay. Instead of relying on merely calcium stains or enzyme stains (i.e. for alkaline phosphatase) investigators are strongly encouraged to first use a dual stain for both alkaline phosphatase and von Kossa (Karp et al., 2006; Purpura et al., 2003) as described in detail above. Only based on this data can a retrospective analysis be used to correctly derive frequencies of recruited osteoprogenitors. Colonies that express APase but not von Kossa may be representative of a variety of cell types (including those of the osteogenic) but without further analysis, these cells cannot be considered as osteogenic, whereas those staining positive for both APase and von Kossa can be "loosely" considered as colony forming unit osteoblasts (CFU-O) (Baksh et al., 2003; Purpura et al., 2003). Although, further analysis is required to justify this (Karp et al., 2006). If calcium stains such as alizarin red are used to stain for mineralized bone nodules, the pattern of staining should mimic the von Kossa staining pattern, otherwise this could indicate dystrophic mineralization or the presence of other cations that may be identified with this stain. Following positive identification of CFU-O, it is useful to examine the produced matrix with FTIR to confirm the mineral is organized in an apatite form (Bonewald et al., 2003). (**Figure 4**). In addition to these methods of analysis, hallmarks of *de novo* bone formation can be examined through ultrastructural studies with electron microscopy (Karp et al., 2003a). This is one of the only methods to date that can be used to detect the cement line matrix. Energy dispersive X-ray analysis is also useful for semi-quantitatively examining the calcium to phosphate ratio, which in hydroxyapatite is 1.67:1.

Osteogenic markers

For a list of human ES cell markers refer to **Chapter 7**. Alkaline phosphatase is also a marker for osteogenic cells, it is important to use other markers to ensure complete differentiation of the human ES cells. In addition to alkaline phosphatase, classical osteogenic markers include:

Osteocalcin (bone Gla protein, OCN)

Is believed to be exclusively found in bone tissue and dentin. It accounts for 10–20% of the non-collagenous protein in bone and contains three residues of gamma carboxy

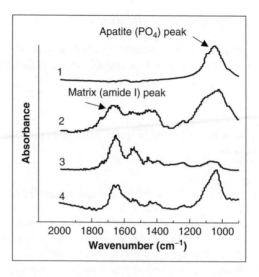

Figure 4 FTIR spectra of (curve 1) hydroxyapatite, (curve 2) human bone, (curve 3) human
ES cells cultured under conditions without dexamethasone, (curve 4) human ES cells cultured under conditions with dexamethasone.
The mineral and matrix peaks from human
bone and hydroxyapatite are comparable to the
extracellular matrix produced by the human ES
cells.

glutamic acid which allows it to bind strongly to bone tissue. This is a late marker of
osteogenesis and its expression typically corresponds with mineralization.

Osteopontin (OPN)

Is a glycoprotein synthesized by a variety of cell types including osteoblasts, hyper-
trophic chondrocytes, macrophages, smooth muscle cells and endothelial cells. Osteo-
pontin is one of the extracellular proteins that constitute the organic component of
bone.

Bone sialoprotein (BSP)

Is a phosphorylated glycoprotein that is expressed almost exclusively in bone and other
mineralized tissues. BSP is believed to be involved in the nucleation of hydroxyapatite
at the mineralization front of bone.

Osteonectin (Secreted protein acidic and rich in cystine — SPARC)

Is a matrix associated glycoprotein that binds to hydroxyapatite and collagen.

Core binding factor alpha1 (Cbfa1 or RUNX2)

Is the osteogenic master gene. It is a transcriptional activator of osteoblast differentiation during embryonic development and is also expressed in differentiated osteoblasts.

Collagen 1

Is the main organic component of bone (**Figure 3 II**) and is also found in other tissues including skin and tendon.

Fourier transform infrared microscopy (FITR)

Through examination of spectral vibrations using FTIR, the nature of the mineral and matrix of bone matrix can be characterized within the same sample without the use of molecular markers. For FTIR analysis of mineralized matrix, there are two main methods which can be used. In one method, dry samples are powdered, mixed with KBr, and pressed into pellets. Alternatively, samples may be analyzed directly within the culture dish which facilitates additional analyses using alternate assays. Regardless of the instrument used, the region of interest spans from $840-1,725$ cm^{-1} as previously reported (Shimko *et al.*, 2004). A single or double band between $900-1,200$ cm^{-1} (depending on the instrument resolution) is characteristic of the presence of a phosphorus based apatite phase which has the general formula $A_5(PO_4)_3$ where A are cations typically consisting of calcium. The PO_4 can be substituted to a limited extent by a carbonate anion or HPO_4^{2-}. Hydroxyapatite, which has the following composition: $Ca_{10}(PO_4)_6(OH)_2$, may be substituted with a variety of ions which may alter the width the bands. For example calcium may be substituted with lead, strontium, sodium, magnesium, potassium, lithium, iron, manganese, zinc and copper. The hydroxyl group may be substituted by fluorine, chlorine or carbonate. One important metric for examination of bone tissue includes the mineral to matrix ratio which can be obtained by integrating the area under the curve between $900-1,200$ cm^{-1} (which includes phosphate v_1 and v_3 absorption bands) and dividing by the area under the curve between 1585 and 1,725 cm^{-1} (collagen amide I band). Although methods exist for quantifying the carbonate fraction (integrating 840 and 890 cm^{-1} (carbonate v_2 absorption band) and dividing by the area under the phosphate absorption region) and the crystallinity (1020-to-1030 ratio), the ability to perform these types of analysis depend strongly on the spatial resolution and signal to noise ratio of the instrument and the purity of the sample.

Troubleshooting

How do I improve the frequency of osteogenic differentiation, can serum be the problem?

There can be significant differences between lots or suppliers of serum and certain lots of serum may not support osteogenic differentiation and production of bone nodules. Therefore, it may be useful to assess the ability of the serum to produce bone nodules with a conventional model such as primary rat bone marrow cells where the frequency of bone nodules (or osteoprogenitors) should be approximately 1 in 500 adherent cells (Purpura *et al.*, 2004). Using a typical rodent cell culture system, osteogenesis *in vitro* has been demonstrated to culminate in the formation of mineralized nodules which are discrete islands of bone that display histological, ultrastructural and immunohistochemical similarities to bone formed *in vivo* (Baksh *et al.*, 2003; Purpura *et al.*, 2004). Therefore, it is typically best to screen serum from a variety of lots or suppliers prior to moving ahead with human ES cells experiments. This normally involves determining frequencies of osteoprogenitor cells with a well defined osteogenic differentiation system — i.e. primary rodent cells. After determining the serum that produces the greatest frequency of osteopro-genitor cells (determined indirectly through counting APase and von Kossa stained nodules), it is advisable to purchase enough serum for all projected experiments. Typically it is best to test samples of serum from various suppliers prior to beginning experiments with human ES cells. After determining the serum that produces the greatest frequency of osteoprogenitor cells (determined indirectly through counting APase and von Kossa stained nodules), it is advisable to purchase an excess of serum required for all projected experiments.

Why do I see so much non-specific staining?

Although a number of publications report supplementation of with beta glycerophos-phate (*β*-GP) at a concentration of 10 mM (Sottile *et al.*, 2003; Bielby *et al.*, 2004) (twice the concentration we suggest), this concentration has been associ-ated with increased levels of dystrophic mineralization (Bonewald *et al.*, 2003) and thus 5 mM is more advisable. Previous work has demonstrated that aberrant min-eralization and cell death may occur when *β*-GP is greater than 6 mM (Gronowicz *et al.*, 1989). We demonstrated that 5 mM is sufficient for development of bone nodules from differentiated human ES cells (Karp *et al.*, 2006). In addition, spe-cific mineralization of osteogenic cultures can be confirmed by demonstrating that mineral stains including tetracycline, alizarin red and von Kossa produce similar staining patterns.

Why do I get an insufficient number of bone nodules from embryoid bodies?

While EB are used as a model for recapitulating the simultaneous formation of multiple tissues during embryonic development, achieving high frequencies of osteoblasts in this system may present a challenge. This may be due to complex cell–cell and cell–matrix interactions in addition to gradients of biomolecules. This creates several microenvironments within each EB where gradients of biomolecules can present different stimuli to the cells. Thus a system devoid of EB may be useful to improve the derivation efficiency of osteogenic cells where one would anticipate more homogenous microenvironments (Karp *et al.*, 2006). When using H1 cells it may be more difficult to differentiate these cells along the osteogenic lineage due to problems associated with EB formation with these cells as has been described in one study (Sottile *et al.*, 2003). We found that H9 cells readily differentiated to osteogenic lineage with increased frequencies when the EB step is skipped. H9 cells are most commonly used for study of osteogenic differentiation of human ES cells to date.

How do I avoid contamination during differentiation experiments?

To avoid contamination, it is imperative that reconstituted antibiotics stored at 4°C are only kept for 2 weeks. Typically it is convenient to make fresh antibiotic and fully supplemented media solutions once per week. If in some cases penicillin/streptomycin is not enough to stop contamination and antifungal agents such as Fungizone (amphotericin) should be added.

Why do I see unusual cell morphology in SEM images?

If the morphology of cells and matrix within the electron micrographs is not of suitable quality, it may be of interest to replace the final preparation step using HMDS with critical point drying as previously described (Karp *et al.*, 2003a). Although the HMDS step significantly reduces processing time, critical point drying is the gold standard method for preparing samples for publication quality images.

How do I determine the level of spontaneous differentiation?

Cultures treated without osteogenic supplements may be used to assess spontaneous differentiation. It may be useful to include ascorbic acid and betaglycerophosphate without dexamethasone (or other differentiation stimulating factors). This condition will thus provide the supplements that osteogenic cells would be able to use to form mineralized collagenous tissue, yet these supplements should not be able to stimulate osteogenic differentiation on their own.

Why do I get variable results from alkaline phosphatase staining?

It is important to know that alkaline phosphatase is both a marker for human ES cells and for osteogenic cells. Given that conventional antibodies for APase cannot differentiate between these two forms, it may be useful to examine APase kinetics during differentiation. One should observe an initial high level of APase followed by a decrease to almost zero and then an increase. This corresponds to a high number of human ES cells that differentiate and lose their APase expression followed by differentiation into osteogenic cells indicated by the re-expression of APase.

How does alkaline phosphatase (APase) and von Kossa (VK) staining work?

The APase/VK protocol is used to stain mineralized nodules in culture (**Figure 5**). Undifferentiated human ES cells exhibit a strong signal for APase, which is a hydrolase enzyme expressed by both human ES cells (Draper *et al.*, 2002) and

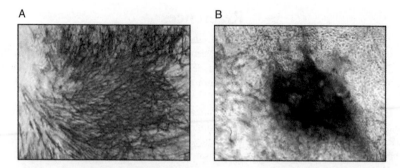

Figure 5 Alkaline phosphatase and von Kossa staining results. (**A**) Unmineralized regions containing colonies of fibroblast will stain red without any black staining. Alternatively, (**B**) mineralized bone nodules will exhibit positive dual staining typically consisting of a black core with red around the boarders.

osteoblasts (Aubin *et al.*, 1995) amongst other cell types. APase is present in all tissues throughout the body, and most concentrated in the liver, bile duct, kidney, bone and placenta and its chemical structure varies depending on where it is produced. It is possible that alkaline phosphatases produced by human ES cells and by osteoblasts likely represent different isoenzymes although this needs to be verified. Furthermore, most antibodies and assays for examining alkaline phosphatase cannot distinguish between these isoenzymes.

Von Kossa has been used to stain mineralized tissues since the 19th century (Kossa 1901). In a standard reaction, calcium is displaced from phosphate ions and replaced by silver ions, generated from the addition of silver nitrate. Therefore, the Von Kossa technique is used to directly demonstrate the anion (e.g. phosphate) and indirectly the cation (e.g. calcium). The reaction is promoted through providing activation energy with ultraviolet light for 20 min (or a 100 W incandescent desk lamp for 1 hr) thereby reducing the silver, which replaced the calcium, to metallic silver which is visualized as black deposits. The deposits appear black just as small metal particles do when they are shaved from a normally shiny metallic surface. Unreacted silver is typically removed with 5% sodium thiosulfate for 5 min.

An alternative protocol that we prefer avoids exposure to light. Instead, the intensity of the black stain (reduction reaction) can be achieved through addition of sodium carbonate formaldehyde as described below. For those interested, von Kossa protocols using UV light induced reduction reactions are readily available on the internet.

Although immunocytochemistry is a useful technique to identify osteogenic cells and related extracellular proteins, we believe it should be a complement to other techniques describe above. It should be noted that cellular markers for osteogenic cells do not necessarily correlate with the ability of those cells to produce bone nodules (Gronowicz *et al.*, 1989). For differentiation experiments from human ES cells, it is important to examine the percentage of human ES cells that remain during differentiation and determine the degree of differentiation by determining the percentage of Oct4 positive cells.

References

Ahn SE, S Kim, KH Park, SH Moon, HJ Lee, GJ Kim, YJ Lee, KH Park, KY Cha and HM Chung. (2006). Primary bone-derived cells induce osteogenic differentiation without exogenous factors in human embryonic stem cells. *Biochem Biophys Res Commun* **340**(2): 403–408.

Alsberg E, KW Anderson, A Albeiruti, RT Franceschi and DJ Mooney. (2001). Cell-interactive alginate hydrogels for bone tissue engineering. *J Dent Res* **80**(11): 2025–2029.

Aubin JE. (1999). Osteoprogenitor cell frequency in rat bone marrow stromal populations: role for heterotypic cell–cell interactions in osteoblast differentiation. *J Cell Biochem* **72**(3): 396–410.

Aubin JE, F Liu, L Malaval and AK Gupta. (1995). Osteoblast and chondroblast differentiation. *Bone* **17**(2 Suppl): 77S–83S.

Baksh D, JE Davies and PW Zandstra. (2003). Adult human bone marrow-derived mesenchymal progenitor cells are capable of adhesion-independent survival and expansion. *Exp Hematol* **31**(8): 723–732.

Barberi T, LM Willis, ND Socci and L Studer. (2005). Derivation of multipotent mesenchymal precursors from human embryonic stem cells. *PLoS Med* **2**(6): e161.

Baron J, Z Huang, KE Oerter, JD Bacher and GB Cutler, Jr. (1992). Dexamethasone acts locally to inhibit longitudinal bone growth in rabbits. *Am J Physiol* **263**(3 Pt 1): E489–E492.

Bhatia M. (2005). Derivation of the hematopoietic stem cell compartment from human embryonic stem cell lines. *Ann N Y Acad Sci* **1044**: 24–28.

Bielby RC, AR Boccaccini, JM Polak and LD Buttery. (2004). In vitro differentiation and in vivo mineralization of osteogenic cells derived from human embryonic stem cells. *Tissue Eng* **10**(9–10): 1518–1525.

Bonewald LF, SE Harris, J Rosser, MR Dallas, SL Dallas, NP Camacho, B Boyan and A Boskey. (2003). von Kossa staining alone is not sufficient to confirm that mineralization in vitro represents bone formation. *Calcif Tissue Int* **72**(5): 537–547.

Cao T, BC Heng, CP Ye, H Liu, WS Toh, P Robson, P Li, YH Hong and LW Stanton. (2005). Osteogenic differentiation within intact human embryoid bodies result in a marked increase in osteocalcin secretion after 12 days of in vitro culture, and formation of morphologically distinct nodule-like structures. *Tissue Cell* **37**(4): 325–334.

Caplan AI. (1991). Mesenchymal stem cells. *J Orthop Res* **9**(5): 641–650.

Caplan AI. (2005). Review: mesenchymal stem cells: cell-based reconstructive therapy in orthopedics. *Tissue Eng* **11**(7–8): 1198–1211.

Cornet F, K Anselme, T Grard, M Rouahi, B Noel, P Hardouin and J Jeanfils. (2002). The influence of culture conditions on extracellular matrix proteins synthesized by osteoblasts derived from rabbit bone marrow. *J Biomed Mater Res* **63**(4): 400–407.

Davies JE. (1996). In vitro modeling of the bone/implant interface. *Anatomical Record* **245**(2): 426–445.

Draper JS, C Pigott, JA Thomson and PW Andrews. (2002). Surface antigens of human embryonic stem cells: changes upon differentiation in culture. *J Anat* **200**(Pt 3): 249–258.

Feng J, AH Melcher, DM Brunette and HK Moe. (1977). Determination of L-ascorbic acid levels in culture medium: concentrations in commercial media and maintenance of levels under conditions of organ culture. *In Vitro* **13**(2): 91–99.

Gronowicz G, FN Woodiel, MB McCarthy and LG Raisz. (1989). In vitro mineralization of fetal rat parietal bones in defined serum-free medium: effect of beta-glycerol phosphate. *J Bone Miner Res* **4**(3): 313–324.

Heng BC, T Cao, LW Stanton, P Robson and B Olsen. (2004). Strategies for directing the differentiation of stem cells into the osteogenic lineage in vitro. *J Bone Miner Res* **19**(9): 1379–1394.

Karp JM, MS Shoichet and JE Davies. (2003a). Bone formation on two-dimensional poly(DL-lactide-co-glycolide) (PLGA) films and three-dimensional PLGA tissue engineering scaffolds in vitro. *J Biomed Mater Res* **64A**(2): 388–396.

Karp JM, MS Shoichet and JE Davies. (2003b). Bone formation on two-dimensional poly(DL-lactide-co-glycolide) (PLGA) films and three-dimensional PLGA tissue engineering scaffolds in vitro. *J Biomed Mater Res A* **64**(2): 388–396.

Karp JM, LS Ferreira, A Khademhosseini, AH Kwon, J Yeh and RS Langer. (2006). Cultivation of human embryonic stem cells without the embryoid body step enhances osteogenesis in vitro. *Stem Cells* **24**(4): 835–843.

Khademhosseini A, R Langer, J Borenstein and JP Vacanti. (2006). Microscale technologies for tissue engineering and biology. *Proc Natl Acad Sci USA* **103**(8): 2480–2487.

Kossa JV. (1901). Uber die in Organismus kunstlich erzeugbaren Verkalkungen. Beitr Pathol Anat Allg Pathol. **29**: 163–202.

Langer R and JP Vacanti. (1993). Tissue engineering. *Science* **260**(5110): 920–926.

Lecoeur L and JP Ouhayoun. (1997). In vitro induction of osteogenic differentiation from non-osteogenic mesenchymal cells. *Biomaterials* **18**(14): 989–993.

Maniatopoulos C, J Sodek and AH Melcher. (1988). Bone formation in vitro by stromal cells obtained from bone marrow of young adult rats. *Cell Tissue Res* **254**(2): 317–330.

Olivier EN, AC Rybicki and EE Bouhassira EE. (2006). Differentiation of human embryonic stem cells into bipotent mesenchymal stem cells. *Stem Cells* **24**(8): 1914–22.

Perlingeiro RC, M Kyba and GQ Daley. (2001). Clonal analysis of differentiating embryonic stem cells reveals a hematopoietic progenitor with primitive erythroid and adult lymphoid-myeloid potential. *Development* **128**(22): 4597–4604.

Purpura KA, JE Aubin and PW Zandstra. (2003). Two-color image analysis discriminates between mineralized and unmineralized bone nodules in vitro. *Biotechniques* **34**(6): 1188–+.

Purpura KA, JE Aubin and PW Zandstra. (2004). Sustained in vitro expansion of bone progenitors is cell density dependent. *Stem Cells* **22**(1): 39–50.

Qu Q, M Perala-Heape, A Kapanen, J Dahllund, J Salo, HK Vaananen and P Harkonen. (1998). Estrogen enhances differentiation of osteoblasts in mouse bone marrow culture. *Bone* **22**(3): 201–209.

Sarugaser R, D Lickorish, D Baksh, MM Hosseini, JE Davies. (2005). Human umbilical cord perivascular (HUCPV) cells: a source of mesenchymal progenitors. *Stem Cells* **23**(2): 220–229.

Shimko DA, CA Burks, KC Dee and EA Nauman. (2004). Comparison of in vitro mineralization by murine embryonic and adult stem cells cultured in an osteogenic medium. *Tissue Eng* **10**(9–10): 1386–1398.

Sodek J and S Cheifetz. (2000). Molecular regulation of osteogenesis. In: *Bone Engineering*, LE Davies, ed. EM Squared, Toronto, p 37.

Sottile V, A Thomson and J McWhir. (2003). In vitro osteogenic differentiation of human ES cells. *Cloning Stem Cells* **5**(2): 149–155.

Thomson JA, J Itskovitz-Eldor, SS Shapiro, MA Waknitz, JJ Swiergiel, VS Marshall and JM Jones. (1998). Embryonic stem cell lines derived from human blastocysts. *Science* **282**(5391): 1145–1147.

13, Part D

Directed differentiation of human embryonic stem cells into hematopoietic *in vivo* repopulating cells

SHANNON McKINNEY-FREEMAN, THORSTEN M. SCHLAEGER, GEORGE Q. DALEY

Division of Hematology/Oncology, Children's Hospital Boston, Department of Biological Chemistry and Molecular Pharmacology, Harvard Medical School, Harvard Stem Cell Institute, 300 Longwood Avenue, Boston, MA 02115, USA

Introduction

Hematopoietic stem cells, the propagators of the hematopoietic system, are most rigorously defined functionally: they are self-renewing cells capable of sustaining all lineages of the peripheral blood for the lifetime of a given organism (Lensch and Daley, 2004). Although their immediate downstream progeny, lineage-restricted progenitors, can be easily assessed functionally via *in vitro* colony-forming assays, the hemapoietic stem cells themselves can only be stringently observed via *in vivo* hematopoietic repopulation. In these assays, to meet the definition of a hematopoietic stem cell, a given cell population must be capable of life-long reconstitution of the entire hematopoietic system of a recipient animal whose own hematopoietic system has been ablated by either irradiation or drug treatment. Furthermore, the cells must be capable of serial transplantation into secondary ablated recipients to provide evidence of their potential for self-renewal. Although it has been thoroughly established by multiple groups that human embryonic stem cells (human ES cell) can yield hematopoietic progenitors after differentiation via embryoid body (EB) formation or co-culture with stromal-derived cell lines (Chadwick *et al.*, 2003; Kaufman *et al.*, 2001; Vodyanik *et al.*, 2005; Zambidis *et al.*, 2005), their ability to yield definitive hematopoietic stem cells has been presumed, but not yet rigorously demonstrated. Thus, as yet, there is no standard protocol for the derivation of definitive hepatopoietic stem cells from human ES cells.

Functional demonstrations of human ES cell-derived hematopoietic stem cells are particularly difficult due to the absence of ideal animal models. As stated, *in vivo* repopulation is the gold standard to define hematopoietic stem cells. Yet, due to its high morbidity and mortality, bone marrow transplantation cannot be justified in human subjects as an assay for putative human hematopoietic stem cells. Thus, researchers are limited to xenotransplantation models using immuno-deficient recipients (Legrand *et al.*, 2006). Even when transplanted with traditional highly purified human hematopoietic stem cell populations, these models yield only limited reconstitution (Legrand *et al.*, 2006). However, despite these limitations, several groups have recently described the capacity of cells ultimately derived from human ES cells to repopulate mice. This chapter will describe and contrast the techniques employed by these groups with the goal of providing a foundation from which future work can build.

Overview of protocols

Using distinct protocols, several groups have recently reported success in coaxing human ES cell-derived cell populations to repopulate the hematopoietic system of ablated mice (**Figure 1, Table 1**) (Galic *et al.*, 2006; Tian *et al.*, 2006; Wang *et al.*, 2005a). One group employs EB differentiation while two others use direct co-culture of human ES cell with stromal cell lines. Here we outline all three protocols.

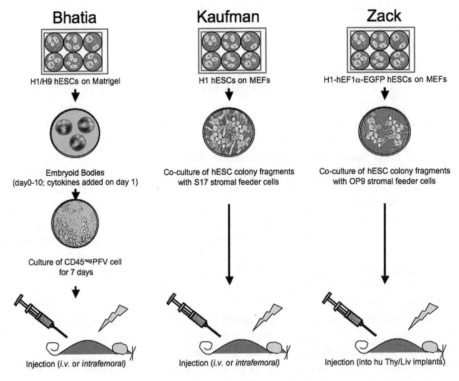

Figure 1 A schematic diagram highlighting the broad differences in the protocols of Bhatia, Kaufman, and Zack.

Materials, reagents and equipment

Table 1 Summary of methods and results of murine hESC-derived repopulation studies

	Bhatia	Kaufman	Zack
human ES cell line(s)	H1, H9	H1	H1 (hEF1a-EGFP transduced)
differentiation medium	KO-DMEM, 20% FCS	DMEM, 20% FCS	aMEM, 10% FCS (non heat-inactivated)
differentiation method	Embryoid Bodies	S17 co-culture (3000 cGy irradiated)	OP9 co-culture
cytokines	SCF, Flt3L, IL3, IL6, G-CSF, BMP4	none added	none added
Time in differentiation culture	cytokines added day 1–10	co-culture for 7–24 days	co-culture for 5–14 days
sorting	$CD45^{neg}CD31^{pos}$ or $CD45^{neg}Flk\text{-}1^{pos}7$ AAD^{neg}	—	—
further *in vitro* maturation/ expansion	7 days on fibronectin in serum-free media with SCF, Flt3L, IL3, IL6, G-CSF	—	—
animal model	NOD/SCID (8-10wk; 3.5 Gy) NOD/SCIDbm2$^{-/-}$ (8-10wk; 3.25 Gy)	NOD/SCID (6-10wk; 300 cGy) +/− anti-ASGM1 treated	SCID, hu. fetal liver/ thymus (300 rads) Rag2$^{-/-}$, hu. fetal liver/thymus (900 rads)
sorting for transplantation	—	—	$CD34^{pos}$ *vs* $CD34^{neg}133^{pos}$
# cells transplanted	$5 \times 10^6 - 1.6 \times 10^7$ (i.v.; unsuccessful) $4 \times 10^4 - 1.5 \times 10^5$ (intrafemural; successful)	$2-4 \times 10^6$ (i.v.) $0.5-3.8 \times 10^6$ (intrafemural)	SCID: 5×10^4 ($CD34^{pos}$); 1×10^6 ($CD34^{neg}133^{pos}$); Rag2: 8×10^5 ($CD34^{pos}$); 1×10^6 ($CD34^{neg}133^{pos}$) plus 2×10^6 mouse BM inj. into hu. fetal liver/thymus implants
route of delivery	1) tail vein (unsuccessful) 2) intrafemural (successful)	1) tail vein 2) intrafemural	
evidence of engraftment	BM PCR/Southern; Survival; Multi-lineage BM FACS analysis	BM+PB PCR; BM hCD45^{pos} FACS	EGFP^{pos}CD45^{pos} thymocytes (incl. SP and DP; CD3^{pos}; CD7^{pos}; CD1a^{pos}; TCR^{pos}; CD127^{pos})
frequency of engraftment	58% (11/19 recipients)	>90% (i.v. and intrafemural)	34% (11/32; SCID); 67% (6/9; Rag2$^{-/-}$)
% BM chimerism	~0.01−4% (PCR and hCD45^{pos} FACS)	~0.16−1.44% (PCR and hCD45^{pos} FACS) 2/12 recipient BM-derived CFUs	*not tested*
% contralat. BM chimerism	very low	≈% ipsilateral chimerism	N/A
% PB, thymus chimerism	*not tested*	very low (PB PCR analysis only)	~0.1−6.2% (SCID; Thy/ Liv implant) ~3.1−24.4% (Rag2; Thy/ Liv implant)

(continued overleaf)

Table 1   (*continued*)

	Bhatia	Kaufman	Zack
duration of engraftment	8 weeks (later time points: *not tested*)	3-6 months (later time points: *not tested*)	3-6 weeks (later time points: *not tested*)
secondary engraftment	*not done*	5×10^6 BM cells/2ndary recipient at 3 mo: 13/28 showed 0.08-0.2% BM hCD45pos	*not done*
teratomas	*unknown*	none observed	*unknown*
comment/specific issues	human ES cell-derived cells cause lethal pulm. emboli upon tail vein injections (aggregation caused by mouse serum; specific to human ES cell-derived blood cells)	hESC-derived CD34pos cells express less HLA class I than UCB-derived cells (NK cell ablation improves engraftment) hESC-derived BM cells express less CD45pos than UCB-derived cells	functional (anti-CD3/CD28-reponsive) hESC-derived thymocytes; lower CD45pos (but equal HLA-A2-) expr. on hESC- *vs* fetal-derived T-cells

Reagents and materials for Protocol I

Table 2   Culture media

Media	Ingredients	Vendor	catalog number
Differentiation medium	Knock-out D-MEM	Invitrogen	10829-018
	20% non-heat-inactivated FBS	Hyclone	SH30070.03E
	0.1 mM non-essential amino acids	Invitrogen	11140-050
	1 mM L-glutamine	Invitrogen	25030-164
	0.1 M β-mercaptoethanol	Sigma	M7522
Differentiation plus medium	Knock-out D-MEM	Invitrogen	10829-018
	20% non-heat-inactivated FBS	Hyclone	SH30070.03E
	0.1 mM Non-essential amino acids	Invitrogen	11140-050
	1 mM L-glutamine	Invitrogen	25030-164
	0.1 mM β-mercaptoethanol	Sigma	M7522
	300 ng/mL Flt-3 ligand	R&D systems	308-FK-025
	10 ng/mL IL-3	R&D systems	203-IL-010
	10 ng/mL IL-6	R&D systems	206-IL-010
	50 ng/mL G-CSF	Amgen	
	50 ng/mL BMP-4	R&D systems	314-BP-010
Inductive medium	9500 BIT media	Stem Cell Technologies	09500
	0.1 mM 2-mercaptoethanol	Sigma	M7522
	300 ng/mL SCF	Amgen	
	50 ng/mL G-CSF	Amgen	
	10 ng/mL IL-3	R&D systems	203-IL-010
	10 ng/mL IL-6	R&D systems	206-IL-010
	300 ng/mL Flt3 ligand	R&D systems	308-FK-025

Table 3 Enzymes and reagents

Reagent	Concentration	Vendor	Catalog number
Collagenase IV	200U/mL	Invitrogen	17104-019
Collagenase B	0.4U/mL	Roche Diagnostics	1-088-087
Cell dissociation buffer		Gibco	13151-014

Table 4 Antibodies and mice

Antibody/mouse	Clone/isotype		Vendor	Catalog number
Anti-human CD45 APC	2D1/mouse IgG1		BD Biosciences	340943
Anti-human CD31/PECAM-1 PE	WM59/mouse IgG1		BD Biosciences	555446
Anti-human VE-cadherin FITC	BV6/mouse IgG2a		Alexis Corporation	210-232F-T100
Anti-human Flk-1/VEGFR-2 (single chain antibody)	ScFvA7		Research Diagnostics, Inc.	RDI-VEGFR2scX
NOD/SCID mice			Taconic NODSC	

NOD/SCID, non-obese diabetic/severe combined immune deficiency

Table 5 Tissue culture plastic and materials

Material	Vendor	Catalog number
Six-well tissue culture plates	BD Biosciences	353046
Glass pipettes	VWR	53047-602
Six-well low attachment plates	Corning	3471
70 μm cell strainer	BD Falcon	352350
Fibronectin-coated plates	BD Biosciences	354402
28G insulin syringe	BD Biosciences	309300

Reagents and materials for Protocol II

Table 6 Culture media

Medium	Ingredients	Vendor	Catalog number
S17 medium	RPMI-1640	Invitrogen	10-404-CV
	10% FCS	Hyclone	SH30088.03
	2 mM L-glutamine	Cellgro/Mediatech	25-005-CI
	0.1 mM 2-mercaptoethanol	Invitrogen	21985-023
	1% non-essential amino acids	Invitrogen	11140-050
	1% penicillin–streptomycin	Invitrogen	15140-122
S17 differentiation medium	RPMI-1640	Invitrogen	10-404-CV
	15% defined FCS	Hyclone	SH30070.03
	2 mM glutamine	Cellgro/Mediatech	25-005-CI
	0.1 mM 2-mercaptoethanol	Invitrogen	21985-023
	0.1 mM Non-essential amino acids	Invitrogen	11140-050
	1% penicillin–streptomycin	Invitrogen	15140-122

Table 7 Enzymes, reagents, cell lines and mice

Reagent	Concentration	Vendor	Catalog number
Collagenase IV	1 mg/mL	Invitrogen	17104-019
Trypsin/EDTA	0.5%	Invitrogen	25300-054
70 μm filter		BD Falcon	352350
S17 cell line		Can be requested from the laboratory of Dr. Kenneth Dorshkind, Department of Cellular & Molecular Pathology, UCLA	
NOD/SCID mice		Taconic NODSC	

Reagents and materials for Protocol III

Table 8 Culture media

Medium	Ingredients	Vendor	Catalog number
OP9 medium	α-MEM	Fisher	10-022-CV
	20% FCS	Hyclone	SH30070.03E
Human ES cell/OP9 medium	α-MEM	Fisher	10-022-CV
	10% FCS	Hyclone	SH30070.03E
	0.1 mM monothioglycerol	Sigma	M6145

Table 9 Enzymes, reagents, cell lines and mice for Protocol III

Reagent	Concentration	Vendor	Catalog number
Collagenase IV	1 mg/mL	Invitrogen	17104-019
Trypsin/EDTA	0.5%	Invitrogen	25300-054
70 μm filter		BD Falcon	352350
Gelatin-coated six-well tissue cultures plates		BD Falcon	356652
T162 flasks		Corning-Costar	3150
OP9 cell line		ATCC	CRL-2749
SCID mice		Taconic	CB17SCRF-M

Protocols

Protocol I: Hematopoietic induction of embryoid bodies (Wang et al., 2005a)

In 2005, Bhatia's group published the first report of *in vivo* repopulating potential from human ES cell-derived cells (Wang *et al.*, 2005a). This report built on earlier work in which they first optimized conditions that promoted the production of detectable hematopoietic progenitors during EB differentiation and then phenotypically and functionally defined a "hemogenic precursor (Wang *et al.*, 2004)." This population, which emerged prior to the appearance of both mature CD45^{pos} cells and hematopoietic progenitor activity during EB differentiation, demonstrated *in vitro* endothelial and hematopoietic progenitor activity and expressed the cell surface antigens CD31/PECAM-1, CD309/VEGF-R2 (=KDR/Flk-1) and CD144/VE-cadherin but not the pan-hematopoietic molecule CD45. These cells were dubbed CD45^{neg}PFV cells.

For their transplantation studies, Wang and colleagues purified CD45^{neg}PFV cells from human ES cells allowed to differentiate for 10 days as EBs and subjected them to *in vitro* serum-free "hematopoietic inductive" culture prior to transplantation into NOD/SCID mice. This culture generally consisted of plating the purified cells on fibronectin-coated plates in 9500 BIT media base supplemented with multiple hemato- poietic cytokines. These inductive conditions were previously shown to promote the modest expansion and maintenance of cord blood-derived hematopoietic stem cells subsequently transplanted into SCID mice (Bhatia *et al.*, 1997). Wang *et al.* also trans- planted CD45^{pos}CD34^{pos}CD38^{neg}Lin^{neg} cord blood-derived cells cultured in these hematopoietic inductive conditions as a positive control for hematopoietic repopulation of their xenotransplantation model. Using this approach, these authors demonstrated that cells capable of repopulating the bone marrow of NOD/SCID mice could be cul- tured from human ES cells. However, the detected engraftment was extremely limited: generally <1% of BM, compared to >10% cord blood-derived BM repopulation. Fur- thermore, the authors only reported engraftment out to eight weeks post-transplant, did not assess serial repopulating potential and did not evaluate the peripheral blood and lymphoid organs for multi-lineage human ES cell-derived reconstitution. Thus, although promising and suggestive, the most rigorous definition of hematopoietic stem cell was not met in these studies. Significantly, the authors noted that human ES cell-derived hematopoietic populations aggregated when mixed with mouse serum and formed pulmonary emboli when injected intravenously (*i.v.*) into mice, causing their death. However, when injected directly into the femur, recipient mice did not succumb to pulmonary emboli and displayed modest human ES cell-derived BM repopulation. In contrast to cord blood derived-hematopoietic stem cell transplanted intrafemorally (*i.f.*), the human ES cell-derived repopulating cells were incapable of reconstituting the contra lateral femur, suggesting a deficiency in their ability to respond to endogenous trafficking cues. Alternatively, migration of the human ES cell-derived hematopoi- etic cells to the contra lateral marrow may be hampered by the still ill understood intravascular aggregation due to exposure to mouse serum (see also below).

Preparation of EBs

a. Harvest a confluent six-well plate of human ES cell in clumps by treatment with collagenase IV and scraping with a glass pipette. Resuspend clumps in *differenti- ation media* and transfer to one six-well low attachment plate. Incubate overnight.

b. The next day, change media to *differentiation plus medium.*

c. Continue culture for 10 days, changing medium every 3 days.

Isolation of CD45^{neg} PFV cells from day 10 EBs

d. Incubate day ten EBs with 0.4U/mL collagenase B for 2 hr at 37°C.

e. Collect EBs and wash with serum-free media base.

f. Incubate EBs with cell dissociation buffer for 10 min at 37°C in water bath. Next, liberate single cells via gentle trituration with pipette and filter with a 70 μm cell strainer.

g. Resuspend single cells at $2–5 \times 10^5$ cells/mL for antibody staining according to manufacturer's instructions.

h. Stain cells with PECAM-1-FITC and CD45-APC at a concentration of 5μg/mL.

i. Also stain a small aliquot of cells with the appropriate isotype controls (IgG1 for both PECAM-1 and CD45).

j. After staining, wash cells according to manufacturer's instructions, collect via centrifugation and resuspend for sorting.

k. Use isotype controls to direct gating for purification of CD45^{neg}PECAM^{pos} cells by flow cytometry of (shown to be nearly homogeneous for VE-cadherin and Flk-1 expression) (Wang *et al.*, 2004).

Inductive culturing and transplantation of CD45^{neg}PFV cells

l. Plate $1 \times 10^5/$ cm^2 CD45^{neg}PFV cells in fibronectin-coated plates.

m. Culture for 7 days in *inductive medium*, refreshing medium every 2 days.

n. Collect cells, resuspend at desired cell dose in total final volume of 25 μL sterile serum-free media base.

o. Subject 8–10 week old NOD/SCID to a sublethal (3.5Gy) dose of irradiation.

p. Deliver cells into the femur of sublethally irradiated NOD/SCID mice via injection with 28G insulin syringe.

q. Eight weeks post-transplant, assess reconstitution of NOD/SCID BM via staining with human-specific antibodies followed by flow cytometry.

Protocol II: Co-culture with S17 cells (Tian et al., 2004; Tian et al., 2006)

Shortly after the first derivation of human ES cell from blastocysts, Kaufman in the laboratory of Thomson reported that when co-cultured directly with the murine bone marrow stromal line, S17, for 17 days in DMEM plus serum, human ES cell differentiated into CD34^{pos}CD38^{neg} hematopoietic cells (Kaufman *et al.*, 2001). When plated in hematopoietic colony-forming assays, cells derived from human ES cells enriched for CD34 expression were also functionally enriched for hematopoietic *in vitro* colony formation. Later work from Kaufman's own laboratory demonstrated that co-culturing human ES cells with irradiated S17 cells under serum-free conditions did not have a negative impact on colony-forming activity, but that this activity does require direct contact with the stromal cell line (Tian *et al.*, 2004). In a recent paper, Kaufman advanced this work by demonstrating that human ES cells co-cultured with irradiated S17 stromal cells for various times are capable of repopulating ablated NOD/SCID mice (Tian *et al.*, 2006). Cells cultured for only 7–10 days repopulated recipient mice slightly better than cells cultured for longer periods. Kaufman saw no difference in engraftment between cells transplanted *i.v.* or *i.f.*, and further reported that cells injected *i.f.* were capable of equal contribution to both femurs. The time period of reconstitution was reportedly 3–6 months post-transplant, and mice injected with BM harvested from primary recipients also displayed human ES cell-derived BM repopulation. Thus, Kaufman's work has met the standards of long-term reconstitution and serial repopulation. However, once again, human ES cell-derived BM engraftment was exceedingly low (<1% on average) and multi-lineage reconstitution of the peripheral blood and lymphoid organs was not demonstrated.

Interestingly, Kaufman and colleagues noted that CD34^{pos} human ES cell-derived cells expressed lower levels of MHC class I molecules than CD34^{pos} cells derived from human CB. These data led them to speculate that natural killer (NK) cell activity might mediate an inhibitory effect on putative hematopoietic engraftment of human ES cell-derived repopulating cells. To address this, experiments were performed in which cells derived from human ES cells were transplanted into NOD/SCID mice either treated or untreated with anti-ASGM1, an antibody that depletes NK cell activity. This treatment resulted in a slight, but statistically significant, increase in human ES cell-derived BM repopulation of NOD/SCID mice, suggesting that low levels of MHC class I expression may hinder human ES cell-derived hematopoietic repopulation.

a. Maintain S17 cells in *S17 medium*, and refresh media weekly. Passage cells 1:2 as needed.

b. Subject confluent S17 cells to 30Gy of irradiation.

c. Passage human ES cells onto confluent irradiated S17 cells.

d. Maintain cells in *S17 differentiation medium* for 7–10 days.

e. Collect single cells via treatment with 1 mg/mL collagenase IV followed by treatment with 0.05% trypsin/EDTA. Pass cells through a 70 μm filter to remove clumps.

f. Subject NOD/SCID mice to sub-lethal irradiation (3Gy) 3–6 hr prior to injection of cells.

g. Inject 2–4 × 10^6 cells *i.v.* via tail vein or 0.5–4 × 10^6 cells *i.f.* per mouse.

h. 8–16 weeks post transplant assess BM of recipients for human ES cell-derived engraftment via flow cytometry with human specific antibodies.

Protocol III: Co-culture with OP9 cells (Vodyanik et al., 2005; Vodyanik et al., 2006)

In 2005 and 2006, Vodyanik and colleagues reported studies in which they also utilized direct culture of human ES cells with a stromal-derived cell line to induce hematopoietic differentiation (Vodyanik *et al.*, 2005; Vodyanik *et al.*, 2006). They focused their work on the calvaria-derived line, OP9, previously shown capable of promoting the expansion of murine ES cell-derived repopulating cells (Kyba *et al.*, 2002; Kyba *et al.*, 2003; Wang *et al.*, 2005b). Their protocol generally involves direct plating of human ES cells on ultra-confluent OP9 cells for 10–14 days in alpha-MEM media base, serum and monothioglycerol. Galic *et al.* recently employed this protocol for the derivation of T cells from human ES cells (Galic *et al.*, 2006). In their studies, they utilized a humanized mouse model in which pieces of human fetal liver and thymus were inserted under the renal capsule of either SCID or Rag-2-deficient mice prior to the transplantation of CD34[pos] or CD133[pos] human ES cell-derived cells into the same site. They found that when analyzed 3–6 weeks post transplantation, human ES cell-derived T cells were generated. They also noted that higher doses of both irradiation and cells transplanted promoted higher levels of T cell engraftment. The focus of this work on the repopulation of a single lineage made assessment of putative hematopoietic stem cell derivation impossible. However, *in vivo* repopulation, even of a single lineage, is noteworthy and suggests that OP9 co-culture of human ES cells may be worth exploring as a resource for the derivation of *bone fide* hematopoietic stem cells.

General OP9 cell-line maintenance

a. OP9 cells are maintained on gelatinized 10 cm tissue culture plates in *OP9 medium*.

b. To avoid adipocyte differentiation, OP9 cells should never reach greater than 80% confluence prior to passaging.

c. OP9 cells are generally passaged 1:3 every 4–5 days during general maintenance.

d. It has been reported that OP9 cells may lose their hematopoietic inductive properties if maintained in culture longer than 1 month post-derivation.

Human ES cell co-culture with OP9 cells

e. Passage OP9 cells into gelatin coated six-well plates.

f. Refresh 50% of media 5 days post plating. Cells should be confluent.

g. Continue OP9 culture for an additional 4 days. Add about 3×10^5 human ES cells in clumps collected by collagenase IV to OP9 cultures in 4 mL of *human ES cell/OP9 medium*.

h. Continue culture on OP9 cells for 10 days, refreshing 50% of medium every 2 days.

i. Prepare a single-cell suspension by subsequent treatments with collagenase IV for 20 min and 0.05% trypsin/EDTA for 15 min at 37°C. Triturate clumps after enzyme incubation to further disperse cells and finally filter through 70 μm filter.

j. Plate cells on T162 flasks for 1 hour at 37°C. This step should allow OP9 cells adhere to the flask.

k. Human ES cell-derived cells containing putative T cell progenitors and/or hematopoietic stem cells should be enriched in the resulting supernatant. Harvest supernatant and collect cells via centrifugation for experimental studies.

Troubleshooting

Are all human ES cell lines equally capable of yielding hematopoietic repopulating cells in the assays described above?

The protocols outlined above were derived from studies that employed the human ES cell lines H1 and H9. The authors of these studies explicitly state that these two lines yielded identical results. A broad, side-by-side survey of the ability of many human ES cell lines to yield cells capable of hematopoietic repopulation has not yet been reported in the literature.

What variables have been shown to impede the ability of human ES cell-derived cells to repopulate the hematopoietic system of mice?

Protocol I: Bhatia's group clearly demonstrated that their putative hematopoietic repopulating cells "clumped" when cultured with mouse serum (Wang *et al.*, 2005a). They also noted that whereas 100% of animals transplanted with cord blood-derived cells survived at least 8 weeks post-transplant, only about 40% of animals transplanted intravenously with human ES cell-derived cells survived to this time point. When these animals were examined, emboli were discovered in the lungs, likely caused by the clumping of injected human ES cell-derived cells. To overcome this problem, Bhatia's group utilized direct injections into the femur of recipient animals. More than 90% of animals injected in this manner survived 8 weeks post-transplant and a significant proportion displayed human ES cell-derived repopulation of the BM. However, they did note that the injected cells were not capable of trafficking through the circulation to engraft contra lateral femurs. Whether this was due to an inherent defect in trafficking or to further clumping phenomenon was not addressed. Interestingly, Kaufman's group did not report a problem with recipient survival and hematopoietic repopulation when injecting cells *i.v.*, and did observe repopulation of contra lateral femurs if cells were transplanted *i.f.* (*Tian et al.*, 2006).

Protocol II: Kaufman's group addressed the issue of potential NK cell-mediated rejection of human ES cell-derived cells due to their observation that human ES cell-derived CD34[pos] cells expressed lower levels of MHC class I than CD34[pos] cord blood-derived cells (Tian *et al.*, 2006). When human ES cell-derived cells were transplanted into NOD/SCID mice treated with anti-ASGM1, an antibody that depletes NK cell activity, they observed slightly higher human ES cell-derived BM repopulation of NOD/SCID mice. These data suggest that NK cell activity may aberrantly impact human ES cell-derived hematopoietic repopulation.

Protocol III: Galic and colleagues examined the ability of cells derived from human ES cells to repopulate T cell lineages using two different humanized mouse models: SCID mice and Rag-2 deficient mice (Galic *et al.*, 2006). Although they saw maximal repopulation using the Rag-2-hu model, they also changed the dose of irradiation and greatly increased the number of CD34^{pos} or CD133^{pos} cells being transplanted. Thus, it is difficult to conclude from these studies whether their increased repopulation was due to the background of the mouse model, the dose of irradiation or the number of cells injected.

What is the minimum number of cells that must be injected to achieve repopulation of the BM?

It is not clear from the current literature what minimum dose of human ES cell-derived cells is required to effect *in vivo* hematopoietic repopulation. The three groups discussed transplanted a wide range of cells, both purified based on cell surface marker expression and unpurified. None of the groups systematically evaluated the effect of cell dose on repopulation. However, data from Bhatia and Kaufman's groups may suggest that higher numbers of cells transplanted *i.f.* result in a higher frequency of engraftment: $4–14 \times 10^4$ cells resulting in 58% of recipients repopulated versus $0.5–3.8 \times 10^6$ cells resulting in>90% of recipients repopulated (**Table 1**) (Tian *et al.*, 2006; Wang *et al.*, 2005a). However, higher numbers of cells injected *i.f.* do not seem to lead to higher level levels of repopulation in the BM.

References

Bhatia M, D Bonnet, U Kapp, JC Wang, B Murdoch and JE Dick. (1997). Quantitative analysis reveals expansion of human hematopoietic repopulating cells after short-term ex vivo culture. *J Exp Med* **186**: 619–624.

Chadwick K, L Wang, L Li, P Menendez, B Murdoch, A Rouleau and M Bhatia. (2003). Cytokines and BMP-4 promote hematopoietic differentiation of human embryonic stem cells. *Blood* **102**: 906–915.

Galic Z, SG Kitchen, A Kacena, A Subramanian, B Burke, R Cortado and JA Zack. (2006). T lineage differentiation from human embryonic stem cells. *Proc Natl Acad Sci USA* **103**: 11742–11747.

Kaufman DS, ET Hanson, RL Lewis, R Auerbach and JA Thomson. (2001). Hematopoietic colony-forming cells derived from human embryonic stem cells. *Proc Natl Acad Sci USA* **98**: 10716–10721.

Kyba M, RC Perlingeiro and GQ Daley. (2002). HoxB4 confers definitive lymphoid-myeloid engraftment potential on embryonic stem cell and yolk sac hematopoietic progenitors. *Cell* **109**: 29–37.

Kyba M, RC Perlingeiro, RR Hoover, CW Lu, J Pierce and GQ Daley. (2003). Enhanced hematopoietic differentiation of embryonic stem cells conditionally expressing Stat5. *Proc Natl Acad Sci USA* **100**(Suppl 1): 11904–11910.

Legrand N, K Weijer and H Spits. (2006). Experimental models to study development and function of the human immune system in vivo. *J Immunol* **176**: 2053–2058.

Lensch MW and GQ Daley. (2004). Origins of mammalian hematopoiesis: in vivo paradigms and in vitro models. *Curr Top Dev Biol* **60**: 127–196.

Tian X, JK Morris, JL Linehan and DS Kaufman. (2004). Cytokine requirements differ for stroma and embryoid body-mediated hematopoiesis from human embryonic stem cells. *Exp Hematol* **32**: 1000–1009.

Tian X, PS Woll, JK Morris, JL Linehan and DS Kaufman. (2006). Hematopoietic engraftment of human embryonic stem cell-derived cells is regulated by recipient innate immunity. *Stem Cells* **24**: 1370–1380.

Vodyanik MA, JA Bork, JA Thomson and Slukvin, II (2005). Human embryonic stem cell-derived CD34+ cells: efficient production in the coculture with OP9 stromal cells and analysis of lymphohematopoietic potential. *Blood* **105**: 617–626.

Vodyanik MA, JA Thomson and Slukvin, II (2006). Leukosialin (CD43) defines hematopoietic progenitors in human embryonic stem cell differentiation cultures. *Blood* **108**: 2095–2105.

Wang L, L Li, F Shojaei, K Levac, C Cerdan, P Menendez, T Martin, A Rouleau and M Bhatia. (2004). Endothelial and hematopoietic cell fate of human embryonic stem cells originates from primitive endothelium with hemangioblastic properties. *Immunity* **21**: 31–41.

Wang L, P Menendez, F Shojaei, L Li, F Mazurier, JE Dick, C Cerdan, K Levac and M Bhatia. (2005a). Generation of hematopoietic repopulating cells from human embryonic stem cells independent of ectopic HOXB4 expression. *J Exp Med* **201**: 1603–1614.

Wang Y, F Yates, O Naveiras, P Ernst and GQ Daley. (2005b). Embryonic stem cell-derived hematopoietic stem cells. *Proc Natl Acad Sci USA* **102**: 19081–19086.

Zambidis ET, B Peault, TS Park, F Bunz and CI Civin. (2005). Hematopoietic differentiation of human embryonic stem cells progresses through sequential hematoendothelial, primitive, and definitive stages resembling human yolk sac development. *Blood* **106**: 860–870.

13, Part E

Directed differentiation of human embryonic stem cells into lymphocytes

PETTER S. WOLL AND DAN S. KAUFMAN

Stem Cell Institute and Department of Medicine, University of Minnesota, Minneapolis, MN 55455, USA

Introduction

Embryonic stem (ES) cells provide a unique starting point to study lineage commitment and differentiation of multiple cell types. Although much progress has been achieved to understand aspects of development with murine ES cells, there are important differences between humans and mice that necessitate a model system that more closely recapitulates human development. The derivation of human ES cells and the potential of these cells to differentiate into the hematopoietic lineage clearly demonstrate these cells as an alternative source of cells to study human hematopoietic ontogeny (Galic *et al.*, 2006; Kaufman *et al.*, 2001; Slukvin *et al.*, 2006; Vodyanik *et al.*, 2005; Wang *et al.*, 2005; Woll *et al.*, 2005).

Mature hematopoietic cells are derived from the hematopoietic stem cell which gives rise to series of lineage restricted progenitor cells. The earliest stage of hematopoietic stem cell differentiation generates myeloid progenitor and lymphoid progenitor cells (Blom and Spits, 2006; Kondo *et al.*, 2003; Laiosa *et al.*, 2006). This chapter focuses on derivation of lymphoid lineage cells from human ES cells. To produce mature blood cells from human ES cells there are at least two steps of differentiation needed. The first step generates hematopoietic precursor/progenitor cells. This process has been successfully achieved by several research groups using different versions of embryoid body (EB) formation or co-culture with stromal cells to promote or support hematopoietic differentiation. Both methods result in a heterogeneous population of cells, which after approximately 2 weeks of culture contains a significant population of hematopoietic progenitor cells (Kaufman *et al.*, 2001; Tian *et al.*, 2004; Vodyanik *et al.*, 2005; Wang *et al.*, 2005; Zambidis *et al.*, 2005). The hematopoietic progenitor cells can be identified and isolated based on expression of specific cell surface antigens such as CD34 and CD45 (Woll *et al.*, 2005). To induce the differentiation of the human ES cell-derived hematopoietic progenitor cells into mature blood cells, a second differentiation step is needed specific for the hematopoietic cell type of interest. The first mature hematopoietic cells generated from human ES cells were of the myeloid lineage, and included erythrocytes, macrophages, granulocytes and megakaryocytes (Kaufman *et al.*, 2001). Recently, derivation of lymphoid cells from human ES cells has also been reported. By engrafting hematopoietic progenitors derived from human ES cells into human thymic tissue in immunodeficient mice (so-called SCID-Hu mice), Galic and colleagues were able to demonstrate development of human ES cell-derived T cells (Galic *et al.*, 2006). Also, CD19$^+$ B cells derived from human ES cells after *in vitro* culture have been demonstrated by phenotypic analysis, but not functional studies (Vodyanik *et al.*, 2005; Vodyanik *et al.*, 2006). However, as the best functionally characterized lymphoid cell lineage derived from human ES cells are natural killer (NK) cells (Vodyanik *et al.*, 2005; Woll *et al.*, 2005), the protocols described here focus on the generation of NK cells from human ES cells.

NK cells are cells of the innate immune system that do not require pre-stimulation in order to perform their effector functions. Like most blood cells, NK cells are produced in the bone marrow, and once mature they enter circulation and peripheral tissues where their main effector functions include lysis of virally-infected and tumor cells. Several studies have demonstrated development of NK cells from human hematopoietic precursors isolated from bone marrow, peripheral blood, or umbilical cord blood (UCB). While stromal-cell-independent differentiation protocols for NK cells have been described (Mrozek *et al.*, 1996), most analyses use a stromal cell line to support NK cell differentiation from hematopoietic progenitors. These stromal cells include AFT024 (from murine fetal liver) and MS-5 (from murine bone marrow) (Miller and McCullar, 2001; Sivori *et al.*, 2003). By supplementing with cytokines know to be important for lymphoid commitment and NK cell differentiation, Flt3-ligand (FL), stem cell factor (SCF), IL-7 and IL-15, both stromal cells are able to support the generation of CD56$^+$ NK cells. However, MS-5 stromal cells also require IL-21 to generate fully mature CD56$^+$ CD16$^+$ KIR$^+$ NK cells, whereas this is not an requirement for AFT024 cells (Miller and McCullar, 2001; Sivori *et al.*, 2003). *In vitro* differentiation of human ES cells into NK cells can be achieved by applying similar stromal cell-mediated protocols to human ES cell-derived hematopoietic progenitors (Woll *et al.*, 2005).

Several NK cell subtypes have been identified based on CD56 expression and cytokine profile (Cooper *et al.*, 2001; Loza and Perussia, 2001), and there is still much to be learned about their developmental intermediates. One unique aspect of the ES cell system is that the differentiation process follows distinct sequential steps of hematopoietic maturation that allow better identification of developmental intermediates during *in vitro* culture. Unlike UCB and bone marrow, human ES cells do not initially contain mature hematopoietic cells, reducing the possibility that contaminating mature cells might obscure *in vitro* analysis of differentiation pathways. The generation of mature NK cells with cytolytic activity and cytokine production allows human ES cells to serve as a model system to investigate human NK cell development and possibly serve as an alternative source of cells for cancer immunotherapy. The clinical potential of lymphocytes to treat human malignancies has been established for two decades now. By adaptive transfer of lymphokine-activated killer (LAK) cells and tumor-infiltrating lymphocytes (TIL), both consisting mainly of CD8$^+$ T lymphocytes, cancer regression can be achieved in some but not all patients (Rosenberg *et al.*, 1987; Rosenberg *et al.*, 1988). This still relatively low success rate is probably related to the requirement of CD8$^+$ T cells to recognize tumor-specific antigens, which remains poorly characterized. In this regard, NK cells have proven an attractive alternative to anti-tumor therapy, as NK cells do not depend on recognition of tumor antigens for activation of anti-tumor activity. Recently, the ability of NK cells to treat human malignancy has also been demonstrated (Miller *et al.*, 2005; Ruggeri *et al.*, 2002). Still, this therapy only works for some but not all patients, and depends on cancer type. Therefore, human ES cell-derived NK cells could offer a novel source of anti-tumor lymphocytes with a broader application compared to conventional sources.

Overview of protocol

The methods described here outline (1) induction of hematopoietic differentiation of undifferentiated human ES cells to generate hematopoietic progenitors, (2) isolation of progenitor cell populations derived from differentiated human ES cells and (3) terminal differentiation of these cells into functional NK cells (**Figure 1**).

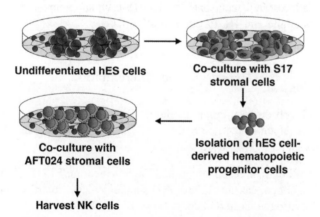

Undifferentiated hES cells

Co-culture with S17 stromal cells

Co-culture with AFT024 stromal cells

Isolation of hES cell-derived hematopoietic progenitor cells

Harvest NK cells

Figure 1   Schematic diagram of the two-step differentiation protocol for generating natural killer cells from human embryonic stem cells. Briefly, undifferentiated human ES cells are induced to differentiate into the hematopoietic lineage by co-culture with inactivated monolayer of S17 stromal cells. This generates hematopoietic progenitor cells that can be isolated based on expression of specific cell surface markers. To generate NK cells, human ES cell-derived progenitors are co-cultured with AFT024 stromal cells in medium supplemented with cytokines supporting NK cell differentiation. After 4 weeks in NK culture mature and functional NK cells can be harvested.

Materials, reagents and equipments

Cells

1. H1 and H9 line human ES cells (WiCell, Madison, WI). Maintained as undifferentiated cells as previously described (Kaufman *et al.*, 2001).

2. S17 stromal cells (kindly provided by Dr. K. Dorshkind, University of California, Los Angeles, CA). Maintained in RPMI-1640 (Invitrogen Corp./Gibco, cat. no. 10-404-CV) supplemented with 10% fetal calf serum (FCS) (Hyclone, cat. no. SH30088.03), 2 mM L-glutamine (Cellgro/Mediatech, cat. no. 25-005-CI), 0.1 mM β-mercaptoethanol (2-ME) (Invitrogen Corp./Gibco, cat. no. 21985-023), 1% MEM non-essential amino acids (NEAA) (Invitrogen Corp./Gibco, cat. no. 11140-050) and 1% penicillin–streptomycin (P/S) (Invitrogen Corp./Gibco, cat. no. 15140-122) at 37°C, 5% CO_2.

3. AFT024 stromal cells (ATCC, cat. no. SCRC-1007). Maintained in Dulbecco's modified Eagle's medium (DMEM) (Invitrogen Corp./Gibco, cat. no. 11965-092) supplemented with 20% FCS, 0.05 mM 2-ME and 1% P/S at 33°C, 5% CO_2.

4. K562 cells (ATCC, cat. no. CCL-243). Maintained in Iscove's modified Eagle's medium (Cellgro/Mediatech, cat. no. 10-016-CV) supplemented with 10% FCS, 2 mM L-glutamine and 1% P/S at 37°C, 5% CO_2.

Hematopoietic differentiation

1. Human ES cell-differentiation medium: RPMI-1640 supplemented with 15% defined FCS (Hyclone, cat. no SH30070.03), 2 mM L-glutamine, 0.1 mM 2-ME, 1% NEAA and 1% P/S. Store at 2–8°C.

2. ES wash medium: Dulbecco's modified Eagle's medium/F12 (DMEM/F12) (Invitrogen Corp./Gibco, cat. no. 11330-032) supplemented with 10% knockout-serum replacer (Invitrogen Corp./Gibco, cat. no. 10828-028).

3. Six-well tissue culture plates (NUNC™ Brand Products, Nalgene Nunc cat. no. 152795).

4. Mitomycin C (American Pharmaceutical Partners, product no. 109020).

5. Distilled phosphate buffered saline (D-PBS) without Ca^{2+} and Mg^{2+} (Cellgro/Mediatech, cat. no. 21-031-CV)

6. Collagenase type IV (Invitrogen Corp./Gibco, cat. no. 17104-019), 1 mg/mL diluted in DMEM/F12. Filter sterilized with a 50 mL 0.22 μm membrane Steriflip (Millipore, cat. no. SCGP00525). Store at 2–8°C.

7. 0.05% Trypsin–EDTA (Cellgro/Mediatech. cat. no. 25-052-CI)

8. Chick serum (Sigma, cat. no. C5405).

9. 70 μm cell strainer filter (Becton Dickinson/Falcon, cat. no. 352350).

10. EasySep® human CD34 positive selection kit (StemCell Technologies, cat. no. 18056) and EasySep® PE positive selection kit (StemCell Technologies, cat. no. 18557).

11. Easy Sep Buffer: DPBS without Ca^{2+} and Mg^{2+} supplemented with 2% FCS and 1 mM EDTA. Store at 2–8°C.

12. EasySep® Magnet (StemCell Technologies, cat. no. 18000).

13. Flow cytometer for sorting (for example FACSAria)

Natural killer cell differentiation

1. NK cell-differentiation medium: DMEM (Cellgro/Mediatech, cat. no. 10-017-CV) mixed 2:1 with Ham's F12 (Cellgro/Mediatech, cat. no. 10-080-CV) supplemented with 20% heat-inactivated human AB serum (Valley Biomedicals, cat. no. HP1022 HI), 5 ng/ml sodium selenite (Sigma-Aldrich, cat. no. S5261), 50 μM ethanolamine (MP Biomedicals, cat. no. 194658), 20 mg/L ascorbic acid (Sigma-Aldrich, cat. no. A-5960), 25 μM 2-ME, 2 mM L-glutamine, 1% P/S, 10 ng/ml IL-15 (Pepro-Tech, cat. no. 200-15)), 5 ng/ml IL-3 (PeproTech, cat. no. 200-03), 20 ng/ml IL-7 (National Cancer Institute bulk cytokine repository), 20 ng/ml stem cell factor (SCF) (PeproTech, cat. no. 300-07), and 10 ng/ml Flt3 ligand (Flt3L) (PeproTech, cat. no. 300-19). Store at 2–8°C and protect from light.

2. 24-well tissue culture plate (Falcon, cat. no. 353047)

3. Gelatin (Sigma, cat. no. G-1890), 0.1% solution made in water and sterilized by autoclaving.

Protocols

Generation of hematopoietic progenitors from human ES cells

There are several methods to generate hematopoietic progenitors from human ES cells. For simplicity, we describe here methods involving co-culture with S17 stromal cells.

1. Inactivate S17 stromal cells by incubating cells with conditioned S17 medium containing 10 μg/mL mitomycin C for 3 hr at 37°C, 5% CO_2. After mitomycin treatment cells are washed twice in PBS and dissociated with 0.05% trypsin–EDTA. To make a confluent S17 cell-layer in a six-well plate, 2.5 mL of inactivated S17

cells resuspended in S17 medium at 1.0×10^5 cells/mL are transferred to each well pretreated with 0.1% gelatin for at least 1 hr.

2. Before undifferentiated human ES cells are transferred to confluent and inactivated monolayer of S17 stromal cells, human ES cell colonies are dislodged and disrupted by incubation in 1 mg/mL collagenase type IV for approximately 7 min at 37°C, followed by scraping colonies off with a 5 mL glass pipette. Cells are washed twice in ES wash medium to remove residual collagenase.

3. Disrupted human ES cell colonies are resuspended in hematopoietic differentiation medium and transferred to S17 plates and cultured at 37°C, 5% CO_2 for 16–20 days, with complete medium changes every 3–4 days.

4. Human ES cells cultured on S17 can be observed to transit from confined colonies to then spread and proliferate after a few days of culture. Later in culture, a heterogeneous population of cells can be observed with a variety of structures **(Figure 2)**.

Isolation of hematopoietic progenitors

Optimal time for differentiation of human ES cells varies some depending on the ES cell line, stromal cell line used and serum lot. However, 5–10% CD34[+] human ES cell-derived cells are typically observed by flow cytometry after 17 days of co-culture with S17 stromal cells and these cells are enriched for hematopoietic progenitors as assayed by colony forming assays in methylcellulose. Further sorting for CD34

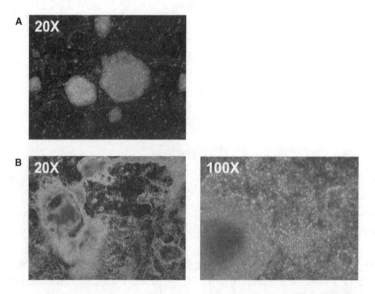

Figure 2 Representative images of undifferentiated and differentiated human ES cells. **(A)** H9 human ES cells goring on mitotically inactivated MEF feeders. **(B)** Human ES cells co-cultured with S17 stromal cells for 14 days demonstrate typical differentiated cell types.

and CD45 double positive cells significantly increases hematopoietic progenitors (Woll et al., 2005).

1. To prepare a single-cell suspension from differentiated human ES cells, aspirate medium from all wells of six-well plate cultures. Add 1.5 mL 1 mg/mL collagenase IV per well and incubate at 37°C for 5–10 min. until S17 stromal cells can be observed to break up. Scrape with a 5 mL glass pipette to disrupt the cells and transfer to a 15 mL conical tube. Add 5 mL of ES wash medium and mix by pipetting up and down to break up colonies. Spin tube at 350 g (~1,100 rpm) for 3 min.

2. Carefully pipette off supernatant, resuspend cell pellet and wash cells by adding 10 mL Ca^{2+} and Mg^{2+}-free DPBS. Mix vigorously by pipetting up and down to break up colonies. Spin tube at 350 g (~1,100 rpm) for 3 min.

3. Carefully pipette off supernatant, agitate the tube to resuspend cell pellet and add 5 mL 0.05% trypsin prewarmed in a 37°C waterbath. Vigorously pipette up and down with 5 mL pipette to break up the cells and add 100 µL chick serum (2%). (Note: chick serum adds proteins to improve cell viability, but does not contain trypsin inhibitors). Place tube in 37°C waterbath and incubate for 10–15 min. until a single cell suspension can be observed. The efficiency of trypsin digestion can be increased by shaking the tube at regular intervals during the incubation.

4. Add 8 mL ES wash medium and mix by pipetting. Spin tube at 450 g(~1,400 rpm) for 5 min in a refrigerated centrifuge (4°C).

5. Aspirate off supernatant and resuspend cells in 3 mL ES wash medium.

6. Eliminate remaining cell clumps by passing the single-cell suspension through a 70 µm mesh. Wash mesh with additional 3 mL ES wash medium to get all cells.

7. Count live cells by tryphan blue staining in hemocytometer.

8. Isolate $CD34^+$ $CD45^+$ human ES cell-derived hematopoietic progenitor cells by:

   a. Enriching for $CD34^+$ cells using Easy Sep® CD34 positive selection cocktail and magnetic separation from CD34 negative cells according to manufacturer's instruction, followed by flow cytometric sorting for $CD34^+$ $CD45^+$ cells.

   b. Alternatively, enrich for $CD34^+$ cells using the Easy Sep® CD34 positive selection cocktail as above, followed by CD45-PE staining of enriched cells and a second enrichment for CD45-PE positive cells using the PE-positive selection cocktail according to manufacturer's instructions.

9. Count live cells by tryphan blue staining in hemocytometer.

Natural killer cell differentiation

This method was adapted from previous studies that demonstrated differentiation of human UCB- and BM-derived progenitor cells into NK cells (Miller and McCullar,

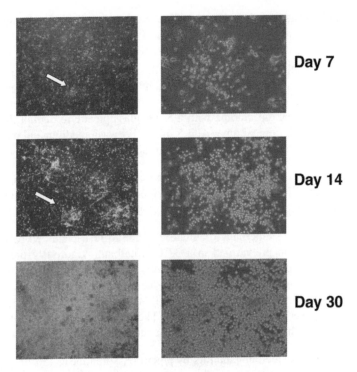

Figure 3   Representative images of human ES cell-derived progenitors cultured in NK cell conditions for 7, 14 and 28 days. Arrows
indicate presence of hematopoietic cell clusters that expand over
time. Left panel is 20× (original magnification) and right panel is
100× (original magnification).

2001). For optimal NK cell differentiation a high quality AFT024-feeder layer and a
high number (>10,000 cells per well) of human ES cell-derived hematopoietic progenitors are required. During culture in NK cell differentiation conditions, CD56⁺
CD45⁺ cells are typically observed as early as after 14 days of culture of human ES
cell-derived hematopoietic progenitor cells. These cells appear as cluster of refractile
cells typical of hematopoietic cells on the AFT024 stromal cells, and will continue
to proliferate and eventually take over the well **(Figure 3)**. At least four weeks of
culture is needed to generate a significant population of mature human ES cell-
derived NK cells. The mature NK cells can be identified by expression of CD56,
killer-cell Ig-like receptors (KIRs) and CD16, ability to secrete cytokines and cytolytic
activity.

1.   Add 0.5 mL 0.1% gelatin to all wells in a 24-well plate. Incubate at RT for at
     least 30 min.

2.   Dissociate AFT024 cells grown in T-75 flask in AFT024 medium at 33°C, 5%
     CO_2 with 0.05% trypsin and count cells.

3. AFT024 cells are plated into gelatinized 24-well plate, allowed to grow to confluency and inactivated by cesium irradiation at 2,000 rad. Irradiated plates are best if used 1 day after irradiation and should not be used 5–6 days after irradiation.

4. Isolated human ES cell-derived hematopoietic progenitors are plated onto inactivated AFT024 cells at 10,000–50,000 cells per well in 1 mL NK cell differentiation medium.

5. Half-medium changes are done every 5 days. The first week the NK cell differentiation medium contains 10 ng/mL IL-3, but after this IL-3 is omitted from the medium.

6. Isolation of mature NK cells with cytolytic activity and cytokine production is typically optimal after 30 days of culture in NK cell conditions. The cells can be isolated by gentle pipetting up and down a few times to mix the cells.

7. Mature CD56$^+$ NK cells can be identified by flow cytometry. *In vitro* function of human ES cell-derived NK cells can be analyzed by measurement of direct cytolytic activity against K562 by a standard ^{51}Cr-release assay or other common immunological assays (Woll *et al.*, 2005).

Troubleshooting

The human ES cells do not differentiate into hematopoietic progenitors during co-culture with S17 stromal cells. What is going wrong?

After too many passages, S17 stromal cells loose their ability to support hematopoietic differentiation. New S17 cells should be thawed and used instead. There are also alternative methods that support hematopoietic differentiation of human ES cells, including other stromal cells (OP9) or embryoid body formation (Vodyanik *et al.*, 2005; Zambidis *et al.*, 2005).

How can one tell if the human ES cells are starting to differentiate when co-cultured with S17 stromal cells?

If human ES cells continue to stay in confined colonies even after several days in co-culture with S17 stromal cells, this indicates a poor differentiation. Typically, human ES cells start to spread out from the initial colonies plated on the S17 stromal cells and a heterogeneous population of cells should appear.

During isolation of hematopoietic progenitor cells many cells are lost due to poor viability and cell clumping. What can be done to minimize this?

For best viability and cell recovery, use cold reagents and equipment unless specified otherwise in the protocol. Media and buffers should be kept at 4°C when not in use. Tubes and magnets should be pre-cooled on ice before use. This reduces cell clumping dramatically. If cell clumping becomes a problem, the EDTA concentration in the Easy Sep Buffer can be increased. Incubation of cells with the CD34+ Positive selection cocktail and magnetic beads works best at room temperature.

Many cells are lost after half-medium changes of NK cell differentiation medium. Any suggestions?

The human ES cell-derived hematopoietic cells are very loosely attached to the AFT024 cells and can be accidentally pipetted off during the medium change. To prevent this, the medium changes can be done with P1000 pipette by carefully aspirating the top layer in each well. While adding medium back to the well, slowly pipette the medium on the side of the well. This reduces the chance of disturbing the AFT024 layer and detaching human ES cell-derived cells.

References

Blom B and H Spits. (2006). Development of human lymphoid cells. *Annu Rev Immunol* **24**: 287–320.

Cooper MA, TA Fehniger and MA Caligiuri. (2001). The biology of human natural killer-cell subsets. *Trends Immunol* **22**: 633–640.

Galic Z, SG Kitchen, A Kacena, A Subramanian, B Burke, R Cortado and JA Zack. (2006). T lineage differentiation from human embryonic stem cells. *Proc Natl Acad Sci USA* **103**: 11742–11747.

Kaufman DS, ET Hanson, RL Lewis, R Auerbach and JA Thomson. (2001). Hematopoietic colony-forming cells derived from human embryonic stem cells. *Proc Natl Acad Sci USA* **98**: 10716–10721.

Kondo M, AJ Wagers, MG Manz, SS Prohaska, DC Scherer, GF Beilhack, JA Shizuru and IL Weissman. (2003). Biology of hematopoietic stem cells and progenitors: implications for clinical application. *Annu Rev Immunol* **21**: 759–806.

Laiosa CV, M Stadtfeld and T Graf. (2006). Determinants of lymphoid–myeloid lineage diversification. *Annu Rev Immunol* **24**: 705–738.

Loza MJ and B Perussia. (2001). Final steps of natural killer cell maturation: a model for type 1–type 2 differentiation? *Nat Immunol* **2**: 917–924.

Miller JS and V McCullar. (2001). Human natural killer cells with polyclonal lectin and immunoglobulinlike receptors develop from single hematopoietic stem cells with preferential expression of NKG2A and KIR2DL2/L3/S2. *Blood* **98**: 705–713.

Miller JS, Y Soignier, A Panoskaltsis-Mortari, SA McNearney, GH Yun, SK Fautsch, D McKenna, C Le, TE Defor, LJ Burns, PJ Orchard, BR Blazar, JE Wagner, A Slungaard, DJ Weisdorf, IJ Okazaki and PB McGlave. (2005). Successful adoptive transfer and in vivo expansion of human haploidentical NK cells in patients with cancer. *Blood* **105**: 3051–3057.

Mrozek E, P Anderson and MA Caligiuri. (1996). Role of interleukin-15 in the development of human CD56+ natural killer cells from CD34+ hematopoietic progenitor cells. *Blood* **87**: 2632–2640.

Rosenberg SA, MT Lotze, LM Muul, AE Chang, FP Avis, S Leitman, WM Linehan, CN Robertson, RE Lee, JT Rubin, CA Seipp, CG Simpson and DE White. (1987). A progress report on the treatment of 157 patients with advanced cancer using lymphokine-activated killer cells and interleukin-2 or high-dose interleukin-2 alone. *N Engl J Med* **316**: 889–897.

Rosenberg SA, BS Packard, PM Aebersold, D Solomon, SL Topalian, ST Toy, P Simon, MT Lotze, JC Yang, CA Seipp, C Simpson, C Carter, S Bock, D Schwartzentruber, JP Wei and DE White. (1988). Use of tumor-infiltrating lymphocytes and interleukin-2 in the immunotherapy of patients with metastatic melanoma. A preliminary report. *N Engl J Med* **319**: 1676–1680.

Ruggeri L, M Capanni, E Urbani, K Perruccio, WD Shlomchik, A Tosti, S Posati, D Rogaia, F Frassoni, F Aversa *et al.* (2002). Effectiveness of donor natural killer cell alloreactivity in mismatched hematopoietic transplants. *Science* **295**: 2097–2100.

Sivori S, C Cantoni, S Parolini, E Marcenaro, R Conte, L Moretta and A Moretta. (2003). IL-21 induces both rapid maturation of human CD34+ cell precursors towards NK cells and acquisition of surface killer Ig-like receptors. *Eur J Immunol* **33**: 3439–3447.

Slukvin II, MA Vodyanik, JA Thomson, ME Gumenyuk and KD Choi. (2006). Directed differentiation of human embryonic stem cells into functional dendritic cells through the myeloid pathway. *J Immunol* **176**: 2924–2932.

Tian X, JK Morris, JL Linehan and DS Kaufman. (2004). Cytokine requirements differ for stroma and embryoid body-mediated hematopoiesis from human embryonic stem cells. *Exp Hematol* **32**: 1000–1009.

Vodyanik MA, JA Bork, JA Thomson and Slukvin II. (2005). Human embryonic stem cell-derived CD34+ cells: efficient production in the coculture with OP9 stromal cells and analysis of lymphohematopoietic potential. *Blood* **105**: 617–626.

Vodyanik MA, JA Thomson and Slukvin II. (2006). Leukosialin (CD43) defines hematopoietic progenitors in human embryonic stem cell differentiation cultures. *Blood* **108**: 2095–2105.

Wang L, Menendez P, F Shojaei F, L Li, F Mazurier, JE Dick, C Cerdan, K Levac and M Bhatia. (2005). Generation of hematopoietic repopulating cells from human embryonic stem cells independent of ectopic HOXB4 expression. *J Exp Med* **201**: 1603–1614.

Woll PS, CH Martin, JS Miller and DS Kaufman. (2005). Human embryonic stem cell-derived NK cells acquire functional receptors and cytolytic activity. *J Immunol* **175**: 5095–5103.

Zambidis ET, B Peault, TS Park, F Bunz and CI Civin. (2005). Hematopoietic differentiation of human embryonic stem cells progresses through sequential hemato-endothelial, primitive, and definitive stages resembling human yolk sac development. *Blood* **106**: 860–870.

13, Part F

Directed differentiation of human embryonic stem cells into myeloid cells

CHANTAL CERDAN AND MICKIE BHATIA

McMaster Stem Cell and Cancer Research Institute, Faculty of Health Sciences, McMaster University, 1200 Main Street West, MDCL 5029, Hamilton, Ontario, Canada, L8N 3Z5

Introduction

Human embryonic stem (ES) cells spontaneously and randomly differentiate into multiple ectodermal, endodermal and mesodermal cell types, in the absence of FGF which sustains their undifferentiated state. These pluripotent cells provide therefore a powerful model system to understand the cellular and molecular basis of human embryonic hematopoietic development. Differentiation methodologies of human ES cells are largely adapted from methodologies used for murine ES cell differentiation and include two main approaches: coculture on supportive stromal cell layers (Kaufman *et al.*, 2001; Vodyanik *et al.*, 2005; Qiu *et al.*, 2005; Tian *et al.*, 2004; Woll *et al.*, 2005; Slukvin *et al.*, 2006; Narayan *et al.*, 2005) and/or formation of embryoid bodies (EBs) (Tian *et al.*, 2004; Chadwick *et al.* 2003; Cerdan *et al.*, 2004; Zambidis *et al.*, 2005; Ng *et al.*, 2005; Cameron *et al.* 2006; Bowles *et al.*, 2006; Wang *et al.*, 2004, Wang *et al.*, 2005a; Zhan *et al.*, 2005; Kim *et al.*, 2005). Despite the different procedures applied in studying hematopoietic development from human ES cells, different groups have achieved considerable common outcomes. First, hematopoietic development from human ES cells displays a spatial and temporal pattern (Kaufman *et al.*, 2001; Vodyanik *et al.*, 2005; Chadwick *et al.*, 2003; Wang *et al.*, 2004; Zhan *et al.*, 2004; Zambidis *et al.*, 2005). Second, during early human ES cell differentiation, hematopoietic cells are derived from CD45 negative precursors that coexpress PECAM-1/CD31 and CD34 surface markers (Kaufman *et al.*, 2001; Vodyanik *et al.*, 2005; Chadwick *et al.*, 2003; Wang *et al.*, 2004; Zambidis *et al.*, 2005). Two groups, including ours, have identified an immature endothelial population as being responsible for giving rise to hematopoiesis from human ES cells (hereinafter CD45^{neg}PFV) (Zambidis *et al.*, 2005; Wang *et al.*, 2004). Through clonal experiments our group (Wang *et al.*, 2004) demonstrated that this rare CD45^{neg}PFV resident population within EBs was able to generate both the endothelial and hematopoietic lineages, suggesting that this population contains cells with "hemangioblastic" properties. These combined findings recapitulate observations from human embryos (Oberlin *et al.*, 2002; Tavian *et al.*, 1999, 2001), further illustrating that human ES cells can be applied as a model for studies of early human development. Third, human ES cell-derived hematopoietic cells have similar colony and cellular morphologies to those derived from committed adult hematopoietic tissues (peripheral blood, cord blood and bone marrow), including macrophage, granulocyte, erythroid, as well as multipotent hematopoietic progenitors, and their hematopoietic progenitor capacity is enriched in the CD34+ subfraction.

Preliminary data from our group (Wang *et al.*, 2005b) and others (Tian *et al.*, 2006; Narayan *et al.*, 2005) suggest that human ES cell-derived hematopoietic cells have hematopoietic stem (HS) cell properties. However, in our hands intravenous transplantation of human ES cell-derived hematopoietic cells caused mortality in recipient mice due to emboli formed from rapid cellular aggregation in response to mouse serum, and bypassing the recipient mouse circulation by direct intra bone marrow transplantation (IBMT) gave rise to very low levels of engraftment. Thus, the derivation of long term engrafting human ES cell-derived HS cells still remains a challenge, and current human ES cell differentiation protocols toward hematopoietic lineages result in heterogeneous cell populations and insufficient yields.

Overview of protocol

The following section describes methodologies that have been successfully applied in our laboratory in the generation and characterization of human ES cell-derived hematopoietic cells *in vitro* and *in vivo*. These involve EB formation, myeloid induction in hematopoietic cytokines and BMP-4 supplemented medium, and finally the isolation and culture of CD45^{neg}PFV hemogenic precursors.

Materials, reagents, and equipment

Cell lines

- Human embryonic stem cell lines: in our hands, this protocol has been successfully used with H1, H9 and CA1 cell lines (WiCell, Madison, WI, www.wicell.org and Nagy, A., www.mshri.on.ca/nagy respectively). These lines are propagated without antibiotics according to standard protocols (described in **Chapter 5** — Extended Protocols), using a Matrigel matrix and mouse embryonic fibroblast conditioned medium (MEF-CM), supplemented with basic fibroblast growth factor (bFGF, 8 ng/mL) (see section **Preparation of media, solutions and growth factors aliquots** for preparation of MEF-CM and bFGF).

Media, solutions, and reagents

- Acetone (500 mL bottle; Fisher; cat. no. A18-500).

- Agarose (100 g bottle; Invitrogen; cat. no.15510-019).

- Ammonium chloride 0.8% solution (500 mL bottle; Stem Cell Technologies; cat. no. 07850).

- Avertin (10 g bottle; Sigma; cat. no. 840-2).

- BIT media 9500, 5 × (100 mL bottle; Stem Cell Technologies, Vancouver, Canada; cat. no. 09500). Aliquot into 4 mL per tube and store at −30°C.

- Buprenorphine (BUPRENEX®) (Pack of 10, 1 mL vial; Burn Veterinary Supply, Inc., Farmers, Branch, TX; cat. no. 254-0251).

- Bovine serum albumin (BSA), 30% solution (50 mL bottle; Sigma, Oakville, Canada; cat. no. A9576).

- Cell dissociation buffer, enzyme-free PBS-based (100 mL bottle; Gibco; cat. no. 13151-014).

- Collagenase B (100 mg bottle; Roche Diagnostics GmbH, Manheim, Germany; cat. no. 1088-807).

- Collagenase IV (1 g bottle; Gibco; cat. no. 17104-019).

- DNAzol reagent (100 mL bottle; Gibco; cat. no. 10503-027).

- D-PBS, Ca^{2+} and Mg^{2+} free (500 mL; Gibco; cat. no. 14190-144).

- Distilled water DNAse, RNAse-free (500 mL; Gibco; cat. no. 10977-015).

- Dulbecco's modified Eagle medium (DMEM) (500 mL; Gibco, Burlington, Canada; cat. no. 11965-092).

- EcoRI restriction enzyme (5,000 U; MBI Fermentas; cat. no. ERO271).

- Ethanol 70% and 100% (standard).

- Eye ointment sterile (from any pharmacy).

- Fetal bovine serum (FBS), heat inactivated and non-heat inactivated (500 mL; Hyclone, Logan, UT; cat. no. 30071-03).

- L-Glutamine 200 mM (100 mL bottle; Gibco; cat. no.15039-027).

- Glycogen 20 mg/mL aqueous solution (1 mL vial; Fermentas; Burlington, Canada; cat. no. R0561).

- Iscove's modified Dulbecco's medium (IMDM) (500 mL; Gibco; cat. no.12200-028).

- Iscove's modified Dulbecco's medium (IMDM) (Powder 10 × 1 L; Gibco; cat. no. 12200-036).

- Knockout DMEM (KO-DMEM) (500 mL; Gibco; cat. no.10829-018).

- Knock Out Serum Replacement (KO-SR) (500 mL; Gibco; cat. no. 10828-028).

- Matrigel (10 mL; BD, Mississauga, Canada; cat. no. 353234).

- β-Mercaptoethanol 14.3 M (100 mL; Sigma; cat. no. M 7522).

- Methanol (1 L bottle; ICN, Montreal, Canada; cat. no. ICN155386).

- Methocult SF H4230, methylcellulose medium (MC) (80 mL, Stem Cell Technologies; cat. no. HCC-4230).

- $MgCl_2$ 50 mM (Invitrogen; cat. no. Y02016).

- MilliQ H_2O: local autoclaved.

- Monothioglycerol (100 mL bottle; Sigma; cat. no. M6145).

- Non-essential amino acid solution, 10 mM, 100 × (100 mL bottle; Gibco; cat. no.11140-050).

- dNTP set 100 mM (Invitrogen; cat. no. 55082-85).

- dNTPs 20 mM mixture (5 mM each): combine 20 μL of each dNTP at 100 mM with 320 μL of distilled water DNAse, RNAse-free water. Store at −30°C.

- PCR buffer 10 × (Invitrogen; cat. no. Y02028).

- PCR primers diluted to 50 μM with sterile distilled water DNAse, RNAse-free water.

- Phenol:chloroform isoamyl alcohol solution 1:1 (vol:vol). Chloroform isoamyl alcohol 24:1 (1 L bottle; Merck; Montreal, Canada; cat. no. CX1054-6). Store at room temperature.

- Proviodine detergent, 0.75% iodine (500 mL bottle; Rougier, Mirabel, Canada; cat. no. DIN00172936).

- Proviodine solution, 1.0% free iodine (500 mL bottle; Rougier; cat. no. DIN00172944).

- Sodium acetate 5 M.

- Sodium bicarbonate $NaHCO_3$ (250 g, Sigma; cat. no. S6297).

- Taq DNA polymerase (Invitrogen; cat. no. 18038-042).

- TE buffer pH=8. To prepare: mix 100 mM Tris-HCl and 10 mM EDTA.

- *tert*-Amyl alcohol (Sigma; cat. no. 246-3).

- Tris-buffered Phenol (100 mL bottle; Invitrogen; Mississauga, Canada; cat. no. 15513-039). Store at 4°C.

- Trypan blue (100 mL bottle; Gibco; cat. no. 1691049).

- 0.25% Trypsin/EDTA (100 mL bottle; Invitrogen; cat. no. 25200-056).

- Wright Giemsa stain solution (500 mL bottle; Sigma; cat. no. WG-16).

- Xylenes (4 L bottle; Sigma; cat. no. 53405-6).

- Basic Fibroblast Growth Factor (bFGF), recombinant human (10 μg vial; Gibco, cat. no. 13256-029).

- Bone Morphogenetic Protein-4 (BMP-4), recombinant human (300 μg bottle; R&D; cat. no. 314-BP).

- Erythropoietin (EPO), recombinant human (5,000 IU/mL; R&D; cat. no. 286-EP).

- Flt-3 ligand (Flt-3 L), recombinant human (250 μg bottle; R&D; cat. no 308-FK/CF).

- Granulo-Colony Stimulating Factor (G-CSF), recombinant human (300 μg/mL bottle; Amgen Inc.; cat. no. 3105100).

- Granulo-Macrophage Colony Stimulating Factor (GM-CSF), recombinant human (10 μg bottle; StemCell Technologies; cat. no. 02532).

- Interleukin-3 (IL-3), recombinant human (50 μg bottle; R&D; Minneapolis, MN; cat. no. 203-IL).

- Interleukin-6 (IL-6), recombinant human (50 μg bottle; R&D; cat. no. 206-IL).

- Stem Cell Factor (SCF), recombinant human (1,000 μg bottle; R&D; cat. no. 255-SC/CF).

- PE-conjugated anti-human CD19 mAb (2 mL vial; Beckman Coulter, Marseille, France; cat. no. IM1285).

- PE-conjugated mouse anti-human CD31/PECAM-1 mAb (2 mL vial; BD; cat. no. 555446).

- PE-conjugated anti-human CD33 mAb (2 mL vial; BD; cat. no. 347787).

- FITC-conjugated anti-human CD34 mAb (2 mL vial; BD; cat. no. 555821).

- FITC-conjugated anti-human CD36 mAb (1 mL vial; Research Diagnostics Inc., Flanders, New Jersey; cat. no. RDI-M1613016).

- APC-conjugated anti-human CD45 mAb (0.5 mL vial; BD; cat. no. 340943).

- Mouse anti-human Glycophorin A-PE mAb (2 mL vial; Beckman Coulter; cat. no. IM2211).

- Single chain anti-human VEGFR2 (KDR/Flk-1) scFvA7 mAb (0.5 mL; Research Diagnostics Inc.; cat. no. RDI-VEGFR2scX).

- HRP-labeled mouse anti-E tag mAb (0.5 mg; Amersham Pharmacia Biotech, Piscataway, NJ; cat. no. 27-9413-01).

- PE-conjugated $F(ab')_2$ fragment donkey anti-mouse IgG (H+L) mAb (1 mL; Research Diagnostics Inc.; cat. no. 715-116-150).

- FITC-conjugated mouse IgG isotype mAb (2 mL; BD; cat. no. 349041).

- PE-conjugated mouse IgG isotype mAb (2 mL; BD; cat. no. 340043).

- APC-conjugated mouse IgG isotype mAb (0.5 mL; BD; cat. no. 340442).

- 7-Amino actinomycin D (7-AAD) viability dye (3 mL ready to use solution; Beckman Coulter; cat. no. IM3422).

Equipment

- 40 μm cell strainer, nylon (case of 50; BD; cat. no. 352340).

- Chamber slides, four wells, Permanox (Lab Tek; Nunc; VWR; cat. no. CA62407-331).

- Cover slides, 24 × 50, thickness #1 (Fisher; cat. no. 12-545F).

- 35 × 10 mm dishes (case of 500; Sarstedt; cat. no. 83-1800-002).

- Filter system, sterile, Corning 0.22 μm cellulose acetate filter, low binding protein (case of 12, 150 mL and 500 mL; Fisher, Ottawa, Canada; cat. no. 09-761-119 and 09-761-5).

- Ultra-free-MC Filter, sterile 0.22 μm, low binding protein (pack of 50; Millipore, Nepean, Canada; cat. no. UFC30GVOS).

- Fume cabinet.

- Gauze, 3 inch × 3 inch, sterile, pack of two (case of 48 trays; Johnson&Johnson, Titusville, NJ; cat. no. 344–2339).

- Hybond-N+ membrane, nylon 20 cm × 3 m (1 roll; Amersham Bioscience; cat. no. RPN203B).

- Microscope slides, 25 × 75 × 1.0 mm (Fisher; cat. no. 12-550-15).

- 16 G needles, 1.5 inch bevel needles (box of 100; BD; cat. no. 305198).

- 27 G needles, 0.5 inch bevel (Box of 100; VWR; cat. no. CABD305109).

- 5 mL pipettes, sterile (case of 200; Fisher; cat. no. CS004487).

- 10 mL pipettes, sterile (case of 200; Fisher; cat. no. CS004488).

- 25 mL pipettes, sterile (case of 200; Fisher; cat. no. CS004489).

- Pipetmen (P2, P10, P20, P100, P200, and P1000 µL) and appropriate filter tips.

- Six-well plates, flat bottom (case of 50; VWR, Mississauga, Canada; cat. no. CA62406-161).

- Six-well plates, ultra-low attachment (case of 24; Fisher; cat. no. CS003471).

- 12-well plates, non TC treated, flat bottom (BD; cat. no. 351143).

- 24-well plates, Fibronectin-coated (pack of 5; VWR; cat. no. CACB354411).

- 96-well plates, Fibronectin-coated (pack of 5; VWR; cat. no. CACB354409).

- 0.3 cm^3 syringes, insulin 28 G (1/2 inch) (case of 200; VWR; cat. no. CABD309300).

- 1 cm^3 syringes, slip tip (pack of 100; BD; cat. no. 309602).

- 3 cm^3 syringes (case of 100; BD; cat. no. CABD309585).

- 1.5 mL microtubes (Sarstedt; cat. no. 72690).

- 15 mL polypropylene conical tubes (case of 500; VWR; cat. no. CA21008-918).

- 50 mL polypropylene conical tubes (case of 500; VWR; cat. no. CA21008-940).

- Fluorescence activated cell sorting (FACS) tubes, 5 mL polystyrene round bottom tubes (BD; cat. no. 352054).

- Electrophoresis equipment.

- FACS sorter.

- Freezers (−80°C, −30°C) and refrigerators (4°C).

- Gel reader system (we use Gel Doc and Quantity One software, BioRad).

- GeneQuant II for quantification of DNA, RNA and oligonucleotides (Pharmacia Biotech).

- Hemacytometer (VWR; cat. no. 15170-208).

Preparation of media, solutions, and growth factors aliquots

Preparation of culture and differentiation media

• BIT medium

The serum-free hematopoietic conducive BIT medium consists of 9500 BIT medium, 2 mM L-glutamine, 0.1 mM β-mercaptoethanol, 300 ng/mL SCF, 50 ng/mL G-CSF, 300 ng/mL Flt-3 L, 10 ng/mL IL-3, 10 ng/mL IL-6.

Notes. Thawed 5 × BIT media can be stored at 4°C for up to 3 months. 1 × BIT medium must be prepared prior to use and can be stored at 4°C for only up to 1 day. Once thawed, the growth factors and β-mercaptoethanol can be stored at 4°C for up to a month, and the L-glutamine can be stored at 4°C for up to 2–4 weeks.

• EB medium

EBs are cultured in KO-DMEM supplemented with 20% of non-heat-inactivated FBS, 1% non-essential amino acids, 1 mM L-glutamine, 0.1 mM β-mercaptoethanol. To prepare 100 mL of medium: combine 20 mL of non-heat-inactivated FBS, 1 mL non-essential amino acids, 0.5 mL L-glutamine and 7 µL β-mercaptoethanol (1.43 M). Filter through a 0.22 µm sterile cellulose acetate membrane (150 mL volume). Store at 4°C no longer than 14 days.

Note. Once thawed, L-glutamine and β-mercaptoethanol can be stored at 4°C as described above. Non-essential amino acids stable at 4°C can be stored up to the date of expiration.

• IMDM medium

Rinse all glassware with distilled water prior to use to ensure that there are no traces of detergent. Add package of IMDM powder (17.67 g) to 200 mL milliQ H_2O, rinse package. Weigh and add remaining ingredients (Sodium Bicarbonate $NaHCO_3$: 3.024 g, 10% Monothioglycerol: 65 µL, see below for preparation). Bring volume up to 1 L with MilliQ H_2O. Mix solution gently to avoid damaging proteins.

Filter sterilize 0.22 µm. Aliquot under 40 mL into 50 mL centrifuge tubes and store at −30°C. After thawing, can be stored at 4°C up to 12 months. Do not use antibiotics.

• Methylcellulose medium (MC)

MC Methocult SF H4230 is supplemented with the following media and growth factors. Thaw one 80 mL bottle overnight at 4°C and add 20 mL of sterile IMDM, 50 ng/mL SCF (3.4 µL at $1.5 × 10^3$ ng/µL), 10 ng/mL IL-3 (40 µL at 25 ng/µL), 3 units/mL EPO (60 µL at 5,000 U/mL) and 10 ng/mL GM-CSF (3.4 µL at 1:10 dilution of

3×10^3 ng/µL). Shake vigorously and leave overnight to settle. Make 5 or 10 mL aliquots and store at $-30°$C. Once thawed, keep at $4°$C for no longer than 2 weeks.

- Mouse embryonic fibroblast conditioned medium (MEF-CM)

MEF-CM is produced by daily collection of feeding medium from irradiated (40Gy) mouse MEF cells within a 7–10 day period. The medium consists of 80% KO-DMEM supplemented with 20% KO-SR, 1% non-essential amino acids, 1 mM L-glutamine, 0.1 mM β-mercaptoethanol, and 4 ng/mL bFGF. To prepare 500 mL of medium: combine 100 mL KO-SR, 5 mL non-essential amino acids, 2.5 mL L-glutamine, 35 µL 1.43 M β-mercaptoethanol and 392.5 mL KO-DMEM. The medium is then filtered through a 0.22 µm sterile cellulose acetate membrane and kept at $4°$C up to 14 days. Immediately prior to use, add 200 µL of bFGF (working solution, 10 ng/µL) to 500 mL of medium. Feed MEF cells and collect medium daily up to 10 days. Pool all collected MEF-CM, filter them through a 0.22 µm sterile cellulose acetate membrane (500 mL volume). Store aliquots in $-80°$C up to 6 months. Store thawed MEF-CM at $4°$C no more than 7 days.

Note. MEF-CF quality is crucial for maintenance of undifferentiated human ES cells and subsequent formation and development of EBs. Since MEF-CM quality varies among different batches of MEF cells and MEF-CM preparations, assessing each batch of MEF-CM prior to use is recommended.

Preparation of solutions

• Avertin

To prepare 50% stock solution: add 20 mL of *tert*-amyl alcohol to dissolve Avertin. Wrap in foil in a glass jar and store at room temperature up to 6 months.

To prepare 2.5% working solution: combine 1 mL of Avertin stock solution with 19 mL D-PBS.

• Collagenase B

To prepare stock solution: calculate the total amount of International Units (IU) in each individual bottle, dilute with KO-DMEM to make 4.0 IU/mL and filter through a 0.22 µm membrane. Aliquot 1 mL stock solution into 15 mL polypropylene tubes and store at $-30°$C. Avoid repeated freeze–thaw cycles.

To prepare working solution (0.4 IU/mL): add 9 mL of KO-DMEM to 1 mL of stock solution. The working solution can be stored at $4°$C up to 1 week.

• Collagenase IV

To prepare stock solution: calculate the total amount of IU in each individual bottle, dilute with KO-DMEM to make 10,000 IU/mL, aliquot under 2 mL and store at $-30°$C.

To prepare working solution: combine 2 mL of stock solution with 98 mL of KO-DMEM. The working solution is then filtered through a 0.22 µm membrane and stored at −30°C. Avoid repeated freeze–thaw cycles. Thawed working solution can be stored at 4°C up to 1 week.

- FACS buffer

To prepare 500 mL of FACS buffer: combine 470 mL of PBS and 30 mL heat-inactivated FBS.

- Matrigel

To prepare stock solution: thaw Matrigel slowly to avoid gel formation (overnight at 4°C). Add 10 mL of cold KO-DMEM (4°C). Put Matrigel bottle and 20 tubes (15 mL polypropylene conical tubes) on ice. Pipette 20 times to mix Matrigel and KO-DMEM medium. Make 1 mL/tube aliquots. Store at −30°C.

To prepare 1:15 working solution: thaw one tube of stock Matrigel at 4°C into a container with cold water to avoid polymerization (which occurs just above 4°C) overnight or within 1–2 hr. Add 5 mL of cold KO-DMEM, pipette ten times. Then, add 9 mL of cold KO-DMEM and pipette ten times. Add 0.75–1 mL of working solution to one well of a six-well plate. Spread evenly without forming bubbles. Wrap the plate with plastic film. Incubate the plate overnight at 4°C or for 2 hr at room temperature. Use coated plates immediately or store at 4°C no longer than 7 days. For 1:6 dilution for the passage of human ES cells 1 week prior to EB formation, proceed the same way as for 1:15 working solution but add only 5 mL of cold KO-DMEM into 1 mL Matrigel aliquots.

Note. Once diluted, Matrigel can be stored at 4°C for up to 1 week.

- β-Mercaptoethanol

To prepare 1.43 M β-mercaptoethanol: combine 2 mL of 14.3 M β-mercaptoethanol with 18 mL of PBS and store at −30°C in 0.25 mL aliquots. Keep thawed 1.43 M β-mercaptoethanol at 4°C up to 1 month.

To prepare 55 mM β-mercaptoethanol: add 10 μL of 1.43 M β-mercaptoethanol to 240 μL IMDM and store at 4°C up to 2 weeks.

- Methanol/acetone 80:20 solution

Combine 80 mL methanol and 20 mL acetone and store at room temperature in a fume cabinet. Refrigerate the bottle at 4°C for 30 min prior to use as a fixative for Giemsa staining.

- 10% Monothioglycerol

Dilute monothioglycerol stock in PBS, 1:10. Filter sterilize 0.22 μm. This stock is stored at 4°C in the dark (wrap tube in aluminium foil).

Preparation of growth factors aliquots

- ### bFGF

To prepare working solution (10 µg/mL): dissolve each vial of bFGF in 1 mL D-PBS containing 0.1% BSA (add 30 µL of 30% BSA into 10 mL D-PBS). Pre-filter Ultra-free-MC 0.22 µm filter with 3 mL of D-PBS containing 10% BSA (combine 1 mL of 30% BSA with 2 mL D-PBS), then filter bFGF working solution. Aliquot 200 µL of filtered bFGF into sterile eppendorf tubes.

- ### BMP-4

To prepare stock solution (1,000 µg/mL): add 300 µL of D-PBS containing 2% heat inactivated FBS to the bottle, pipette and aliquot 50 µL/tube.

To prepare working solution (50 µg/mL): add 950 µL of D-PBS containing 2% heat inactivated FBS to 50 µL of stock solution, pipette and aliquot 50 µL/tube.

- ### EPO

Pipette, aliquot 70 µL/tube and use as it is at 5U/µL.

- ### Flt-3L

To prepare working solution (250 µg/mL): add 1 mL of D-PBS containing 2% heat inactivated FBS to the bottle, pipette and aliquot 100 µL/tube.

- ### G-CSF

Pipette, aliquot 50 µL/tube and use as it is at 300 ng/µL.

- ### GM-CSF

To prepare stock solution (3 µg/mL): add 3.333 mL of D-PBS containing 2% heat inactivated FBS to the bottle, pipette and aliquot 100 µL/tube.

- ### IL-3

To prepare working solution (25 µg/mL): add 2 mL of D-PBS containing 2% heat inactivated FBS to the bottle, pipette and aliquot 100 µL/tube.

- ### IL-6

To prepare working solution (50 µg/mL): add 1 mL of D-PBS containing 2% heat inactivated FBS to the bottle, pipette and aliquot 50 µL/tube.

- ### SCF

To prepare working solution (1,500 µg/mL): add 660 µL of D-PBS containing 2% heat inactivated FBS to the bottle, pipette and aliquot 50 µL/tube.

Note for growth factors solutions

Store both stock and working solutions at $-30°C$ no longer than 3 months. Keep thawed aliquots at 4°C no longer than 1 month. Avoid repeated freeze–thaw cycles.

Protocols

Embryoid body formation

1. Approximately 1 week before forming EBs, coat each well of six-well plates with 0.75–1 mL of Matrigel at 1:6 dilution, as described in section Preparation of solutions/Matrigel/working solution. Refer to **Figure 1** for exemplary outline of the protocols and microscopic photographs of differentiating EBs.

2. Remove the excess of Matrigel, rinse each well of six-well plate with 2 mL KO-DMEM, aspirate, and seed undifferentiated human ES cells in the 1:6 Matrigel pre-coated wells.

3. Feed cells daily with 3 mL MEF-CM supplemented with fresh bFGF (8 ng/mL) to each well of six-well plates.

4. At day 2, after addition of MEF-CM+bFGF to the human ES cell culture, add 0.5 mL of 1:6 diluted Matrigel to each well, and gently swirl the plate to mix media.

5. At day 7, aspirate the medium and add 0.5 mL of pre-warmed Collagenase IV (200 IU/mL, working solution) to each well.

6. Incubate at 37°C for 3–10 min until colony edges slightly pull away from the well.

7. Aspirate the Collagenase IV.

8. Wash the residual Collagenase IV with 2 mL of pre-warmed KO-DMEM or EB medium per well and aspirate.

9. Add 2 mL of EB medium to each well.

10. Using a 10 mL pipette, scrape gently the bottom of each well in "strips".

11. Triturate gently (ten times) using a 10 mL pipette.

12. Transfer the content of two wells of undifferentiated human ES cells to one well of a low attachment plate (ratio 2:1). This ratio can be 1.5/1. Total volume of EB medium per well is 4 mL.

13. Incubate overnight at 37°C, 5% CO_2 to allow for EB formation; change EB medium 18–24 hr later.

A   Experimental approach

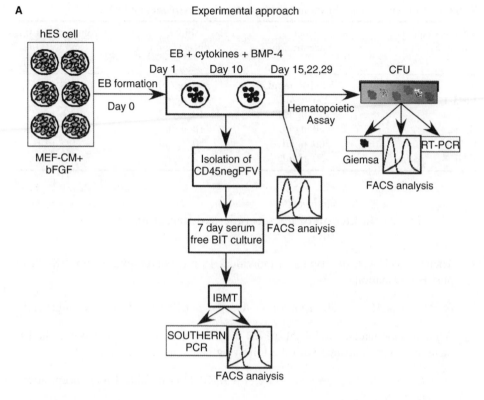

B

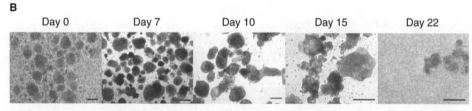

Figure 1 (A) Exemplary outline of hematopoietic differentiation from human ES cells (see protocols for details of each protocol). (**B**) Development of EBs in the presence of hematopoietic cytokines and BMP-4. Debris and single cells can be seen after EB formation (day 0). Cystic EBs can be observed after day 7. Bright single cells found at day 22 are CD45+ cells that have been confirmed by FACS analysis. Bar = 50 μm.

Notes. This step is considered as day 0 of EB differentiation. This procedure has been tested successfully in six-wells but not 24-wells of ultra-low attachment plates. When dissociation of human ES cells to form EBs, determine the appropriate incubation time with collagenase IV by examining the colonies under microscope, as incubation time will vary between different batches of collagenase IV as well as human ES cell lines and passage numbers. In addition, avoid vigorous pipetting.

Table 1 Concentrations and volumes of hematopoietic cytokines and BMP-4

Recombinant human cytokines	Stock concentration	Final concentration	Volume for 4 mL (one well of six-well plate)
SCF	1.5×10^3 ng/μL	300 ng/mL	0.8 μL
Flt-3 L	250 ng/μL	300 ng/mL	4.8 μL
IL-3	25 ng/μL	10 ng/mL	1.6 μL
IL-6	50 ng/μL	10 ng/mL	0.8 μL
G-CSF	300 ng/μL	50 ng/mL	0.7 μL
BMP-4	50 ng/μL	25 ng/mL	2 μL

Induction of hematopoietic differentiation

1. Transfer each well of EBs to an individual 15 mL polypropylene tube 18–24 hr post EB formation.

2. Centrifuge at 150 g/~800 rpm for 5 s to separate EBs from debris and single cells.

3. Aspirate supernatant. Add 4 mL of EB medium to each tube. Prior to use, the EB medium is supplemented with the following hematopoietic cytokines :

   a. SCF + Flt3L + IL-3 + IL-6 + G-CSF + BMP-4* (see **Table 1** for concentrations and volumes).

   b. Only SCF + Flt3L + BMP-4*.

4. Transfer each tube of EBs to the previous well. Culture EBs in a 37°C and 5% CO_2 humidified incubator.

5. Change EB medium, hematopoietic cytokines and BMP-4 every 4–5 days or earlier if medium changes color (orange/yellow). Usual kinetic for a total of 15 days of differentiation = days 1, 5, 10.

*Notes.** No statistical difference between these two treatments in our hands. This step is considered as day 1 of EB differentiation. EB differentiation can be maintained up to one month, with regular change of EB medium and cytokines.

Derivation of hematopoietic cells from CD45^{neg} PFV precursors (Wang *et al.*, 2004)

A subpopulation of primitive endothelial-like cells purified from day 10 EBs that express PECAM-1, Flk-1, VE-cadherin, but not CD45 (CD45^{neg}PFV) is uniquely responsible for hematopoietic development (Wang *et al.*, 2004). To generate hematopoietic cells from EB-derived CD45^{neg}PFV precursors: (1) dissociate day 10 EBs to produce single cells; (2) isolate CD45^{neg}PFV precursors from these single cells; (3) derive hematopoietic cells by culture of CD45^{neg}PFV precursors in a hematopoietic conducive serum-free medium — BIT medium.

Dissociation of day 10 EBs*

1. Transfer each well of EBs to an individual 15 mL polypropylene tube.

2. Centrifuge at 100 g (~250 rpm) for 3 min. Aspirate supernatant.

3. Add 4 mL of Collagenase B (working solution, 0.4 IU/mL). Transfer to the previous well. Let EBs dissociate in a 37°C incubator for 2 hr.

4. Transfer the cells to a 15 mL polypropylene tube. Centrifuge for 10 min at 450 g/~1,500 rpm.

5. Aspirate the Collagenase B supernatant.

6. Add 2 mL of cell dissociation buffer and incubate for 10 min in a 37°C water bath.

7. Centrifuge for 10 min at 450 g/~1,500 rpm. Aspirate the supernatant and depending on the number of EBs, resuspend in 200–500 μL of :

    a. IMDM for CFU plating or *in vivo* experiments.

    b. FACS buffer for flow cytometry analysis.

8. To achieve single cell suspension, gently triturate (50–70 times) with a 200 μL pipettor (set at 100 μL) or a 1,000 μL pipettor (set at 200–450 μL) depending on the number of cells.

9. Filter through a sterile 40 μm cell strainer.

10. Count live and dead cells by trypan blue exclusion using a hemacytometer.

Notes. *May vary between days 7 and 11. This protocol can be used for EB dissociation into single cells at any time of the differentiation period.

Isolation of hematopoietic precursor CD45^{neg}PFV cells by FACS sorting

Isolation of CD45^{neg}PECAM1+Flk1+, CD45^{neg}PECAM1+, or CD45^{neg}Flk1+ cells from day 10 EBs demonstrates that VE-cadherin segregates with either of these populations and produces identical results, indicating that any of these sorting strategies isolate a functionally identical subpopulation in which close to 90% of the cells are

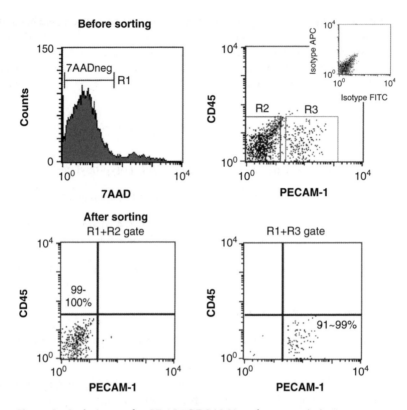

Figure 2 Isolation of CD45^{neg}PECAM1+ hematopoietic precursors. CD45^{neg}PECAM1+ cells are isolated from day 10 EBs by a FACS Vantage SE. Dead cells are excluded by 7AAD positive staining. Sorting purities are 99–100% for CD45^{neg}PECAM1- (R1+R2), and 91–99% for CD45^{neg}PECAM1+ (=CD45^{neg}PFV+) (R1+R3) subpopulations, as determined using the same sorting gate settings.

CD45^{neg}PFV (Wang *et al.*, 2004). Therefore, isolation of CD45^{neg}PECAM1+ cells represents a simpler and accurate strategy for purification of CD45^{neg}PFV precursors. To purify CD45^{neg}PECAM1+ cells (refer to **Figure 2** for sorting gates):

1. Resuspend single cells dissociated from day 10 EBs in sterile FACS buffer (2 × 10^6 cells/mL).

2. Pre-coat a few tubes (5 mL FACS tubes) with 0.5 mL of heat-inactivated FBS for 30 min for collection of sorted cells.

3. Add 20 μL of anti-human PECAM-1-PE mAb and 20 μL of anti-human CD45-APC mAb to 1 mL of cell suspension. Add 2 μL of mouse IgG-PE and 2 μL IgG-APC mAbs into isotype control tubes containing 1 × 10^5 cells/200 μL. Incubate for 30 min at 4°C.

4. Centrifuge the tubes at 450 g (~1,500 rpm) for 5 min. Aspirate supernatant. Wash twice with 3 mL of FACS buffer.

5. Resuspend cells in FACS buffer at a concentration of 2×10^6/mL. Add 20 μL of 7-AAD to 1 mL of cell suspension and stain for 10 min at room temperature to exclude dead cells.

6. Filter through a 40 μm cell strainer just prior to sort to avoid clogging of the sorting nozzle.

7. Set sorting gates, including histogram markers and dot plot quadrants, by use of respective IgG isotype controls. Sort CD45^{neg}PECAM-1+ subpopulation on a FACS Vantage SE (BD).

8. Determine purity immediately after sorting using the same sorting gate settings.

9. Count live and dead cells by trypan blue exclusion using a hemacytometer.

Hematopoietic culture of EB-derived CD45^{neg}PFV cells

To derive hematopoietic cells from CD45^{neg}PFV precursors:

1. Centrifuge CD45^{neg}PECAM1+ (containing the majority of CD45^{neg}PFV cells) cells at 453 g for 5 min. Aspirate supernatant and add 2 mL of 1 × BIT medium. Centrifuge at 453 g for 5 min. Repeat this wash step once to get rid of residual FBS.

2. Resuspend cells in 1× BIT medium supplemented with cytokines and reagents as described in **Table 2**. Make a final cell concentration of 2.5×10^5 cells/mL. *Note.* This medium must be prepared prior to use.

Table 2 To prepare 10 mL of 1 × BIT medium, follow the right-hand column of the table

Media/cytokines/ reagents	Stock concentration	Final concentration	Volume for 10 mL
BIT medium	5×	1×	2 mL
L-Glutamine	200 mM	2 mM	100 μL
β-Mercaptoethanol	55 mM	0.1 mM	18.2 μL
SCF	1.5×10^3 ng/μL	300 ng/mL	2 μL
Flt-3 L	250 ng/μL	300 ng/mL	12 μL
IL-3	25 ng/μL	10 ng/mL	4 μL
IL-6	50 ng/μL	10 ng/mL	2 μL
G-CSF	300 ng/μL	50 ng/mL	16.7 μL (1:10 dilution)
IMDM			Up to 10 mL

3. Seed CD45^{neg}PECAM1+ cells in fibronectin-coated plates. Add 200 µL/well for a 96-well plate, 1 mL/well for a 24-well plate. Make a final cell density of 5×10^4 cells/cm^2.

4. At days 3 and 5, gently replace half the volume of medium with freshly pre-pared 1× BIT medium containing the hematopoietic cytokines. Single bright round hematopoietic cells emerge around day 3. Avoid discarding these cells during medium replacement.

5. At day 7, transfer medium from each well to a 15 mL polypropylene tube. Add 0.25% Trypsin/EDTA to each well (50 µL for 96-well plate, 200 µL for 24-well plate) and incubate at room temperature to dissociate the adherent cells. Examine the cells under the microscope until they round up (requires 1–5 min). Neutralize trypsin immediately with 5-fold volume of IMDM containing 10% FBS.

6. Gently pipette the dissociated cells. Transfer the cells to a 15 mL polypropylene tube. Centrifuge at 450 g (~1,500 rpm) for 5 min.

7. Aspirate supernatant. Wash once with IMDM medium.

8. Count live and dead cells using a hematocytometer.

9. Prepare 5,000–100,000 viable cells/FACS tube in FACS buffer for FACS analysis, 10,000–50,000 viable cells in IMDM for CFU plating, and 500,000–1,000,000 viable cells in IMDM for *in vivo* transplant experiments, all in sterile eppendorf tubes.

Notes. At the end of the *in vitro* culture (may be longer than 7 days) of CD45^{neg}PFV hemogenic precursors, we generally observe a twofold cell expansion as compared with the number of cells originally seeded. Pre-warm all media required at 37°C prior to use. DO NOT warm unnecessary media, since certain components in the media might be sensitive to temperature. This protocol is optimized for 96- or 24-well plates. However, if you have enough cells you may scale-up the procedure by using 48- to six-well plates as long as you proportionally increase all the volumes and maintain the concentration of all different cytokines and reagents.

Characterization of human ES cell-derived hematopoietic cells *in vitro*

Flow cytometric analysis of cell surface markers

One-step staining with fluorochrome conjugated mAbs

This method is used for detection of CD45, CD34, CD36, CD33, CD19, CD31/ PECAM-1 and Glycophorin A on human ES cell-derived hematopoietic cells. Refer to **Figure 3** for representative flow cytometric analysis of both erythroid and myeloid colonies harvested from MC.

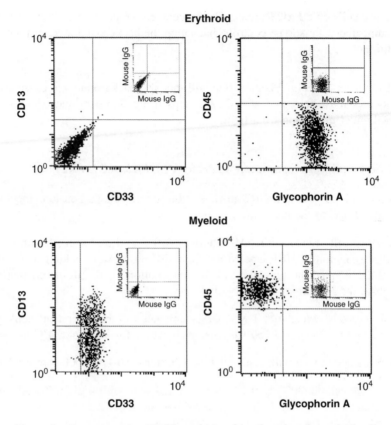

Figure 3 Representative FACS analysis of erythroid and myeloid cells comprising day 15 CFUs derived from day 15 EBs. Cells comprising the erythroid colonies express erythroid marker Glycophorin A, but lack pan-leukocyte marker CD45 and myelomonocytic markers CD33 and CD13. Cells from the macrophage or granulocyte colonies express CD45, CD33 and CD13, but lack Glycophorin A.

1. Resuspend the cells in 200 µL of FACS buffer containing 50,000 to 100,000 cells.

2. Add 3–5 µL of mAbs conjugated with APC, FITC or PE fluorochromes to 200 µL of cell suspension and mix briefly by pipetting or vortexing. Incubate for 30 min at 4°C in the dark.

3. In the same conditions, stain 50,000 cells with 3 µL of the appropriate fluorochrome-conjugated matched isotype control mAbs.

4. Wash the cells twice with 2 mL of FACS buffer (453 g/~1,500 rpm for 5 min).

5. Resuspend cells in 300 µL of FACS buffer and stain with 4 µL of 7-AAD at room temperature for 10 min in the dark to exclude dead cells.

6.   Analyze cells on a FACScalibur. The percentages of positive cells are determined as compared with isotype controls that set up the background level of non-specific staining.

Notes. Do not fix the cells. Fixation step will increase background and cause artefacts. All antibodies must be stored at 4°C. All steps using fluorochromes should also be performed under protection from light.

Three-step staining to detect VEGFR2 (KDR/Flk1)

1.   Add 2.5 μL of anti-VEFGR2 mAb to 200 μL of cell suspension. Incubate for 30 min at 4°C in the dark.

2.   Wash the cells twice with FACS buffer (450 g (~1,500 rpm), 5 min). Prepare working solution of HRP-labeled mouse anti-E tag mAb by adding 1 μL of mAb to 19 μL of FACS buffer. Add 2 μL of this solution to 200 μL of cell suspension. Incubate for 30 min at 4°C in the dark.

3.   Wash the cells twice. Add 1 μL of PE-conjugated F(ab')$_2$ fragment donkey anti-mouse IgG to 200 μL of cell suspension. Incubate for 30 min at 4°C in the dark.

4.   Wash the cells twice with 2 mL of FACS buffer (450 g/~1,500 rpm for 5 min).

5.   Resuspend cells in 300 μL of FACS buffer and stain with 4 μL of 7-AAD at room temperature for 10 min in the dark to exclude dead cells.

6.   Analyze cells on a FACScalibur. The percentages of positive cells are determined as compared with isotype controls that set up the background level of non-specific staining.

Morphological characterization of EB-derived hematopoietic colonies

CFU assay in MC

1.   Collect and count cells from dissociated EBs (See EB dissociation) or hematopoietic cells derived from CD45^{neg}PFV (see section **Hematopoietic culture of EB-derived CD45^{neg} PFV cells**). Plate 10,000–50,000 cells per well of a 12-well non-TC-treated plate.

2.   Thaw MC supplemented with the hematopoietic cytokines at room temperature.

3.   Resuspend the total number of cells to be plated in a final volume of 100 μL IMDM in a 1.5 mL eppendorf tube.

4.   Mix cell suspension by vortexing.

5. Add 0.6 mL of MC supplemented with the cytokines using a 1 mL syringe coupled to a 16 G needle.

6. Vortex the tube for a few seconds to completely mix the cells and MC.

7. Let the tube stand for 5 min to allow bubbles to move up.

8. Using a 1 cm^3 syringe coupled to a 16 G needle, transfer 0.5 mL of the preparation to one well of 12-well non-TC treated plates. Move the plate slowly in order to distribute MC evenly in the well and to avoid bubbles.

9. Add sterilized water to the surrounding empty wells to maintain moisture.

10. Incubate the plate in a 37°C and 5%CO$_2$ humidified incubator.

11. According to the standard morphological criteria, identify and count the different colonies (CFU) under an inverted microscope at two different time points after the day of plating (between days 7–10 and at day 14). You can count/examine the CFU up to day 40.

Cytospin preparation and Giemsa staining

1. Under a microscope, pluck individual colonies between 14 and 21 days using a P20 pipetman. Either use large single colonies or pool several smaller colonies with similar morphology.

2. Deposit cells in 1.5 mL eppendorf tubes containing 1 mL of D-PBS. Gently pipette to release the cells from MC.

3. Wash the cells twice with 1 mL of D-PBS.

4. Resuspend the cells in 100 μL of D-PBS.

5. Position a slide, a filter card, and a sample chamber to the slide holder. Carefully align the hole on the filter card with the hole on the bottom of the chamber. Load it to a cytocentrifuge rotor.

6. Add 100 μL of cell suspension to the bottom of chamber.

7. Centrifuge at 150 g/~800 rpm for 5 min. To avoid extrusion of the nucleus and distortion of cells' morphology, DO NOT use higher speed or longer centrifugation time.

8. Gently separate the slide with the filter card and chamber. Leave slide at room temperature for 10 min.

9. Fix slide in cold methanol:acetone (80:20 vol/vol) for 5–10 min.

10. Place slide on a horizontal staining rack over a sink. Flood the slide with Wright Giemsa solution and incubate for 1 min.

11. Add buffer solution (D-PBS diluted ten times with distilled water, pH6.8) in sufficient quantity to dilute staining 2–3 times and incubate for 3 min.

12. Wash off carefully with tap water, stand the slide on end to drain and air dry at room temperature.

13. Clear in xylene for 2 min. Cover with a cover slide and examine under a microscope.

Characterization of human ES cell-derived hematopoietic cells *in vivo* (Wang *et al.*, 2005b)

Despite *in vitro* analysis, bona fide hematopoietic stem cells can only be functionally defined by sustained multi-lineage *in vivo* reconstitution upon transplantation. Experimentally, the NOD/SCID xenotransplant assay has provided a powerful tool to functionally define candidate human hematopoietic stem cells, defined as SCID-repopulating cells (SRC) (Bhatia, 2003). However, conventional intravenous transplantation of human ES cell-derived hematopoietic cells causes mouse mortality due to emboli formed from rapid cellular aggregation in response to mouse serum (Wang *et al.*, 2005b). Bypassing the circulation by direct intra bone marrow transplantation (IBMT) allows recipient survival and functional *in vivo* assessment of hES cell-derived hematopoietic cells (Wang *et al.*, 2005b).

IBMT of NOD/SCID or NOD/SCID β2m-/- mice

1. Collect $CD45^{neg}$PFV-derived hematopoietic cells by FACS sorting as previously described. Resuspend the cells in IMDM medium.

2. Transfer $5 \times 10^5 — 1 \times 10^6$ cells to 1.5 mL eppendorf tubes. Prepare one tube for each mouse IBMT. Centrifuge the tubes for 2 min at 453 g (~1,500 rpm). Carefully discard supernatant without disturbing the cell pellet.

3. Resuspend the cells in 25 μL of IMDM medium, avoiding bubbles.

4. Sublethally irradiate (3.25Gy) 8 to 12-week old mice.

5. Weigh the animal. Prior to injection, shake Avertin (2.5% working solution) vigorously since the solution is water insoluble.

6. Inject Avertin intra-peritoneally at a dose of 0.014–0.018 mL/g of body weight. Use lower dosage for small females and higher dosage for large males. Leave the animal undisturbed until anesthetic plane has been attained.

7. Fill a 3 cm^3 syringe coupled to a 27 G needle with 2 mL D-PBS. Rinse a 28 G insulin syringe (1/2 inch, 0.3 cm^3) with IMDM medium to fill the dead space, then carefully fill the barrel with the cell suspension.

8. Put the mouse on a sterile gauze. Gently place a small amount of sterile eye ointment on the mouse's eyes using sterile gauze.

9. Sterilize the knee with Proviodine detergent, alcohol wipe and Proviodine disinfectant. Start sterilization at the incision site and move towards the periphery. Allow final solution to dry in order to create a bacteriostatic barrier.

10. Flex the leg in order to move the femur and tibia to a vertical position. Brace the femur with your index finger, the tibia with your thumb and use your middle finger for additional stabilization where needed. Make sure that your index finger can feel the femur and guide the direction.

11. Gently insert in a twisting motion a 27 G needle vertically into the femoral cavity through the patellar tendon to form a channel (do not push the needle straight in).

12. Gently withdraw the 27 G needle and keep the leg position unmoved. Insert a 28 G insulin syringe into the channel made by the 27 G needle. Inject 25 µL of cell suspension.

13. Gently remove the needle while straightening the leg and holding it in that position for several seconds to stop leakage of the injected cells.

14. Subcutaneously inject 0.3 mL of buprenorphine (0.05–0.1 mg/kg of body weight) and 1 mL of 0.9% D-PBS into the scruff.

15. Wrap the mouse in a sterile gauze. Return the mouse to the cage and keep it in a warm (under a heating lamp) and quiet environment until full recovery.

16. Administer the second dose of buprenorphine (0.3 mL) 24 hr post-IBMT. Monitor the mice daily for the first 3 weeks.

Note. The IBMT technique requires practice. Most failures come from: (a) inserting the needle when your index finger fails to identify the femur orientation; (b) moving the 27 G needle under the skin prior to "drilling" the femur (poor alignment of the entrance of skin and bone) causing the 28 G needle to miss the bone entrance of the femur channel; (c) changing the femur position during operations.

Collect bone marrow of mice post IBMT for analysis of SRC

1. Euthanize the mouse 8 weeks after IBMT. Dip the mouse in 70% ethanol and dissect the mouse bones using sterile scissors and forceps.

2. Place injected femur, contra-lateral femur, or two tibias and two iliac crests separately into 35 mm dishes containing 2 mL of DMEM.

3. Flush bone marrow into a FACS tube with an insulin syringe containing 1 mL of DMEM.

4. For analysis of human white cells, lyse red cells with cold 0.8% ammonium chloride solution for 5 min at 4°C. Wash the remaining cells twice in FACS buffer.

FACS analysis of engrafted human hematopoietic cells in the bone marrow of mice

1. Resuspend the cells in FACS buffer at a concentration of 2.5×10^6 cells/mL. Distribute 200 µL of cell suspension (5×10^5 cells) to each FACS tube. To analyze human myeloid and lymphoid lineages, add 5 µL of FITC-conjugated mAb against human CD45 and 5 µL of PE-conjugated mAbs against either human CD19 (lymphoid) or CD33 (myeloid) to each tube. To analyze human erythroid lineage, add 5 µL of PE-conjugated mAb against human glycophorin-A and 5 µL of FITC-conjugated mAb against human CD36. Simultaneously set up corresponding IgG isotype mAb controls.

2. Incubate the tube for 30 min at 4°C in the dark.

3. Add 3 mL of FACS buffer to each tube. Centrifuge at 453 g for 5 min. Aspirate the supernatant.

4. Repeat the wash step once. Leave 0.2 mL of supernatant after aspiration. Add 4 µL of 7-AAD. Incubate at room temperature for 10 min in the dark.

5. Analyze cell surface markers by FACS.

Southern blot and PCR analysis to detect human DNA in mouse bone marrow

1. In parallel, perform southern blot and PCR analyses to detect human DNA in mouse bone marrow.

2. Isolate high-molecular-weight DNA using phenol/chloroform extraction or DNA-zol reagent, according to the manufacturer's instructions.

3. Digest 1 µg of DNA with EcoRI restriction enzyme at 37°C overnight and separate on a 1.0% agarose gel.

4. Transfer the DNA to Hybond-N+ nylon membrane and hybridize with a ^{32}P-labeled human chromosome 17-specific alpha-satellite probe.

5. The level of human cell engraftment is quantified using a phospho-imager and ImageQuant software (Molecular Dynamics, Sunnyvale, CA) by comparing the characteristic 2.7 kb (kilobase) band to human: mouse DNA mixture controls (limit of detection, approximately 0.1% of human DNA).

6. If the level of human DNA is less than 0.1%, perform PCR to detect the human-specific chromosome 17-specific α-satellite. Forward primer

5′-ACACTCTTTTTGCAGGATCTA-3′ and reverse primer 5′-AGCAATGTGAAA CTCTGGGA-3′ are used to amplify a 1171bp sequence (40 cycles, 94°C–30 s, 60°C–30 s, 72°C–15 s+72°C–10 min). The PCR products are separated on 1.0% agarose gels.

Note. The criteria for mouse engraftment are the presence of human DNA in both the transplanted femur and non-transplanted bone marrows.

Troubleshooting

I am getting little or no hematopoietic differentiation of EBs. What can I do?

The efficiency of human ES cell differentiation toward hematopoietic lineages may vary from one cell line to another (for instance, H9 is superior to H1 in our hands) and from one experiment to another. We have observed that cell density of undifferentiated human ES cell cultures is critical to differentiation response during EB formation. Accordingly, it may take longer than 7 days for undifferentiated human ES cells to be passed, since they must be 80–90% confluent, densely packed, while maintaining the adequate ratio of undifferentiated to fibroblast-like cells. If longer than 10 days are required, then passage the cells onto fresh Matrigel at a 1:1 or 2:1 ratio. Passaging the undifferentiated cells onto 1:6 dilution of Matrigel 1 week prior to EB formation enhances the thickness of the colonies. Other important factors are the size of EBs (large EBs produce more hematopoietic CD45+ cells than small EBs, our unpublished observations) and the batch of FBS (we recommend to test different batches for differentiation efficiency and choose the best performer for subsequent differentiation experiments).

Preparation of MEF-CM is time-consuming, can I use other media for the maintenance/propagation of undifferentiated human ES cells ?

We have tried to culture undifferentiated human ES cells in two different standard media: SR media supplemented with 40 ng/mL bFGF (Wang *et al.*, 2005 c) or XVivo-10 (Li *et al.*, 2005) media supplemented with 80–100 ng/mL bFGF.

In our hands, cultures of human ES cells in SR media grow slower than cultures in MEF-CM, so we recommend not to pass the cells beyond 1:2 ratio, and also to pass them less frequently (every 8 to 10 days). In addition, the output of hematopoietic differentiation during EB formation (measured by frequency and total numbers of CD45+ cells) was not as consistent and reproducible from SR-cultures as from MEF-CM-cultures (unpublished observations). Cultures of human ES cells in XVivo-10 maintained an undifferentiated phenotype and the ability to differentiate toward hematopoietic lineages for only 10–12 weeks (unpublished observations).

The CD45^{neg} PFV precursors die after isolation. Why?

CD45^{neg}PFV cells are very fragile and require *gentle* manipulations (for sorting, prefer enrichment to purity mode, use low pressure with 100 μm nozzle, avoid pipetting and keeping the cells too long at room temperature). We have also experienced that isolated CD45^{neg}PFV precursors do not like to be repeatedly collected and transferred to new fibronectin-coated wells, resulting in a notable cell death. It is sometimes possible to harvest the cells with no need of enzymatic digestion by 0.25% Trypsin/EDTA, because they grow loosely attached to the fibronectin.

References

Bhatia M. (2003). The ultimate source of human hematopoietic stem cells: thinking outside the marrow. *Cloning Stem Cells* **5**: 89–97.

Bowles KM, L Vallier, JR Smith, MR Alexander and RA Pedersen. (2006). HOXB4 overexpression promotes hematopoietic development by human embryonic stem cells. *Stem Cells* **24**: 1359–1369.

Cerdan C, A Rouleau and M Bhatia. (2004). VEGF-A165 augments erythropoíetic development from human embryonic stem cells. *Blood* **103**: 2504–2512. Epub 2003 Dec 2504.

Chadwick K, L Wang, L Li, P Menendez, B Murdoch, A Rouleau and M Bhatia. (2003). Cytokines and BMP-4 promote hematopoietic differentiation of human embryonic stem cells. *Blood* **102**: 906–915.

Cameron CM, WS Hu and DS Kaufman. (2006). Improved development of human embryonic stem cell-derived embryoid bodies by stirred vessel cultivation. *Biotechnol Bioeng* **94**: 238–248.

Kaufman DS, ET Hanson, RL Lewis, R Auerbach and JA Thomson. (2001). Hematopoietic colony-forming cells derived from human embryonic stem cells. *Proc Natl Acad Sci USA* **98**: 10716–10721.

Kim SJ, BS Kim, SW Ryu, JH Yoo, JH Oh, CH Song, SH Kim, DS Choi, JH Seo, CW Choi, SW Shin, YH Kim and JS Kim. (2005). Hematopoietic differentiation of embryoid bodies derived from the human embryonic stem cell line SNUhES3 in co-culture with human bone marrow stromal cells. *Yonsei Med J* **46**: 693–699.

Li Y, S Powell, E Brunette, J Lebkowski and R Mandalam. (2005). Expansion of human embryonic stem cells in defined serum-free medium devoid of animal-derived products. *Biotechnol Bioeng* **91**: 688–698.

Narayan AD, JL Chase, RL Lewis, X Tian, DS Kaufman, JA Thomson and ED Zanjani.. (2005). Human embryonic stem cell-derived hematopoietic cells are capable of engrafting primary as well as secondary fetal sheep recipients. *Blood* **107**: 2180–2183.

Ng ES, RP Davis, L Azzola, EG Stanley and AG Elefanty. (2005). Forced aggregation of defined numbers of human embryonic stem cells into embryoid bodies fosters robust, reproducible hematopoietic differentiation. *Blood* **106**: 1601–1603.

Oberlin E, M Tavian, I Blazsek and B Peault. (2002). Blood-forming potential of vascular endothelium in the human embryo. *Development* **129**: 4147–4157.

Qiu C, E Hanson, E Olivier E, M Inada, DS Kaufman, S Gupta and EE Bouhassira (2005). Differentiation of human embryonic stem cells into hematopoietic cells by coculture with human fetal liver cells recapitulates the globin switch that occurs early in development. *Exp Hematol* **33**: 1450–1458.

Slukvin II, MA Vodyanik, JA Thomson, ME Gumenyuk and KD Choi. (2006). Directed differentiation of human embryonic stem cells into functional dendritic cells through the myeloid pathway. *J Immunol* **176**: 2924–2932.

Tavian M, MF Hallais and B Peault. (1999). Emergence of intraembryonic hematopoietic precursors in the pre-liver human embryo. *Development* **126**: 793–803.

Tavian M, C Robin, L Coulombel and B Peault. (2001). The human embryo, but not its yolk sac, generates lympho-myeloid stem cells: mapping multipotent hematopoietic cell fate in intraembryonic mesoderm. *Immunity* **15**: 487–495.

Tian X, JK Morris, JL Linehan and DS Kaufman. (2004). Cytokine requirements differ for stroma and embryoid body-mediated hematopoiesis from human embryonic stem cells. *Exp Hematol* **32**: 1000–1009.

Tian X, PS Woll, JK Morris, JL Linehan and DS Kaufman. (2006). Hematopoietic engraftment of human embryonic stem cell-derived cells is regulated by recipient innate immunity. *Stem Cells* **24**: 1370–1380.

Vodyanik MA, JA Bork, JA Thomson and Slukvin II. (2005). Human embryonic stem cell-derived CD34+ cells: efficient production in the coculture with OP9 stromal cells and analysis of lymphohematopoietic potential. *Blood* **105**: 617–626; Epub 2004 Sep 2016.

Wang L, L Li, F Shojaei, K Levac, C Cerdan, P Menendez, T Martin, A Rouleau and M Bhatia. (2004). Endothelial and hematopoietic cell fate of human embryonic stem cells originates from primitive endothelium with hemangioblastic properties. *Immunity* **20**: 31–41.

Wang L, P Menendez, Shojaei, L Li, F Mazurier, JE Dick, C Cerdan, K Levac and M Bhatia. (2005b). Generation of hematopoietic repopulating cells from human embryonic stem cells independent of ectopic HOXB4 expression. *J Exp Med* **201**: 1603–1614.

Wang J, HP Zhao, G Lin, CQ Xie, DS Nie, QR Wang and GX Lu. (2005a). In vitro hematopoietic differentiation of human embryonic stem cells induced by co-culture with human bone marrow stromal cells and low dose cytokines. *Cell Biol Int* **29**: 654–661.

Wang L, L Li, P Menendez, C Cerdan and M Bhatia. (2005c). Human embryonic stem cells maintained in the absence of mouse embryonic fibroblasts or conditioned media are capable of hematopoietic development. *Blood* **105**: 4598–4603.

Woll PS, CH Martin, JS Miller and DS Kaufman. (2005). Human embryonic stem cell-derived NK cells acquire functional receptors and cytolytic activity. *J Immunol* **175**: 5095–5103.

Zambidis ET, B Peault, TS Park, F Bunz and CI Civin. (2005). Hematopoietic differentiation of human embryonic stem cells progresses through sequential hematoendothelial, primitive, and definitive stages resembling human yolk sac development. *Blood* **106**: 860–870.

Zhan X, G Dravid, Z Ye, H Hammond, M Shamblott, J Gearhart and L Cheng. (2004). Functional antigen-presenting leucocytes derived from human embryonic stem cells in vitro. *Lancet* **364**: 163–171.

14, Part A

Directed differentiation of human embryonic stem cells into forebrain neurons

EMILY A. DAVIS

Department of Cellular and Molecular Medicine, University of California, San Diego, La Jolla, CA 92093, USA

and

LAWRENCE S.B. GOLDSTEIN

Howard Hughes Medical Institute, Department of Cellular and Molecular Medicine, University of California, San Diego, 9500 Gilman Drive, La Jolla, CA 92093, USA

Introduction

The *in vitro* culture of human embryonic stem (ES) cell lines has opened the door to many new research and therapeutic approaches to human biology and disease. One major goal is the controlled differentiation of these cells into any neuron found throughout the nervous system. For example, Parkinson's disease, amyotrophic lateral sclerosis (ALS), and many other diseases could, in principle, etc. be alleviated by the replacement of lost or damaged neurons with healthy cells. Parkinson's disease is characterized by death of midbrain dopaminergic neurons, which could potentially be replaced by implantation of dopaminergic neurons derived from human ES cells. Human ES cells can also be genetically modified, which leads to additional therapeutic possibilities. For example, neurons engineered to express disease specific or injury-response proteins could overcome enzyme deficiencies or spinal cord injury.

Neurons derived from human ES cells can also serve as unique models of human neurodegenerative diseases. Currently, few sources of human neurons are available for research, and obtaining samples to study a disease throughout its progression is often difficult. Furthermore, other model systems currently used to study disease cannot replicate the detailed features of a human cell. For example, mice are often used to study Alzheimer's disease, even though they do not develop typical disease naturally or when genetically manipulated. To overcome these problems, human basal forebrain cholinergic and other defined types of neurons typically affected in Alzheimer's

Table 1 Summary of selected *in vitro* differentiation methods in chronological order

Method	Efficiency	Neurotransmitter types	Reference
High density cultures grown for 3 weeks. Collect PS-NCAM$^+$ rosettes, form neurospheres, plate and continue differentiation	unknown	Glutamate, GABA	Reubinoff *et al.*, 2000
Induced embryoid bodies (EBs) for 4 days, added FGF-2 5 days, collected neural precursors (NPs), grow as neurospheres, remove FGF-2, grow > 14 days	95% nestin$^+$ NPs, no mature quant.	Glutamate, GABA	Zhang *et al.*, 2001
EBs induced w/or w/o retinoic acid, plated in selective media with growth factors. Also, used magnetic beads to sort for PS-NCAM$^+$ and A2B5$^+$ cells	30% Map2$^+$ cells	3% TH$^+$, GABA, glutamate, glycine	Carpenter *et al.*, 2001
Induced EBs, added retinoic acid and βNGF to EB media	76%NF-H$^+$	Serotonin, dopamine	Shuldiner *et al.*, 2001
Collected NPs from high density cultures, expanded progenitor cells into neurospheres, plated neurospheres for final differentiation	unknown	15% glutamate, 35% GABA serotonin<1% TH$^+$ < 1%	Reubinoff *et al.*, 2001
Cultured on MS5 stromal cells, then sequentially exposed to SHH, FGF8 then ascorbic acid and BDNF	50% Tuj−1$^+$	70% dopamine, 5% serotonin, 1% GABA	Perrier *et al.*, 2004
Induced EBs with FGF or RA, then added BDNF or TGF-alpha	75% NF200$^+$	Up to 20% TH$^+$ and GABA	Park *et al.*, 2004
Co-cultured on PA6 stromal cells for ~3 weeks	92% colonies Tuj−1$^+$	87% of colonies are TH$^+$, additional ChAT	Zeng *et al.*, 2004
Co-cultured with PA6 (+GDNF), and added factors from embryonic striatum	unknown	Dopamine	Buytaert-Hoefen *et al.*, 2004
Suspension culture in HepG2 conditioned medium (epithelial hepatocyte carcinoma line) + BMP-4 pulse	unknown	75% dopamine, adrenergic, noradrenergic	Schulz *et al.*, 2004
Co-cultured with PA6 in BHK-21/Glasgow medium	50% Tuj−1$^+$	30% TH$^+$, 15% peripherin$^+$	Pomp *et al.*, 2005
See #2, added retinoic acid to neuroectodermal cells, then SHH, BDNF, GDNF, IFG1	"majority" Map2$^+$	Cholinergic	Li *et al.*, 2005

Table 1 *(continued)*

Method	Efficiency	Neurotransmitter types	Reference
4 days as EBs, 14–16 days in adherent in F12 media. Added FGF8 and SHH sequentially, then NPs harvested and differentiated fully	Unknown	~30% dopamine, some GABA	Yan *et al.*, 2005
3 weeks as floating aggregates +noggin to derive NPs, dissociated and plated	37% Tuj−1$^+$	GABA, glutamate, serotonin, TH$^+$	Itsykson *et al.*, 2005
2 weeks on feeders, removed MEFs, 3 days to develop rosettes, into Neurobasal/B27 + LIF (to make NPs). 7 days w/SHH and RA, 14 more days of differentiation	Unknown	20–30% ChAT, Tuj-1 and Islet1$^+$ cells	Shin *et al.*, 2005
Seeded human ES cells onto human amniotic membrane matrix, grow for up to 42 days	40% Tuj−1$^+$	31% TH$^+$, 5% GABA, < 1% serotonin	Ueno *et al.*, 2006
Use HepG2 cell conditioned media or a F12/N2 media to derive neuroepithelial cells. Propagate cells in neurobasal media, B27 and bFGF, then differentiate further	70% nestin and musashi−1$^+$	unknown	Shin *et al.*, 2006

disease, could be derived from human ES cells and serve as a more realistic model. Human ES cell lines have the potential to be genetically manipulated to introduce, i.e., "knock-in", known mutations that cause familial Alzheimer's disease, so true neuronal models of familial Alzheimer's disease can be studied at the earliest stages of the disease process. Such cells can also be used to answer specific biochemical and cellular questions about Alzheimer's disease as well as testing proposed hypotheses of disease. For example, details of Abeta processing dynamics, tau pathology, the role of apolipoprotein E, and many other issues can be investigated. Similarly, disorders such as Huntington's disease, Down's syndrome, injury response, and a host of other conditions can be studied in cultured human neurons. Finally, questions in basic neurobiology can be answered using neurons derived from human ES cells. Neurotransmitter dynamics, synapse formation, electrical impulses, cell–cell interactions, and neuronal development questions can all be studied using human neuronal cells in culture.

There are currently two general approaches for inducing neurons to differentiate from pluripotent human ES cells. One approach involves treatment of human ES cells to form neurons *in vitro*, while a second approach implants undifferentiated, (or partially differentiated) cells into animals, resulting in *in vivo* formation of neurons.

There are currently a number of protocols for differentiation into neuron-like cells *in vitro*. **Table 1** summarizes a number of the methods published and their efficiency and neurotransmitter types. Most of these protocols attempt to induce formation of either general neuron-like cells, or a more specific type of neuron, most often dopaminergic or motor neurons, which are discussed in following chapters. Three general types of protocol are used: (1) formation of embryoid bodies (EBs) with all three developmental layers, which are then pushed towards neuronal differentiation, (2) co-culture of stem cells with neuron-inducing feeder layers, or (3) induction of neural precursors or neuroectoderm cells which are subsequently plated in defined media for terminal differentiation into neurons.

Overview of protocols

Our laboratory has found that reliable generation of neuron-like cells can be achieved by co-culture of stem cells with PA6 stromal cells. We have successfully used PA6 stromal cells to differentiate the HSF6, HUES9, and H1 human ES cell lines. The methods are adapted from Zeng *et al.*, 2004. We also include at the end of the section a synopsis of selected *in vitro* methods from each of the three general approaches toward neuronal differentiation.

Materials, reagents, and equipment

Culture media

PA6 stromal cell maintenance media

DMEM	GIBCO Invitrogen	11965-092	450 mL
Fetal bovine serum (FBS) (heat inactivated)	Mediatech	35-011-CV	50 mL
L-Glutamine (200 mM)	GIBCO Invitrogen	25030-081	5 mL
Pen/Strep (10,000 u/mL)	GIBCO Invitrogen	15140-122	5 mL

Medium for PA6 differentiation

GMEM	GIBCO Invitrogen	11710-035	450 mL
Knockout Serum Replacement (KSR)	GIBCO Invitrogen	10828-028	50 mL
Sodium pyruvate (100 mM)	GIBCO Invitrogen	11360-070	5 mL
Non-essential amino acids (10 mM)	GIBCO Invitrogen	11140-050	5 mL
Pen/Strep (10,000 u/mL)	GIBCO Invitrogen	15140-148	5 mL
β-Mercaptoethanol (55 mM)	GIBCO Invitrogen	21985-023	500 μL

Other materials

PA6 stromal cells (RIKEN BioResource Center Cell Bank, obtain through a Material
   Transfer Agreement)
Phosphate buffered saline (PBS) w/o Ca^{2+}, Mg^{2+} (Invitrogen, #14190-144)
0.25% Trypsin + EDTA   (Invitrogen, #25300-054)
10 cm sterile culture dishes (coated w/0.1% gelatin) (Fisher, #08−772-E)

Protocol

Following is a detailed protocol for differentiation of human ES cells by PA6 co-culture.
Figure 1 (see below; color version in color plate section) shows immunofluorescence
staining of a typical sample that has been cultured for 48 days on PA6 cells.

A. Maintenance of PA6 stromal cells

a.   PA6 cells are grown in PA6 maintenance medium at 37°C in 5% CO_2

b.   To passage:

   1.   When PA6 cells are ~80% confluent (every 4−6 days), remove culture
       medium.

   2.   Wash once with PBS, then add 1 mL 0.25% trypsin and incubate at 37°C for
       1−2 min.

   3.   When cells have detached, neutralize trypsin with 9 mL of maintenance
       medium.

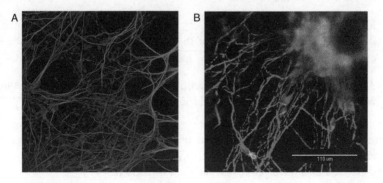

Figure 1   Human ES cells on PA6 for 48 days fixed and stained with (A)
Tuj-1 (green) and dapi (blue), and (B) Tuj-1 (green) and HuC/D (red).
Scale bars: (A) 205 μm; (B) 110 μm. This image is best seen in color;
please refer to plate section to see the color image.

4. Pellet cells at 200 g/~700rpm, 4°C for 5 min. Re-suspend the pellet in mainte-
    nance media, and plate cells onto 0.1% gelatin-coated dishes at approximately
    a 1:4 ratio.

B. Culturing human ES cells onto PA6 feeder layers

a. 1 day before passage of human ES cells, PA6 cells should be plated onto 0.1%
    gelatin-coated dishes in PA6 maintenance media at a density of approximately
    6.4×10^3 cells/cm^2. The PA6 cells continue to divide as you culture, so starting
    plates will be sparse.

b. Seeding of human ES cells is dependent on cell line — in this case, we are using
    HUES9 cells. If using cell lines that require cutting or scraping, passage as usual
    until resuspension of cell pellet.

    1. Wash out PA6 maintenance media from previously seeded PA6 plates, rinse
        once with PBS, and replace with PA6 differentiation media.

    2. When human ES colonies are well-bordered, remove maintenance media and
        rinse once with PBS.

    3. Add 1 mL 0.05% trypsin (we dilute 0.25% trypsin 1:5 into PBS), and incubate
        at 37°C for 1–2 min or until cells appear to separate from each other.

    4. Neutralize trypsin with 9 mL of PA6 differentiation media. Centrifuge for
        5 min at 200 g/~700rpm at 4°C. Resuspend cells in PA6 differentiation media.

    5. Add approximately 8×10^4 cells onto 1–10 cm plate previously seeded with
        PA6 cells. For other human ES cell lines, more cells may be needed. For
        example, the Hsf6 cell line requires about 1.1×10^5 cells, as their growth rate
        is much slower than HUES9.

    6. Maintain plates at 37°C in 5% CO_2 for at least 30 days. Change media when
        it becomes orange to yellow. For the first 7–14 days, media may be changed
        every 2–3 days. After ~14 days, media will need to be changed daily.

    7. After 30 or more days of differentiation, cells will form a very thick mass
        with many Tuj-1 positive cells throughout.

A number of other protocols are also described in the literature. We summarize
below and in Table 1 some of the more clearly defined approaches. These methods are
well-characterized and have published detailed directions.

For differentiation through embryoid body formation, cells grown in suspension culture aggregate to form EBs. In this method (Carpenter *et al.*, 2001) EBs were induced
by growing cells in suspension culture for 4 days to form aggregates. The resulting
EBs were then transferred to adherent plates and grown for 3 days with a variety of

neurotrophic factors. 87% of cells at this stage of differentiation stained with the neural precursor markers A2B5 or PS-NCAM. Adherent cells derived from the EBs were then passaged into a serum-free neurobasal media mixture plus additional factors and grown for 14–16 days to produce neurons. At the termination of the differentiation protocol, 30% of differentiated cells were stained with Map2 antibody, a marker for post-mitotic neurons. Neuron-like cells also stained for a variety of neurotransmitters. In addition to immunohistochemistry, neuronal maturity was confirmed by calcium imaging.

A second method of differentiation employs culturing cells on a feeder layer. In one well-characterized case, MS5 stromal cells were used to induce neural rosettes after 4 weeks of culturing. (Perrier et al., 2004) NCAM, Pax6 and Sox1, indicators of early neuronal differentiation, were found within the rosettes. Following stromal culture, rosettes were selected and transferred to new media plus factors in order to induce precursors into midbrain dopaminergic type neurons. After 7–10 days media was changed for final differentiation into neurons. After 50 days of differentiation, 30–50% of cells were Tuj-1 positive, and thus thought to be neurons. Tuj-1 is β-tubulinIII, a microtubule that is thought to be neuron-specific, and expressed early in differentiation as well as in mature neurons. Immunofluorescence and RT-PCR analysis indicated 64–79% of the Tuj-1 positive cells were positive for tyrosine hydroxylase (TH), indicating possible dopaminergic neurons, while serotonin and GABA-positive (5% and 2% respectively) neurons were also found.

Another commonly used feeder-layer method differentiated stem cells by co-culture with PA6 stromal cells. Undifferentiated human ES cells were plated onto a PA6 monolayer, and left to differentiate for 3 weeks (Zeng et al., 2004). 92% of the colonies contained Tuj-1+ cells, and 87% of colonies contain post-mitotic, TH positive cells, although in our hands the percentage of TH+ cells seems to be much lower. Analysis of PA6 differentiated neurons also included K+ depolarization, RT-PCR of transcripts for neurons and stem cells, and focused micro-arrays.

A final method for differentiation of human ES cells in vitro derives and subsequently subcultures a neuroprecursor, neuroepithelial (NEP) or neural progenitor (NP) line of cells. Human ES cells were grown in ES medium on mouse embryonic fibroblasts for 7 days (Shin et al., 2006). The cells were then differentiated in precursor induction media 7 days until rosettes formed. Cells positive for the markers nestin and musashi-1, markers for CNS progenitor cells, precursors to markers such as PS-NCAM and A2B5, were selected from the rosettes and cultured in media conditioned on HepG2 (an epithelial hepatocyte carcinoma line). Clonal NEP cells were then fully differentiated for 14 days in Neurobasal media plus factors. PS-NCAM and A2B5 were detected early in terminal differentiation, while Hu C/D (a marker of post-mitotic neurons) and Tuj-1 were detected after 14 days, as well as glial and oligodendrocyte markers. Although the efficiency of creating neurons was not measured, 70–80% of human ES cells expressed musashi-1 and nestin after NEP differentiation. These cells were selected for subculture and retained their expression levels after extended time in culture, and may represent a promising avenue for future differentiation experiments.

Many techniques are emerging for the formation of neurons from human ES cells, but there are still many points that need to be addressed. In particular, standards for the characterization of the differentiated cultures need to be defined. We recommend three criteria for analysis of the cells derived through differentiation protocols: identification of the general cell types formed and their frequency, determination of neuron types, and confirmation of maturity.

Since none of the current methods for differentiation produce a population that is 100% neuronal, identification and characterization of cells in this mixed population is critical to evaluate the efficiency of new techniques. We suggest immunofluorescence or quantitative PCR for markers of cell types, including neurons, glia, oligodendrocytes and neural precursors. For neurons in particular, markers such as NeuN and Hu C/D should be used in addition to Tuj-1, as Tuj-1 is expressed in neuronal lineage cells along their developmental timeline.

To determine the type of neurons formed, a combination of methods is probably suitable. Commonly, immunofluorescence for neurotransmitter type is used. Also, some groups depolarize cells and assay the neurotransmitters that are released into the media, usually by HPLC. Which neuron types are in the population is also a lingering question. Are the neurons similar to CNS or PNS? Do they resemble neurons from certain brain regions? Morphology analysis, as well as more markers, can answer these questions.

For confirmation of neuronal maturity, a number of methods can be used, and include patch clamping and calcium imaging, as well as immunofluorescence for mature neuronal markers, polarity and synapse formation. Electrophysiology is a functional, rather than morphological, confirmation of maturity, although it may not be practical for every researcher. In any case, steps should be taken to detect whether the neurons in the cultures are truly mature. Information on all of these important aspects of neuronal differentiation and characterization will be necessary for the use of these neurons in experiments, and more crucial for future therapeutic endeavors.

Troubleshooting

My cultures are very thick and have large masses of cells. Is this normal?

These cultures become very thick and dense when cultured for long periods. You will see many different structures in the cultures, and they will vary between different culture plates. The development of neurons and other cells is difficult to observe by brightfield microscopy because of this density. Also, expect large amounts of cell death in your cultures, even when changing media every day. This is not unusual and does not indicate your cultures are dying. If, however, the media stops changing to yellow after extended time in culture, the cells in your samples may have died.

I would like to use immunofluorescence on these cultures, but they are so thick.

We suggest adding 0.1% Triton-X100 to the buffer for your primary antibody step in addition to a separate permeabilization step. Also, be very gentle and avoid moving the coverslips excessively, as the cells are likely to detach from the coverslip after being fixed. Finally, the staining is easiest to image on a confocal microscope, due to the thickness of the sample.

The cells are peeling off the plates. What do I do?

Unfortunately, we have not found a solution to the cells detaching from the plates. We recommend plating multiple backup plates and suggest you be very careful when changing media. Coating culture dishes with different substrates does not seem to improve this phenomenon, probably because the cells separate from the plate after such long periods in the culture dish. One possible way to rescue your samples is to trypsinize and replate your cells when they begin to detach. We have successfully replated these differentiating cells, but have not yet thoroughly characterized the tryspinized cells.

References

Buytaert-Hoefen A, E Alvarez and CR Freed. (2004). Generation of tyrosine hydroxylase positive neurons from human embryonic stem cells after coculture with cellular substrates and exposure to GDNF. *Stem Cells* **22**: 669–674

Carpenter MK, MS Inokuma, J Denham, T Mujtaba, CP Chiu and MS Rao. (2001). Enrichment of neurons and neural precursors from human embryonic stem cells. *Exp Neurol* **172**: 383–397.

Itsykson P, N Ilouz, T Turetsky, RS Goldstein, MF Pera, I Fishbein, M Segal and BE Reubinoff. (2005). Derivation of neural precursors from human embryonic stem cells in the presence of noggin. *Mol Cell Neruosci* **30**: 24–36.

Li XJ, ZW Du, ED Zarnowska, M Pankratz, LO Hansen, RA Pearce and SC Zhang. (2005). Specification of motoneurons from human embryonic stem cells. *Nat Biotechnol* **23**: 215–221.

Park S, KS Lee, YJ Lee, HA Shin, HY Cho, KC Wang, YS Kim, HT Lee, KS Chung, EY Kim and J Lim. (2004). Generation of dopaminergic neurons in vitro from human embryonic stem cells treated with neurotrophic factors. *Neurosci Lett* **359**: 99–103.

Perrier AL, V Tabar, T Barberi, ME Rubio, J Bruses, N Topf, NL Harrison and L Studer. (2004). Derivation of midbrain dopamine neurons from human embryonic stem cells. *Proc Natl Acad Sci USA* **101**: 12543–12548.

Pomp O, I Brokhman, I Ben-Dor, B Reubinoff and RS Goldstein. (2005). Generation of peripheral sensory and sympathetic neurons and neural crest cells from human embryonic stem cells. *Stem Cells* **23**: 923–930.

Reubinoff BE, MF Pera, C-F Fong, A Trounson and A Bongso. (2000). Embryonic stem cell lines from human blastocysts: somatic differentiation *in vitro*. *Nat Biotechnol* **18**: 399–404.

Reubinoff BE, P Itsykson, T Turetsky, MF Pera, E Reinhartz, A Itzik and T Ben-Hur. (2001). Neural progenitors from human embryonic stem cells source: *Nat Biotechnol* **19**: 1134–1140

Schulz TC, SA Noggle, GM Palmarini, DA Weiler, IG Lyons, KA Pensa, ACB Meedeniya, BP Davidson, NA Lambert and BG Condie. (2004). Differentation of human embryonic stem cells to dopaminergic neruons in serum-free suspension culture. *Stem Cells* **22**: 1218–1238.

Shin S, S Dalton and S Stice. (2005). Human motor neuron differentiation from human embryonic stem cells. *Stem Cells Devel* **14**: 266–269.

Shin S, MM Mitalipova, SA Noggle, D Tibbitts, A Venable, R Rao and SL Stice. (2006). Long-term proliferation of human embryonic stem cell-derived neuroepithelial cells using defined adherent culture conditions. *Stem Cells* **24**: 125–138.

Ueno M, M Matsumura, K Watanabe, T Nakamura, F Osakada, M Takahashi, H Kawasake, S Kinoshita and Y Sasai. (2006). Neural conversion of ES cells by an inductive activity on human amniotic membrane matrix. *Proc Natl Acad Sci USA* **103**: 9554–9559.

Yan Y, D Yang, ED Zarnowska, Z Du, B Werbel, C Valliere, RA Pearce, JA Thomson and S-C Zhang. (2005). Directed differentiation of dopaminergic neuronal subtypes from human embryonic stem cells. *Stem Cells* **23**: 781–790.

Zhang S-C, M Wernig, ID Duncan, O Bruestle and JA Thomson. (2001). In vitro differentiation of transplantable neural precursors from human embryonic stem cells. *Nat Biotechnol* **19**: 1129–1133

Zeng X, J Cai, J Chen, Y Luo, Z-B You, E Fotter, Y Wang, B Harvey, T Miura, C Backman, G-J Chen, MS Rao and WJ Freed. (2004). Dopaminergic differentiation of human embryonic stem cells. *Stem Cells* **22**: 925–940.

14, Part B

Directed differentiation of human embryonic stem cells into dopaminergic neurons

JAN PRUSZAK AND OLE ISACSON

Neuroregeneration Laboratories, Center for Neuroregeneration Research, Harvard Medical School, McLean Hospital, Belmont, MA 02478, USA

Introduction

The idea of replacing cells to compensate for degeneration and thereby restore function has entered several medical fields, including neurology (Isacson, 2003). Clinical studies have clearly demonstrated that transplantation of fetal dopaminergic (DA) cells into the brain of Parkinson's disease (PD) patients can alleviate disease-associated motor deficits for years to over a decade (Mendez *et al.*, 2005). However, the use of fetal tissue is associated with many problems including low yield, unpredictable clinical outcome and ethical debate about the use of aborted fetuses as donor tissue. An alternative cell source to fetal midbrain tissue are human embryonic stem (ES) cells, which can generate large amounts of cells, and which have the capacity to mature into DA neurons (Sonntag *et al.*, 2005). Several cell culture protocols have been explored, and continuous discoveries further elucidate the mechanisms that underlie DA specification in normal midbrain development and during ES cell differentiation. However, it has to be stressed that current human ES cell differentiation protocols result in heterogeneous cell populations and that, as of now, sufficient yield of the appropriate therapeutic DA cell has not been achieved (Ben Hur *et al.*, 2004; Brederlau *et al.*, 2006; Park *et al.*, 2005; Perrier *et al.*, 2004; Pruszak *et al.*, 2007; Roy *et al.*, 2006; Schulz *et al.*, 2004; Yan *et al.*, 2005).

The following section provides examples of concepts and detailed methodologies that have been applied in the *in vitro* generation of dopaminergic neurons for cell-therapeutic studies of Parkinson's disease.

Overview of protocol

In our hands, a protocol based on Perrier *et al.* (2004), modified by the addition of Noggin (Sonntag *et al.*, 2006), efficiently generates DA neurons for both studies of midbrain development *in vitro* and neurotransplantation studies in the rodent and primate. Neuroectodermal differentiation ("rosette formation") of human ES cells is induced by coculture on bone marrow-derived stromal feeder cells or alternatively in feeder-free conditions on poly-ornithine/laminin coated dishes. In both cases, neuroectodermal induction is enhanced by supplementing 300 ng/mL mouse-recombinant Noggin to the medium for the first 21 days in culture (Sonntag *et al.*, 2006). Medium is changed every other day and cells are differentiated toward the dopaminergic phenotype, in principle as described previously (Perrier *et al.*, 2004), in the presence of the growth factor combinations as depicted in **Figure 1**.

Materials, reagents, and equipment

- Human embryonic stem cells; in our hands this exemplary protocol has been used with H1, H7, H9 cell lines (WiCell, Madison, WI, www.wicell.org); those are propagated, in principle, according to standard protocols (compare **Chapters 5 and 13**), with the modification of using mitotically inactivated human fibroblasts (e.g. Detroit 551, ATCC, www.atcc.org) as a feeder cell layer.

- Murine stromal feeder cells, e.g. MS5-Wnt1 (Dr. L. Studer, Sloan-Kettering Institute, New York, NY (Perrier *et al.*, 2004)), MS5 (Kirin Pharmaceutical Research Laboratory, Gunma, Japan, www.kirin.co.jp/english) or potential alternatives such as PA-6 stromal feeder cells (Riken, Tsukuba, Japan, www.riken.go.jp/engn)

- Knockout DMEM (Gibco #10829, www.invitrogen.com/Gibco)

- Knockout serum replacement (Gibco #10828)

- DMEM/F12 (1:1) + L-glutamine + 15 mM HEPES (Gibco #11330)

- alpha-MEM (Gibco #12571)

- MEM non-essential amino acid solution, 10 mM, 100× (Gibco #12383)

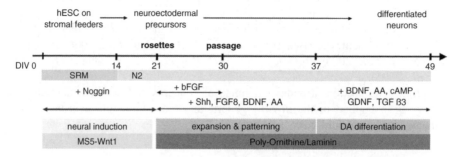

Figure 1   Exemplary protocol of dopaminergic differentiation *in vitro*. Human ES cells are differentiated using growth and signaling factors as indicated in a protocol, in principle according to Perrier *et al.* (2004), modified by the addition of Noggin (see protocol for details).

- L-Glutamine 200 mM, 100× (Gibco #25030)

- Collagenase (Gibco #17104)

- β-Mercaptoethanol 1,000× (Gibco #21985)

- Penicillin–streptomycin, 100× (Gibco #15140)

- 0.05% trypsin–EDTA 1× (Gibco #25300)

- Trypan blue stain 0.4% (Gibco #15250)

- Basic fibroblast growth factor (bFGF; Invitrogen #13256)

- Mitomycin-C (Sigma #M0503, www.sigmaaldrich.com) [photosensitive, *toxic*]

- Laminin (Sigma #L-2020)

- poly-L-Ornithine, 0.01% solution (Sigma #P4957)

- Dibutyryl-cAMP (Sigma #D0260)

- Glial cell line-derived neurotrophic factor (GDNF; Sigma #G1777)

- Ascorbic acid (sodium L-ascorbate; Sigma #A4034) [photosensitive]

- N2-A supplement (100×) (Stem Cell Technologies #07152, www.stemcells.com)

- Heat-inactivated defined fetal bovine serum (FBS; e.g. HyClone #SH30070.03)

- Noggin (R&D Systems #1967-NG, www.rndsystems.com)

- Sonic hedgehog (Shh; R&D Systems #1314-SH)

- FGF-8b (R&D Systems #423-F8)

- Brain-derived neurotrophic factor (BDNF; Peprotech #450–02, www.peprotech.com)

- TGF beta3 (Calbiochem #PF073; via www.emdbiosciences.com or www.merckbiosciences.com)

- Ca^{2+}, Mg^{2+}-free D-phosphate buffered saline (PBS)

- Distilled (DI) water

- Six-well tissue culture dishes

- 24-well tissue culture dishes

- T75, T175 flasks for feeder cell culture

- Filter units 0.22 μm for sterile preparation of stock solutions and media

- Cutting utensil for rosette harvesting (e.g. a 1 mL syringe with a 25 to 27.5 G needle attached)

- Microscope allowing for colony picking (e.g. Zeiss Axiovert 40CFL, 2.5× or 5× objective)

Protocol

Preparation of stock solutions and aliquots (Table 1)

Table 1   Preparation of stock solutions and aliquots

	Concentration of Stock Solution	Final Concentration	Storage [temperature/ max. duration]
Noggin	10 ng/µL	300 ng/µL	4°C/2 weeks
Brain-derived neurotrophic factor (BDNF)	50 ng/µL	20 ng/mL	4°C/2 weeks
Ascorbic acid	200 mM	200 µM	−20°C/2months
Sonic hedgehog	50 ng/ µL	200 ng/mL	4°C/2 weeks
Fibroblast growth factor 8b	50 ng/µL	100 ng/mL	4°C/2 weeks
Dibutyryl-cyclic adenosine monophosphate (dcAMP)	200 mM	500 µM	−20°C/2months
Tumor growth factor beta 3 (TGF β 3)	2 ng/µL	1 ng/mL	−20°C/2months
Glial cell line-derived neurotrophic factor (GDNF)	10 ng/µL	10 ng/mL	4°C/2 weeks
Collagenase (1 mg/mL)	1 mg/mL	1 mg/mL	4°C/4 weeks
Mitomycin-C solution	1 mg/mL	10 µg/mL	−20°C/2months

Note: all solutions prepared in PBS except collagenase (dissolved in KO-DMEM).

Preparation of media (Table 2)

Table 2   Preparation of media

	Volume
Serum replacement medium (SRM)	
Knockout-DMEM	415 mL
MEM non-essential amino acids	5 mL
L-Glutamine	5 mL
β-Mercaptoethanol	0.5 mL
Knockout serum replacement	75 mL

Table 2   (*continued*)

	Volume
Stromal feeder cell culture medium (SFCM)	
alpha-MEM	445 mL
L-Glutamine	5 mL
Penicillin-streptomycin	5 mL
Heat-inactivated FBS	50 mL
N2 medium	
DMEM/F12	500 mL
N2-A	5 mL
BASF medium	
N2 medium	50 mL
BDNF 50 µg/mL	20 µL
Ascorbic acid 200 mM	50 µL
Sonic hedgehog 50 µg/mL	200 µL
FGF-8b 50 µg/mL	100 µL
BCT-GA medium	
N2 medium	50 mL
BDNF 50 µg/mL	20 µL
cAMP 200 mM	125 µL
TGF beta3 2 µg/mL	25 µL
GDNF 10 ng/µL	50 mL
Ascorbic acid 200 mM	50 µL

Note: store media at 4°C for up to 2 weeks. SRM and SFCM
media should be sterile filtered prior to usage.

Neural induction of human ES cells on stromal feeder cells

1.   MS5 cell preparation

- Propagate the MS5 stromal feeder cells in T-175 or T-75 flasks with stromal feeder cell medium (SFCM); split every 3 to 5 days; *avoid overgrowth of cells*, as this may affect the inductive properties.

- Choose a flask at ~80% confluency for mitotical inactivation. Incubate the feeder cells with Mitomycin-C at a final concentration of 10 µg/mL for 2.5 to 3 hrs (compare chapter #13, cardiogenesis protocol).

- Wash 2× with PBS.

- Add 0.05% trypsin–EDTA and incubate for 2–3 min; tap the walls of the flask gently to detach the cells.

- Add SFCM to the flask and triturate. Dissociate any clusters or chunks of stromal feeder cells.

- Transfer the contents of the flask to a tube, and centrifuge at 600 g ($\sim$1,600 rpm) for 3 min.

- Discard the supernatant, and re-suspend the cells with SFCM medium.

- If there are still clumps of MS5 cells, pipette them out.

- Determine the cell number by visual cell counts (dye exclusion assay using trypan blue).

- Plate the mitotically inactivated stromal feeder cells in SFCM medium onto plates coated with 0.1% gelatin at a density of 75,000 cells/cm^2 with SFCM medium (e.g. 750,000 cells per well in a six-well dish); ensure even distribution of cells by carefully tilting the dish; do not swirl, to avoid congregation of the cells in the center of the well.

- Culture over night prior to plating human ES cells for neural induction.

2. *Plating of human ES cells on stromal feeder cells*

- Before beginning, check the Mitomycin-C treated MS5 plates to make sure that the cells are of the appropriate confluency, that they are not detaching at the edge of the well, have not been proliferating and are not forming clusters or aggregates.

- Human ES cells should be plated onto MS5 cells within 1 week after the most recent human ES cells passage. They should be of good quality, and *not* be starting to differentiate.

- For every well of a six-well dish plan to plate ca. one human ES cell colony (1 colony is approximately defined as the size of one visual field at a 10× magnification under the microscope).

- Using a 1 mL syringe and needle, cut and pipette out any portions of human ES cell colonies that have already differentiated or may contain dying cells.

- Count how many colonies you have and determine how many dishes, prepared in the method outlined above, you can plate them into.

- Wash each well 1× with PBS (3 mL), add collagenase solution (1 mL).

- Incubate for ca. 10 to 15 min, then gently harvest the cells with a cell scraper.

- Transfer the cells into a 15 mL conical tube using a P1000 pipette. *Do not triturate.*

- Wash 2× with SRM (2 mL), transferring the cells each time into a 15 mL conical tube using a P1000 pipette. *Do not triturate.*

- Centrifuge the tube at ca. 180 g (~600 rpm) for 5 min.

- Meanwhile, change the medium of the MS5 plates from SFCM to SRM and add Noggin in a final concentration of 300 ng/mL.

- After the tube has been centrifuged, take off all but 1 mL of the supernatant; using a P1000, and vigorously triturate ca. 10 to 20 times to re-suspend the pellet.

- Any remaining chunks in the suspension are likely to be fibroblast feeders; do not worry about pipetting those out, as they will not grow and will be discarded with the first media change.

- Using a P1000 pipette, divide the 1 mL suspension evenly into the number of six-wells you determined can be plated; between plating different plates, triturate gently ca. 3 to 5 times.

- When placing the plates into the incubator ensure that the human ES cells are evenly distributed by carefully tilting the plates from side to side.

3. *Monitoring neural induction and harvesting neuroepithelial precursor cells (rosette structures)*

- The first visible human ES-derived colonies will usually not arise before 7 to 10 days *in vitro* on stromal feeder cells (**Figure 2A**). Media changes are every other day using SRM and adding 300 ng/mL Noggin. After 14 days in culture, conditions are switched to N2 medium with the addition of 300 ng/mL Noggin.

- After about 14 to 21 days in culture, cell colonies with neuroepithelial morphology should have arisen (**Figure 2B**).

- Using a sharp glass pipette, or a slightly bent needle attached to a syringe, carefully detach the areas containing neuroepithelial rosette-like structures, isolating those from the surrounding stromal feeder cells and non-neural cell clusters. Then, collect the floating rosette chunks with a P200 pipette, and *gently* triturate. Ideally, after this step, you have single-rosette colonies, but not a single-cell suspension, that will be plated for subsequent differentiation. See Karki *et al.* (2006) for a video article demonstrating these procedures in detail.

DA patterning and differentiation

- The picked rosette structures are plated on coated tissue culture dishes coated with poly-ornithine (15 µg/mL) and laminin (1 µg/mL) at a density of approximately one rosette colony per cm^2 in medium containing BDNF, ascorbic acid, sonic hedgehog and FGF-8b ("BASF medium") with the addition of 20 ng/mL bFGF.

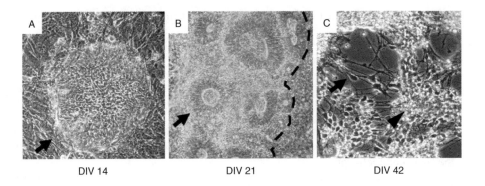

	maturation stages			
	ES cell propagation	neural induction	midbrain patterning	DA differentiation
typical markers	Oct–4 SSEA–3, –4 Tra–1–60, –81	Nestin SSEA–1, Sox–1–2, Doublecortin	Otx–2, En 1/2, Lmx1–a, –b Raldh–1	Nurr–1, Pitx–3, tyrosine hydroxylase AADC DAT GIRK–2, Calbindin

Figure 2 Morphological features of dopaminergic *in vitro* differentiation. (**A**) After plating of dissociated human ES cell colonies on MS-5 feeder cells, new colonies (arrow) arise in the dish from ca. day 10 *in vitro* on. (**B**) After 14 to 21 days *in vitro* of neural induction, neuroepithelial rosette-like structures (arrow) are formed, which are carefully selected by mechanically detaching those areas, separating those from the adjacent stromal feeder and non-neural cells (dashed line). (**C**) The final stage of dopaminergic neuronal differentiation is characterized by the presence of process-bearing neuronal cells (arrow), and some remaining clusters of proliferative cells (arrowhead). (**D**) Specifics of dopaminergic differentiation *in vitro*. Proliferative capacity and expansive potential of the differentiating cells decrease over time. In parallel, the commitment toward the DA phenotype increases during *in vitro* development from the undifferentiated ES cell toward the post-mitotic DA neuron. In this or other human ES cell differentiation protocols, appropriate DA differentiation should be closely monitored by observing the expression of characteristic markers.

- After 9 days in culture, the expanded cells are gently trypsinized, washed, spun, counted and re-cultured on poly-ornithine/laminin coated dishes at a density of approximately 125,000 cells/cm^2 in BASF medium *without* bFGF for another 7 days in culture.

- Following these expansion and patterning steps, the medium is switched to the differentiation conditions using medium containing BDNF, cAMP, TGF beta3, GDNF and ascorbic acid ("BCT-GA medium"). Cells can be cultured at this stage over several weeks for further maturation (**Figure 2C**). We have differentiated human ES cells in this protocol for more than 70 days in culture.

Monitoring DA differentiation *in vitro*

- Neural induction is characterized by a down-regulation of ES cell markers such as Oct-4, SSEA-4, SSEA-3 Tra-1-60, Tra-1-81. This is accompanied by the expression of transcription factors such as Sox-1, Sox-2 or the intermediate filament nestin, which indicate differentiation toward the neural lineage. (**Figure 2D**)

- In the context of therapeutic studies of Parkinson's disease (PD), one should highlight the need for using midbrain-specific dopaminergic (DA) markers to monitor the generation of DA neurons by immunocytochemical and RNA expression analyses. Patterning toward midbrain phenotype is initially characterized by expression of Otx-2, En 1/2, Raldh-1, followed by the occurrence of dopaminergic markers such as Nurr1, Pitx3, TH, AADC, DAT, and GIRK-2, an A9-specific marker. (**Figure 2D**)

Summary and future directions

It should be noted that this protocol allows for various modifying and optimizing steps. For example, neural precursor cells may be further expanded over multiple passages by continued culture in BASF + bFGF medium. In addition, other factors may be tested to optimize the generation of midbrain-type DA neurons, and genetic engineering of transcription factors and cell selection steps may be integrated into the protocol.

Obviously, there still is a need to gain more insight into the transcriptional profile of DA neurons and the development of midbrain-specific DA phenotypes *in vitro* and *in vivo*. The ES cell-based experiments may actually help to further elucidate these steps of neurodevelopment, and also bring us closer to the therapeutic goal of a cell-based therapy of Parkinson's disease.

Troubleshooting

Only few rosette structures occur after 21 days of neural induction. How can I improve this?

> This may be caused by suboptimal quality of both, stromal feeder or human ES cells. Ensure that the stromal feeder cells are of appropriate confluency, and that also prior to plating them as a feeder cell layer, they have been passaged regularly: overgrowth and irregular passaging can reduce their inductive properties. Also, insufficient mitomycin-C inactivation can result in feeder cell growth and affect the neural induction. With regard to plating human ES cells, you might want to harvest single human ES cell colonies, omitting the collagenase, to optimize the trituration procedure. As stated above, only human ES cell colonies of good quality should be selected for plating.

Harvested rosettes do not attach. What has happened?

This may be due to suboptimal poly-ornithine/laminin coating. Particularly the laminin should be handled ice-cold and rapidly applied after reconstitution. Also, growing cells on plastic instead of glass coverslips might facilitate attachment.

Note that infections, for example with Mycoplasma ssp. can lead to various problems in cell differentiation protocols. Preventative measures should be taken, such as thorough screening of incoming cell lines and freshly isolated MEFs for feeder preparation (e.g. using a Mycoplasma detection kit, e.g. Cambrex #LT07-118, www.cambrex.com). For eradication, enrofloxacin (in a concentration of 25 µg/mL; Sigma #17849) may be used on a short term (<14 days) basis, followed by retesting.

Cells die during passaging. Why?

As neural cells are fragile, gentle trituration is crucial. As alternatives, you may omit the trypsin digestion, and just use mechanical trituration, e.g. in HEPES-buffered HBSS, or test other enzymes such as TrypLE (Gibco #12605) or Accutase (www.innovativecelltechnologies.com).

The differentiation does not consistently generate neuronal cells or the fraction of tyrosine hydroxylase positive (TH+) neurons is low. What can be done?

Harvesting the neuroepithelial cells after neural induction is a critical step of this protocol, and suboptimal selection may lead to overgrowth with unwanted cell types. Furthermore, plating density at the different stages has a strong influence, and there may be variation between different batches of cells (and possibly human ES cell lines). As an orientation, in our hands this protocol efficiently yields DA neurons in a fraction of about 20% TH+ cells over total cells. Stock solutions and media should be freshly prepared, particularly the BASF and BCT-GA media should be used between 1 to 3 days after reconstitution.

References

Ben Hur T, M Idelson, H Khaner, M Pera, E Reinhartz, A Itzik and BE Reubinoff. (2004). Transplantation of human embryonic stem cell-derived neural progenitors improves behavioral deficit in Parkinsonian rats. *Stem Cells* **22**(7): 1246–1255.

Brederlau A, AS Correia, SV Anisimov, M Elmi, G Paul, L Roybon, A Morizane, F Bergquist, I Riebe, U Nannmark, M Carta, E Hanse, J Takahashi, Y Sasai, K Funa, P Brundin, PS Eriksson and JY Li. (2006). Transplantation of human embryonic stem cell-derived cells to a rat model of Parkinson's disease: effect of in vitro differentiation on graft survival and teratoma formation. *Stem Cells* **24**(6): 1433–1440.

Isacson O. (2003). The production and use of cells as therapeutic agents in neurodegenerative diseases. *Lancet Neurol* **2**(7): 417–424.

Karki S, J Pruszak, O Isacson and KC Sonntag. (2006). ES cell-derived neuroepithelial cell cultures. *J Visualized Exp* Issue 1: Nov 30, 2006, www.myjove.com.

Mendez I, R Sanchez-Pernaute, O Cooper, A Vinuela, D Ferrari, L Bjorklund, A Dagher and O Isacson. (2005). Cell type analysis of functional fetal dopamine cell suspension transplants in the striatum and substantia nigra of patients with Parkinson's disease. *Brain* **128**(7): 1498–1510.

Park CH, YK Minn, JY Lee, DH Choi, MY Chang, JW Shim, JY Ko, HC Koh, MJ Kang, JS Kang, DJ Rhie, YS Lee, H Son, SY Moon, KS Kim and SH Lee. (2005). In vitro and in vivo analyses of human embryonic stem cell-derived dopamine neurons. *J Neurochem* **92**(5): 1265–1276.

Perrier AL, V Tabar, T Barberi, ME Rubio, J Bruses, N Topf, NL Harrison and L Studer. (2004). Derivation of midbrain dopamine neurons from human embryonic stem cells. *Proc Natl Acad Sci USA* **101**(34): 12543–12548.

Pruszak J, KC Sonntag, MH Aung, R Sanchez-Pernaute and O Isacson. (2007). FACS analysis with neuronal cell selection methods and markers for human embryonic and neural stem cell applications, *submitted*.

Roy NS, Cleren C, Singh SK, Yang L, Beal MF and Goldman SA. (2006). Functional engraftment of human ES cell-derived dopaminergic neurons enriched by coculture with telomerase-immortalized midbrain astrocytes. *Nat. Med.* **12**(11): 1259–1268.

Schulz TC, SA Noggle, GM Palmarini, DA Weiler, LG Lyons, KA Pensa, AC Meedeniya, BP Davidson, NA Lambert and BG Condie. (2004). Differentiation of human embryonic stem cells to dopaminergic neurons in serum-free suspension culture. *Stem Cells* **22**(7): 1218–1238.

Sonntag KC, J Pruszak, T Yoshizaki, J van Arensbergen, R Sanchez-Pernaute and O Isacson. (2006). Noggin enhances the yield of dopaminergic neurons from human embryonic stem cells. *Stem Cells* Oct 12, 2006 [Epub ahead of print].

Sonntag KC, R Simantov and O Isacson. (2005). Stem cells may reshape the prospect of Parkinson's disease therapy. *Brain Res Mol Brain Res* **134**(1): 34–51.

Yan Y, D Yang, ED Zarnowska, Z Du, B Werbel, C Valliere, RA Pearce, JA Thomson and SC Zhang. (2005). Directed differentiation of dopaminergic neuronal subtypes from human embryonic stem cells. *Stem Cells* **23**(6): 781–790.

Web resources

See Karki *et al.* (2006) for a detailed video article on inducing and harvesting human ES-derived neuroepithelial cells. See www.neuroregeneration.org for related manuscripts using this protocol, as well as for updates on modified versions.

14, Part C

Directed differentiation of human embryonic stem cells into spinal motor neurons

BAO-YANG HU AND SU-CHUN ZHANG

Departments of Anatomy and Neurology, School of Medicine and Public Health, Waisman Center, Wisconsin Stem Cell Research Program, WiCell Institute, University of Wisconsin, Madison, WI 53705, USA

Introduction

Neural sub-type specification from stem cells depends on coordination between the intrinsic cellular program of precursors and temporally and spatially available extrinsic factors. For generation of spinal motor neurons, naïve neuroepithelial cells need to be patterned to ventral spinal cord progenitors in response to a caudalizing molecule retinoic acid (RA) and a ventrally derived morphogen sonic hedgehog (SHH) (Jessell, 2000). We have developed a chemically defined, adherent colony culture system that can efficiently direct human embryonic stem (ES) cells to neuroepithelial cells (Zhang *et al.*, 2001). Along the differentiation we have identified a transient primitive neuroepithelial cell stage in which cells exhibit columnar morphology, organize radially into rosettes, and express neuroectodermal transcription factors such as Pax6 and anterior homeodomain proteins such as Otx2 and Bf-1. Importantly, these primitive neuroepithelial cells can be specified to midbrain or spinal cord progenitors and differentiate to dopamine neurons or spinal motor neurons (Yan *et al.*, 2005, Li *et al.*, 2005). When the primitive neuroepithelial cells are treated with RA and SHH, Olig2-expressing spinal motor neuron progenitors appear at around 4 weeks and post-mitotic HB9-expressing motor neurons appear at around 5 weeks (Li *et al.*, 2005). Motor neurons produced using this protocol express hox genes, choline acetyltransferase, and induce acetylcholine receptor clustering in myocytes and muscle contraction (Li *et al.*, 2005). The following protocol is optimized for a more efficient (50%) generation of HB9-expressing motor neurons among the differentiated population.

Materials, reagents and equipment

Media (Table 1)

Table 1   Media

Material/reagent	Vendor	Catalog number
1.Human ES cell medium (500 mL)		
389 mL DMEM-F12	GIBCO Invitrogen	11330-032
100 mL knockout serum replacement	GIBCO Invitrogen	10828-028
5 mL non-essential amino acids	GIBCO Invitrogen	11140-050
2.5 mL L-glutamine (200 mM)	Sigma	G7513
3.5 µL 14.3 M β-mercaptoethanol	Sigma	M7522
Filter and store at 4°C for up to 7–10 days.		
2. Neural induction medium (500 mL)		
489 mL DMEM/F12	GIBCO Invitrogen	11330-032
5 mL N2 supplement	GIBCO Invitrogen	17502-048
5 mL non-essential amino acids	GIBCO Invitrogen	11140-050
1 mL heparin (1 mg/mL)	Sigma	H3149

Stock solutions

1. Basic fibroblast growth factor bFGF (R&D; cat. no. 233-FB): 10 µg/mL, in sterilized PBS with 0.1% bovine serum albumin (BSA) (Sigma; cat. no. A-7906).

2. Dispase (1 mg/mL): 10 mg of dispase (Gibco-BRL; cat. no. 17105-041) in 10 mL F12/DMEM at 37°C for 15 min. Filter with a 50 mL Steri-flip (Fisher Scientific; cat. no. SCGP00525).

3. Heparin (1 mg/mL): dissolve 10 mg heparin (Sigma; cat. no. H3149) in 10 mL DMEM medium.

4. Laminin from human placenta (Sigma; cat. no. L6274).

5. Poly-2-hydroxyethylmethacrylate (Poly-HEME) (Sigma; cat. no. P3932).

6. Brain-derived neurotrophic factor (BDNF, PeproTech Inc; cat. no. 450-02), glial cell line-derived neurotrophic factor (GDNF, PeproTech Inc; cat. no. 450-10), insulin-like growth factor I (IGF-1, GDNF, PeproTech Inc; cat. no. 100-11): 100 µg/mL, in sterilized distilled water.

7. Ascorbic acid (AA, 200 µg/mL): 2 mg ascorbic acid (Sigma; cat. no. A-4403) in 10 mL PBS.

8. Cyclic adenosine monophosphate (cAMP, 1 mM): 10 mg dibutyryl cAMP sodium salt (Sigma; cat. no. D-0260) in 20 mL sterilized distilled water.

Make aliquots and store all the stock solutions at −80°C.

Antibodies

1. For neuroepithelial cell identity:

   Pax6 (monoclonal, Developmental Studies Hybridoma Bank-DSHB Pax6)

   Sox2 (monoclonal, R&D systems MAB2018)

   Sox1 (rabbit polyclonal, Chemicon AB-5768).

2. For motor neurons and progenitors:

   ChAT (affinity purified polyclonal antibody from goat, Chemicon AB144P)

   Olig2 (Rabbit IgG, Santa Cruz SC-19969)

   MNR2 (also known as HB9, monoclonal antibody, DSHB 81.5C10)

   βIII-tubulin (Rabbit IgG, Covance PRB-435P)

Table 2 list supplies.

Table 2   Supplies

Supplies	Vendor	Catalog number
T25 and T75 flasks	Nunc	136196 and 178891
500 mL filter units	Corning	430513
24-well plates	Nunc	143982
15 mL and 50 mL conical tube	BD Biosciences	52095 and 352073
Sterile plastic pipettes 5, 10 and 25 mL	Fisher Scientific	13-678-11D
		13-678-11E and 13-678-11

Protocols

A. Neuroepithelial differentiation (formation of neural tube-like rosettes)

1. Add 1 mL of fresh collagenase (1 mg/mL) or 1 mL of fresh dispase (1 mg/mL) to each well of a six-well plate and incubate the cultures at 37°C for 5–10 min.

2. When the human ES cell colony edges curl, it is ready to remove the colonies from the dishes. Aspirate the enzyme off, add 1 mL of human ES cell medium to each well, gently swirling the plate and/or pipetting the colonies off.

3. Collect the colonies into a 15 mL conical tube and let the tube stand for 1–2 min. The human ES cell colonies should settle down. Aspirate the supernatant gently.

4. Resuspend the colonies with the human ES cell medium and let the colonies settle down for 1–2 min, aspirate the supernatant away. Alternatively, spin down at 50 g/~150 rpm for 2 min.

5. The human ES cell colonies are then plated in a T25 or T75 culture flask in human ES cell medium without bFGF for 4 days with a medium change every day. Plate cells from 2 wells to one T25 flask in 10 mL of medium.

6. Round human ES cell aggregates (often termed embryoid bodies) will be forming in the medium while the feeder cells will attach to the flask (if there are feeder cells carried over). By transferring the aggregates into a new flask, the feeder cells are automatically removed.

7. After 4 days, the human ES cell aggregates are fed with the neural induction medium for another 2 days.

8. Add 300 μL neural induction medium containing 20 μg/mL laminin to each well of a six-well plate. Do not let the medium drain to the edge of the well. Incubate the plate at 37°C for 1 hr.

9. Collect the human ES cell aggregates to a 15 mL conical tube, centrifuge at 50 g/~150 rpm for 2 min. Aspirate off medium and re-suspend human ES cell aggregates in 10 mL neural induction medium. Transfer the aggregates to laminin-coated six-well plate at 20–25 clusters per well in 300 μL of medium. Be careful to let the aggregates distribute evenly.

10. The cell aggregates will attach to the culture surface overnight. Once the aggregates attached, feed the culture with the neural induction medium every other day.

11. 3–4 day after attachment (around 10 days from human ES cell differentiation), columnar cells in the colony center start to form rosettes (**Figure 1C**). At this stage, the cells express Pax6 but not Sox1. Columnar epithelial cells proliferate quickly and form multilayered, neural-tube like rosettes in another 4 days (**Figure 1D**).

B. Differentiation of spinal motor neurons

1. At day 10 of differentiation, replace the medium and feed the cells every other day with fresh neural induction medium supplemented with RA at the final concentration of 0.1 μM.

2. At day 14, multilayered, neural tube-like rosettes are obvious. Rinse the rosette culture with PBS once. Detach the clusters by gently triturating the central part of the attached clusters. The central part of the clusters will easily detach from the culture surface while the peripheral flat cells remain attached.

3. Collect the rosette clumps into a 15 mL centrifuge tube. Triturate the clumps with a 5 or 10 mL serological pipette up and down twice, but not to break up the

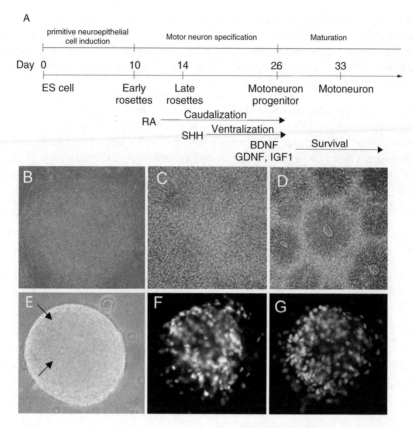

Figure 1 Directed differentiation of spinal motor neuron from human ES cells. (**A**) Motor neurons are specified in a two-step manner as is shown in the strategic scheme. (**B**) Phase contrast image showing a human ES cell colony grown on the mouse embryonic fibroblasts feeder layer. (**C**) The human ES cells are first differentiated to early neuroectodermal cells characterized by early rosettes around 10 days. (**D**) They are then treated with RA for neural induction and caudalization. Typical neural tube-like rosettes form around 14 days. (**E**) The detached rosettes form neuroepithelial clusters in neural induction medium supplemented with B27, SHH and RA. (**F, G**) The neuroepithelial clusters are adhered to the laminin substrate (around day 20) in the presence of RA and SHH for further neuronal differentiation. Olig2+ motoneuron progenitors (**F**) peak around 3.5 weeks, and HB9+ motoneurons (**G**) around 4–5 weeks after differentiation from ES cells.

clumps. Centrifuge at 50 g (approx 600–800 rpm for a small benchtop centrifuge) for 2 min at room temperature.

4. Aspirate off the supernatant, re-suspend the cluster in 5 mL neural induction medium containing B27, sonic hedgehog (SHH) (100 ng/mL) and all-trans retinoic acid (RA) (0.1 µM), and transfer the culture to a T25 flask (cells from 3 wells may be added to one T25 flask).

5. Feed the cultures by replacing half to 2/3 of the medium with the neural induction medium containing B27 and SHH/RA every other day. Spherical neuroepithelial clusters form after one or two days, typically 100–200 μm in diameter (**Figure 1E**). If the spheres grow bigger than 200 μm in diameter, break them using a Pasteur pipette.

6. On day 22 of human ES cell differentiation, Olig2-expressing motor neuron progenitors appear (**Figure 1F**). Plate the neuroepithelial clusters onto laminin-coated Petri dishes. For immunostaining, the cells may be plated onto glass coverslips coated with polyornithine and laminin (2–4 clusters/coverslip). Feed the attached cells with neural induction medium supplemented with RA (0.1 μM), SHH (100 ng/mL), cAMP (1 μM), ascorbic acid (200 ng/mL), and laminin (1 μg/mL).

7. On day 24 (or 2 days after plating the cells onto the growth substrate), neurites are extending out of the cluster.

8. Continue feeding the attached cells with neural induction medium supplemented with BDNF, GDNF, IGF1 (10 ng/mL, respectively), cAMP (1 μM) and AA (200 ng/mL). Differentiated motor neurons extend long axons from the cluster and express βIII-tubulin. Hb9 positive cells appear at 28 days and peak at 35 days from human ES cell differentiation (**Figure 1G**).

9. The cells can be maintained in the same medium but with low concentration of RA and SHH (0.01 μM and 10 ng/mL, respectively). More mature ChAT positive motor neurons appear around 6 weeks after human ES cell differentiation (day 40–42).

Troubleshooting

Human embryoid bodies attach and spontaneously differentiate. How do I prevent this?

This is usually resulted from too many contaminated feeder fibroblasts. If there are a lot of contaminating fibroblasts, they will attach and reform a sparse feeder layer, thus resulting in attachment of the human EBs. This can be solved by transferring the aggregates to a new flask once the fibroblasts have adhered to the plastic, which usually takes place within 24 hr.

Human EBs do not attach to plastic surface in the 'serum free' medium at day 6. What should I do?

If the aggregates do not attach, they may be plated onto the plate coated with laminin (20 μg/mL). Alternatively, addition of 10% FBS into the culture overnight will promote the attachment of the aggregates. Use serum as short time as possible as serum inhibits neural induction.

No typical rosettes formed. Why?

The most common cause is that the human ES cells are partially differentiated or that there is too much damage done to cells during aggregation or embryoid body formation. Also, avoid adding RA before day 8 as RA inhibits neuroepithelial induction, which contrasts neural differentiation from mouse ES cells.

I see a significant amount of non-neural cell contamination. How do I prevent this?

The above procedure typically yields 90–95% neuroepithelial cells based on fluorescence activated cell sorting (FACS) analysis using Pax6 as a marker. The most common source is partially differentiated human ES cells which will produce "bad colonies". Using a object marker in a phase contrast scope, mark partially differentiated human ES cell colonies and/or the "bad colonies" during neural differentiation, then scratch them with a curved Pasteur pipette. Additionally, when removing neural rosettes, add fresh neural induction medium with dispase at a final concentration of 0.2–0.5 mg/mL. Incubate the culture for 15–30 min in a CO_2 incubator until the clump of rosettes retract while the surrounding flat cells are still attached. Gently swirl the flask to release rosette clumps.

I observe poor survival of motor neuron progenitors during preparation for neuronal differentiation. How might this be avoided in repeat efforts?

Enzymatic disaggregation of neuroepithelial spheres can result in significant cell death, particularly of the motor neuron lineage. Addition of B27 and SHH/RA and plating at a high density (30,000 cells/11 mm coverslip) will help. We usually plate small clusters (100–200 µm) for differentiation, which will attach and form monolayer culture after a couple of days. Partial dissociation of progenitor clusters with accutase (for 3–5 min) before plating will facilitate formation of monolayer cells which is helpful for quantification purpose.

References

Jessell TM. (2000). Neuronal specification in the spinal cord: inductive signals and transcriptional codes. *Nat Rev Genet,* **1**: 20–29.

Li XJ, ZW Du, ED Zarnowska, M Pankratz, LO Hansen, RA Pearce and SC Zhang. (2005). Specification of motoneurons from human embryonic stem cells. *Nat Biotechnol* **23**: 215–221.

Yan Y, D Yang, ED Zarnowska, Z Du, B Werbel, C Valliere, RA Pearce, JA Thomson and SC Zhang. (2005). Directed differentiation of dopaminergic neuronal subtypes from human embryonic stem cells. *Stem Cells* **23**: 781–790.

Zhang SC, M Wernig, Duncan, O Brustle and JA Thomson. (2001). In vitro differentiation of transplantable neural precursors from human embryonic stem cells. *Nat Biotechnol* **19**: 1129–1133.

15, Part A

Gene targeting in human embryonic stem cells: Knock out and knock in by homologous recombination

THOMAS P. ZWAKA

Center for Cell and Gene Therapy and the Department of Molecular and Cellular Biology, Baylor College of Medicine, One Baylor Plaza, Houston, TX 77030, USA

Introduction

Gene targeting takes advantage of a cellular DNA repair process known as homologous recombination, which allows a stretch of DNA altered *in vitro* to be precisely inserted into a defined point in the target genome. The introduced modifications could include the removal of exons (knock outs) and/or the introduction of reporter genes or other cDNA sequences (knock ins). The modified DNA must be flanked by DNA that is identical (homologous) to sequences in the target gene. Numerous studies have shown that the length of the homologous region is a critical factor for determining the effectiveness of homologous recombination; targeting was found to require at least 500 base pairs (bp) of sequence, and targeting efficiency increased exponentially as the homologous sequence increased from 2 to 10 kilobase pairs kb, although further increases were less dramatic with regions longer than 14 kb (Bronson, 1994; Thomas and Capecchi, 1987).

Once the desired sequences have been generated *in vitro*, transfection is used to introduce the recombinant DNA into the cell, where it recombines with the homologous genomic region, replacing the normal genomic DNA with recombinant DNA containing the desired genetic modifications. Homologous recombination is a very rare event in cells, and thus a powerful selection strategy must be used to identify recombinant cells (Vasquez *et al.*, 2001). Most commonly, the introduced construct contains an additional gene encoding antibiotic resistance, allowing antibiotic selection of cells that have incorporated the recombinant DNA. However, antibiotic resistance only reveals that the cells have taken up the recombinant DNA and incorporated it somewhere in the genome. To facilitate selection of cells in which homologous recombination has occurred, the end of the recombination construct often includes the thymidine kinase (TK) gene from the herpes simplex virus. This is eliminated during homologous recombination, allowing ganciclovir to be used for negative selection against randomly integrated clones.

However, clonal propagation of human embryonic stem (ES) cells is difficult, which makes gene targeting in these cells more challenging (Thomson *et al.*, 1998). Techniques like gene targeting by homologous recombination rely heavily on the identification of extremely rare clones, which must be propagated, expanded and analyzed. In the case of human ES cells, this strategy requires extensive cell culture experience and a good strategy for identifying homologous recombinant clones. Previous experience with homologous recombination in murine ES cells is a bonus, but keep in mind that human ES cells grow at a much slower rate than murine ES cells, meaning that the entire procedure will take significantly longer (months instead of weeks) (Urbach *et al.*, 2004; Zwaka and Thomson, 2003).

Overview of the protocol

Gene targeting in human ES cells involves: (a) design of the targeting vector and establishment of a screening PCR (polymerase chain reaction) protocol for identifying targeted clones; (b) electroporation of the cells with the gene targeting construct; (c) drug selection and picking of individual clones; (d) expansion of clones and identification of targeting events; (e) characterization of targeted cells.

Materials, reagents, and equipment

Human embryonic stem cells (this technique has not been tested on any lines other than WiCell lines (H1.1 and H9)

BioRad Genepulser II System (Biorad)

0.4 cm gap electrocuvettes (Biorad)

10 cm tissue culture dishes (Nunc)

Six-well tissue culture dishes (Nunc)

24-well tissue culture dishes (Nunc)

28-well tissue culture dishes (Nunc)

Antibiotic resistant mouse embryonic fibroblasts (Open Biosystems)

Antibiotic stocks (G418: Sigma)

Gancyclovir (Sigma)

Phosphate buffered saline (Invitrogen/Gibco #70013-032)

0.05% trypsin (Invitrogen/Gibco #25300-054)

Human ES cell medium (WiCell protocol, www.wicell.org)

Protocol

A. Design of the targeting vector and establishment of a screening PCR protocol for identifying targeted clones

1. Amplify short (1.5–2 kb) and long (5–10 kb) homologous arms directly from genomic DNA isolated from the target human ES cell line.

2. Subclone these fragments into a PCR cloning vector (e.g. the TOPO vector from Invitrogen), sequence the inserts and choose correct clones.

3. Transfer the inserts into the gene targeting vector such that the arms flank the selection cassette in the proper orientation, with the tyrosine kinase TK cassette at the end of the long arm (see **Figure 1**).

4. Amplify a "test short arm" fragment that is slightly longer than the targeting short arm (2–3 kb) and clone this fragment into gene targeting vector; this will be used as the test vector for establishment of the screening PCR (see **Figure 1**).

5. Transfect human cells with the test vector using the method of your choice (e.g. electroporation, see below) and establish stable clones.

6. Isolate genomic DNA from one of these clones and mix with wild type genomic DNA at different ratios (1:1, 1:4, 1:8, 1:16 and 1:32).

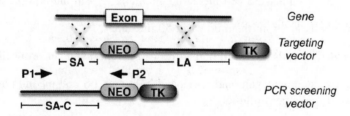

Figure 1 Conventional gene targeting vector contains short homologous arm (SA; 1.5–2 kb) and long homologous arm (LA; 5–10 kb). These arms flank the selection cassette (NEO) in the proper orientation, with the TK (negative selection) cassette at the end of the long arm. A "test short arm" (SA-C) fragment that is slightly longer than the targeting short arm (2–3 kb) cloned into gene targeting backbone vector can be used as the test vector for establishment of the screening PCR (screening PCR primers P1 and P2).

7. Choose a primer that anneals in the selection cassette and a reverse primer that anneals outside the "real" short arm but within the test short arm, and perform PCR amplification. You should be able to generate PCR products from undiluted genomic DNA isolated from the stable clones. Use 50–100 ng template DNA and optimize the PCR procedure such that specific PCR products are generated from the 1:8 or 1:16 dilutions; the more sensitive the PCR, the more effective your screening will be later on.

B. Electroporation with gene targeting construct

1. Expand human ES cells in ten to fifteen 10 cm dishes on murine feeders with daily feedings; the cells are ready for electroporation when many mid-size colonies are visible per dish.

2. Prepare drug-resistant mouse feeders in 10 cm dishes (at least three dishes) the day before electroporation.

3. 4 hr before electroporation, feed the human ES cells with fresh medium and change the medium in the feeder dishes to human ES cell medium (10 mL medium per 10 cm dish).

4. Prepare linearized targeting vector, resuspend 0.4 mg linearized plasmid (dissolved in Tris, no EDTA) in 3 mL phosphate buffered saline (PBS), filter through 0.2 μm filter and store at 4°C until use.

5. Wash cells once with PBS, add 3 mL of freshly thawed and pre-warmed trypsin and incubate at 37°C for 5 min (or until a single cell suspension is obtained). This step is very critical; do not over-trypsinize, but make sure you have a single cell suspension.

6. Add 7 mL of human ES cell medium containing trypsin inhibitor (Sigma).

7. Pool the cells and spin them down in 50 mL tubes.

8. Resuspend the cells in 10 mL human ES cell medium and perform cell counting.

9. Spin the cells down again and resuspend in human ES cell medium at a density of 6×10^7 cells per milliliter.

10. Mix 0.5 mL of these human ES cells and 300 μL of plasmid in PBS (from step 4 above) in an electroporation cuvette; prepare as many cuvettes as you have cells, and incubate the cuvettes for 10 min at room temperature.

11. Electroporate cells with a single 320 V 200 μF pulse and then incubate cells for an additional 10 min at room temperature.

12. Carefully transfer the electroporated cells to the plates containing the drug-resistant mouse feeders (pipette 800 μL directly from the cuvette onto the medium in the 10 cm dish).

13. Return cells to the incubator.

C. Drug selection and picking of individual clones

1. Change medium 24 hr after electroporation and daily thereafter.

2. Begin G418 selection (typically 50 μg/mL in medium on day 2, then double the concentration of G418 1 week post-electroporation.

3. Begin selection with freshly-prepared ganciclovir (typically 1 mM) on day 4 and continue this selection with freshly-prepared ganciclovir until the colonies are picked.

4. Colonies should be evident after 2–3 weeks, and cells are typically ready for picking by the fourth week.

5. Count the number of colonies (you should see between 10–50 colonies per 10 cm dish) and prepare drug-resistant MEFs in a corresponding number of wells in 48-well dishes; on the day of picking, change the medium in the 48-well dishes to human ES cell medium.

6. Change the medium in the 10 cm dishes to fresh human ES cell medium

7. Transfer the 10 cm dishes to a laminar air hood containing a dissecting microscope.

8. Using the tip of a P200 pipette, carefully (and separately) scrape each colony until it consists of multiple small clumps. Suck the clumps into the pipette tip and transfer each colony to a single prepared well; repeat this process until all clones are transferred from the plate.

9. Feed the cells with G418-containing human ES cell medium 24 hr post-plating and daily thereafter.

D. Expansion of clones and identification of targeting events

1. Wait until most wells show a significant number of human ES cells (usually ~10 days after colony picking).

2. On the day before splitting, prepare 24-well plates with drug-resistant mouse feeders.

3. Wash the human ES cells with PBS, add collagenase IV (1 mg/mL) containing DMEM and incubate at 37°C for 15 min.

4. Remove the collagenase-containing PBS and use a 1 mL glass pipette to scrape the cells from each separate well into individual 2 mL Eppendorf tubes. Spin the cells down and resuspend them in 0.5 mL of human ES cell medium.

5. Remove the MEF medium from the 24-well plates and separately pipette the clones from the tubes into individual wells.

6. Feed the cells with G418-containing human ES cell medium 24 hr later and daily thereafter.

7. Wait until most wells show a significant number of human ES cells (usually ~7 days).

8. On the day before splitting, prepare 12-well plates with drug-resistant feeders; prepare twice as many wells as you have clones.

9. Wash the cells with PBS, add collagenase IV (1 mg/mL) containing DMEM and incubate at 37°C for 15 min.

10. Remove the collagenase-containing medium and use a 1 mL glass pipette to scrape the cells from each well into individual 2 mL Eppendorf tubes. Spin the cells down and resuspend them in 2 mL of human ES cell medium.

11. Remove the MEF medium from the 12-well plates and pipette the individual clones directly into separate wells, one clone per well.

12. Feed the cells with G418-containing human ES cell medium 24 hr later and daily thereafter.

13. Wait until most wells show a significant number of human ES cells (usually ~7 days).

14. Harvest cells from one well per clone, pool the clones based on the sensitivity of your screening PCR, and isolate genomic DNA.

15. Perform PCR and identify positive pools.

16. Split individual clones from positive clones into six-well plates, keep clones under drug selection and feed daily.

17. When the six-well plates are ready, re-split them again into six-well plates; transfer half of the cells from each clone into separate Eppendorf tubes and isolate genomic DNA from these cells.

18. Perform screening PCR and identify the positive clones.

19. Expand and freeze these positive clones.

E. Characterization of targeted cells

1. Make sure to freeze down at least three vials of low passage cells for each positive clone.

2. Isolate genomic DNA from each clone and perform genomic Southern blotting with the following three probes:

   a. one in the selection cassette to confirm a single integration event;

   b. one upstream of the targeting vector; and

   c. one downstream of the targeting vector.

3. Since these new cell lines have been propagated from individual cells, they must be fully characterized as human ES cell line subclones. The characterization should include:

   a. karyotyping (see **Chapter 8**);

   b. teratoma formation assays (see **Chapter 9**); and

   c. *in vitro* differentiation assays e.g. embryoid body formation (See **Chapters 10 and 11**).

Troubleshooting

I am having difficulty cloning the targeting vector. Are there other cloning options besides long-distance genomic PCR?

As it is crucial that you choose the right primer combination, you should test several different primer combinations before seeking an alternative cloning method. However, there is currently no evidence that isogenic (from the same cell line) genomic DNA is essential to the success of homologous recombination in human ES cells. Therefore, as an alternative to long-distance PCR, you may use a bacterial artificial chromosome (BAC) clone containing the region of interest as a PCR template, or you may even use BAC fragments directly for cloning of the targeting vector. The latter option has the advantage of not introducing mutations in the homologous arm, which may be a complication of PCR-based methods See **Part C** of this chapter for further details of BACS transgenics in human ES cells.

My screening PCR only detects clones when they are not diluted with wild type DNA. Can I still perform the electroporation?

Yes, you can perform the electroporation step without pooling the clones. The use of pools for DNA isolation and screening is intended to reduce the number of clones during the initial screening rounds. Human ES cells are difficult to maintain, and therefore an effective and robust method for screening positive events is helpful. However, the screening can be performed on individual, un-pooled clones.

After electroporation, I have only a very few clones. What should I do?

Make sure that you are using the correct concentration of antibiotics. Even different passage numbers of the same human ES cell line can have different levels of antibiotic resistance, so it may be necessary for you to optimize the antibiotic concentration by calculating "kill curves". Also, some human ES cell lines are simply more difficult to electroporate than others. Try different cell lines and passage numbers in an effort to identify cell lines yielding a decent number of clones.

I get a lot of colonies, but most of them die after picking.

Optimize your picking protocol by practicing multiple times. With some experience, you will be able to achieve a clone survival rate of more than 70%. Perform the entire procedure with single colonies plated at limited dilutions. Prior to targeting, you should establish the screening PCR using test clone containing a longer fragment of the short arm; this is an ideal opportunity to collect experience with the picking procedure. It may also be helpful to incubate the cells with collagenase IV for 10–15 min and then switch back to human ES cell medium. This treatment makes the colonies "softer" and easier to pick.

We were able to obtain clones and to propagate them, but none had undergone homologous recombination.

Homologous recombination is a very ineffective method and only a very small proportion of cells will be positive. In murine ES cells, sometime several hundred clones must be screened in order to identify positive clones. The efficiency in human ES cells is even lower. Be careful not to overgrow cells prior to electroporation; they must be very healthy and undifferentiated. Beyond that, it may be necessary to screen many clones, and you may find it helpful to perform several electroporation experiments in a row, thus generating a large number of initial colonies. Alternatively, the introduced transgene may be toxic or the introduced mutation may be lethal.

References

Bronson SK and O Smithies. (1994). Altering mice by homologous recombination using embryonic stem cells. *J Biol Chem* **269**: 27155–27158.

Thomas KR and MR Capecchi. (1987). Site-directed mutagenesis by gene targeting in mouse embryo-derived stem cells. *Cell* **51**: 503–512.

Thomson JA, J Itskovitz-Eldor, SS Shapiro, MA Waknitz, JJ Swiergiel, VS Marshall and JM Jones. (1998). Embryonic stem cell lines derived from human blastocysts. *Science* **282**: 1145–1147.

Urbach A, M Schuldiner and N Benvenisty. (2004). Modeling for Lesch–Nyhan disease by gene targeting in human embryonic stem cells. *Stem Cells* **22**: 635–641.

Vasquez KM, K Marburger, Z Intody and JH Wilson. (2001). Manipulating the mammalian genome by homologous recombination. *Proc Natl Acad Sci U S A* **98**: 8403–8410.

Zwaka TP and JA Thomson. (2003). Homologous recombination in human embryonic stem cells. *Nat Biotechnol* **21**: 319–321.

15, Part B

RNA interference in human embryonic stem cells

M. WILLIAM LENSCH, ASMIN TULPULE AND HOLM ZAEHRES
Division of Hematology/Oncology, Children's Hospital Boston, Boston, MA, USA
Department of Biological Chemistry and Molecular Pharmacology, Harvard Medical School, Boston, MA, USA
Harvard Stem Cell Institute, Cambridge, MA, USA

Introduction

Human embryonic stem (ES) cells present unique and powerful opportunities to study aspects of early development including cellular lineage specification and genomic imprinting (Lensch and Daley, 2006). Additionally, human ES cells are promising for the creation of cellular therapeutics and also as platforms for the refined, preclinical interrogation of drugs in a human-specific system. Work in human ES cells has lagged behind comparable efforts in non-human species due to a lack of traditional systems for evaluating genetic gain and loss-of-function studies (Lensch and Daley, 2004). Of particular relevance to this chapter, an ability to evaluate the impact of reduced gene expression in the human ES cell system would indeed prove valuable. Homologous recombination (HR) is a tenable undertaking in human ES cells (Zwaka and Thomson, 2003) but has proven challenging due to the low cloning efficiency of human ES cells and difficulties in electroporation. However, when optimized, HR will no doubt prove to be a widely used strategy for the modification of human ES cells. However, one thing that is lacking in HR systems is an ease of use.

While HR constructs are capable of being developed in many different ways, none would be considered particularly rapid and all require downstream selection or screening in order to identify the rare, recombination events that yield the properly modified cell of interest. As an additional caveat, if the gene of interest impacts the maintenance of the pluripotent state in target ES cells, an investigator could be hard pressed to capture the fruits of HR in a timeframe that also facilitates their rigorous study (if at all). A system capable of being assembled in a relatively short span of time, of reasonable target specificity, durability, and importantly, affecting a large proportion of cells if not the majority within a given population so as to reduce or eliminate the need for selection or screening, would be preferable in certain applications compared to HR. RNA interference or RNAi is such a system.

Double-stranded RNA molecules of certain sequences, termed small interfering RNAs (siRNAs), have been discovered that allow specific silencing of genes in many biological systems from *Caenorhabditis elegans* to mammalian cells (Fire *et al.*, 1998; Elbashir *et al.*, 2001). Though transfection of chemically-synthesized siRNAs has proven to be quite valuable for evaluating gene function in human ES cells (Vallier *et al.*, 2004; Hyslop *et al.*, 2005), in our hands, lentiviral-based RNA interference (RNAi) is preferable (Zaehres *et al.*, 2005). Advantages of lentiviral delivery include the durability of the knockdown (as the construct integrates genomically), the high viral titers achievable that permit infection of a large majority of the cellular population under study, and the ability to more accurately gauge the copy number in cellular target by adjusting the multiplicity of infection (MOI). Such systems use the U6 promoter; one of several different RNA polymerase III-based promoters and yield siRNAs as short hairpin RNAs (shRNA) (Brummelkamp *et al.*, 2002; Lee *et al.*, 2002; Yu *et al.*, 2002; Paddison *et al.*, 2002; Paul *et al.*, 2002; Sui *et al.*, 2002; Miyagishi and Taira, 2002).

Methodologies for the generation and use of lentiviral systems have been described elsewhere [such as (Tiscornia *et al.*, 2006) and on the website of the Broad Institute's RNAi Consortium http://www.broad.mit.edu/genome_bio/trc/publicProtocols.html]. In short, when properly pseudotyped with an appropriate coat protein (such as the glycoprotein of vesicular stomatitis virus or VSVg) lentivirions are capable of being highly concentrated via ultracentrifugation to produce supernatants with titers approaching or exceeding 1×10^9 infectious units per milliliter. Also, the use of lentivirus is preferable to retroviral constructs in certain applications due to an increased resistance to silencing in differentiating cells as well as a capacity for genomic integration in non-dividing targets (Cherry *et al.*, 2000; Lois *et al.*, 2002; Pfeifer *et al.*, 2002).

Materials, reagents and equipment

We obtain our DNA oligonucleotides from Integrated DNA Technologies (www.idtdna. com) and order the 100 nmol scale with PAGE purification and $5'$ phosphorylation. Lentivirus vectors are as described under Web Resources. **Table 1** gives reagents.

Protocol

A scheme highlighting the major steps of the protocol is shown in Figure 1. Tuschl and colleagues have shown that the anti-viral, interferon response that cells develop when exposed to long, double-stranded RNAs is obviated by using shorter duplexes of 21 to 23 nucleotides including 3-prime overhangs of two nucleotides at both ends (Elbashir *et al.*, 2001). Online search engines are available for mining transcript sequences in order to identify appropriate duplexes (in both coding as well as untranslated regions). These include the siRNA Selection Program at the Whitehead Institute for Biomedical Research (Cambridge, Massachusetts): (http://jura.wi.mit.edu/siRNAext/) (Yuan *et al.*, 2004). The sequence of the final core siRNA duplex should be $AAGN_{18}TT$. Three distinct siRNA sequences are preferable for each mRNA target though experiments using fewer remain worthwhile (see below in troubleshooting).

Table 1 Reagents

293T cells	ATCC CRL-11268
Annealing buffer	
Potassium acetate	Sigma P1190–100g
HEPES	Sigma H4034–25g
Potassium hydroxide	Sigma P5958–250g
Magnesium acetate	Sigma M2545–250g
Reagents for transduction/maintenance of human ES cells	
Protamine sulfate	Sigma P3369
Matrigel	BD Bioscience 354234
Basic FGF	Invitrogen PHG0023
PBS (Ca and Mg-free)	Invitrogen 14190250
Collagenase type-IV	Invitrogen 17104-019
Trypsin (0.05%)	Invitrogen 25300-054

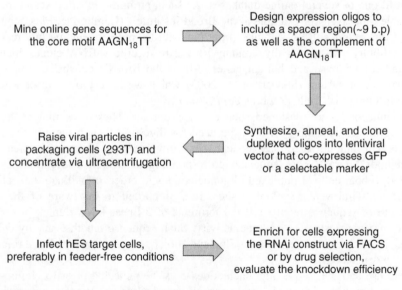

Figure 1 Flowchart of the overall scheme for the production of lentivirally-delivered RNAi as described in the text.

Descriptions of the design of siRNA expression systems have been previously published [such as: (Li and Rossi, 2005; Sano *et al.*, 2005)]. We have used both the Lentilox (Rubinson *et al.*, 2003) and Lentihair (Stewart *et al.*, 2003) systems with good results in hESCs. Each of these systems makes use of the U6 promoter; to produce siRNAs as short hairpin RNAs (shRNA). In shRNAs, both the sense and anti-sense siRNAs are produced as one transcript from a single, common promoter and contain an intervening loop of non-annealing sequence (we routinely use loops of nine nucleotides, see below).

The transcript anneals as a stem-loop structure capable of being processed to its final, active form (Brummelkamp *et al.*, 2002; Paddison *et al.*, 2002; Paul *et al.*, 2002; Sui *et al.*, 2002; Yu *et al.*, 2002). In addition, both systems make use of an independently transcribed marker such as green fluorescent protein (GFP) to allow for the purification of infected cells by fluorescence activated cell sorting (FACS) and a rough titration of siRNA levels.

The overall design of the oligonucleotides to be expressed from the U6 promoter is:

Sense oligonucleotide:	5'-T(GN$_{18}$)(TTCAAGAGA)(N$'_{18}$C)TTTTTT-3'
Note:	N$'_{18}$ = the complement of the proximal N$_{18}$ sequence
Anti-sense oligonucleotide:	Complement of the overall sense oligonucleotide

The 3'-poly-thymidine sequence (at least four nucleotides) is the terminator signal. The loop sequence shown above (5'-TTCAAGAGA-3') is taken from (Brummelkamp *et al.*, 2002). Naturally-occurring loop sequences have proven effective in siRNA applications though variable results are not uncommon (Sano *et al.*, 2005).

It is possible that an siRNA construct corresponding to a gene of interest is already present in one of several online databases. As such, an initial *in silico* search may be a valuable timesaver, for example at the Broad Institute's (Cambridge, Massachusetts) RNAi Consortium shRNA Library (http://www.broad.mit.edu/genome_bio/trc/). The Broad's library will eventually contain 1.5×10^5 specific shRNA clones including those targeting mouse and human genes. All of the Broad's constructs employ the Lentihair vector system (Stewart *et al.*, 2003) which we have proven to be effective for use in human ES cells (Zaehres *et al.*, 2005).

Oligonucleotides are designed such that appropriate cohesive termini are present following duplex annealing without the need for digestion though there is the caveat that a certain amount of duplex concatamer formation may be present when using this method. Alternatively, sequences corresponding to desired restriction sites may be added to both ends of duplexed oligonucleotides in order to facilitate their cloning into the lentiviral vector's cloning sites. It is important to be aware of the minimal length of sequence required at the terminus of a linear DNA duplex in order to achieve maximal restriction enzyme activity. Such sequence lengths vary by enzyme and the website of the company New England Biolabs may be consulted as a reference (http://www.neb.com).

The sense and anti-sense oligonucleotides must be annealed prior to cloning (and restriction digestion if needed). Add each oligonucleotide to a final concentration of 6 μM (60 pmol/μL) in annealing buffer (100 mM K-acetate, 30 mM HEPES-KOH pH 7.4, 2.0 mM Mg-acetate). Add mixed oligonucleotides to a microcentrifuge tube and heat to 95°C for 10 min in a heating cell with a removable aluminum block. Remove the microcentrifuge tube and mix quickly by flicking the tube prior to replacing it into the heating block. Remove the entire aluminum block containing the microcentrifuge tube from the heating cell and set it aside to cool slowly to room temperature (~30 min). Allow the block to slowly cool in an area free of drafts (such as from nearby centrifuges). Alternatively, one may anneal oligos via gradual ramp-down of temperature using a PCR machine.

Ligate the prepared oligonucleotide duplex into the lentiviral vector of choice and transform bacteria in order to raise an appropriate amount of material for use in virus production. Several colonies should be picked for verification using DNA sequencing. Though rare, mutations may be introduced during construct amplification in bacterial hosts. Certain mutations are more deleterious than others [see (Sano et al., 2005)] and it is best to be validate the sequence of any clone as an incorrect siRNA sequence may drastically reduce the efficiency of the desired gene knockdown.

The methodology used to produce and harvest viral particles using mammalian host cells (such as human embryonic kidney-293T) has been extensively described elsewhere and will not be revisited here. Viral titering and the replication competence test should be performed on 293T cells prior to the infection of human ES cells. For details, please see (Tiscornia et al., 2006) or the Broad Institute's RNAi Consortium (http://www.broad.mit.edu/genome_bio/trc/publicProtocols.html).

It is recommended that transduction of human ES cells be performed in feeder-free conditions (on Matrigel, BD Biosciences, Rockville, Maryland) with conditioned media from mouse embryonic fibroblasts supplemented with bFGF, as described in (Daheron et al., 2004; Zaehres et al., 2005). As the majority of human ES lines exhibit poor replating efficiency as single cells and are passaged in large clumps either mechanically or after treatment with collagenase IV, it can be difficult to accurately assess human ES cell number for the calculation of the multiplicity of infection (MOI). For transduction, we recommend harvesting the human ES cells in small clumps (50–200 cells) with collagenase, followed by trypsin treatment of a small aliquot to accurately assess cell number. For trypsin-adapted human ES lines that demonstrate high replating efficiency as single cells, trypsinization to a single cell suspension followed by cell counting is easily achieved. To start, prepare six-well Matrigel plates (three wells per virus) and plate 100,000 human ES cells per well. After 8–12 hr, add lentiviral particles at MOI 10, 50, and 100 in three separate wells (MOI 10 corresponds to ten viral particles per target human ES cells, here 1×10^6 virions for 1×10^5 human ES cells) along with 6 µg/mL protamine sulfate to improve the infection efficiency. Incubate at 37°C overnight. The following day, wash the human ES cells two times with PBS and bleach both the original media and the washes, disposing of them as BL-2+ waste. The cells can then be treated as non-BL2+.

Knockdown validation should be performed in a variety of ways. We suggest expanding the infected cells to obtain sufficient numbers for FACS analysis (Figure 2) and/or sorting (infection efficiencies can vary greatly) followed by immediate lysis to analyze protein levels via Western blotting. The use of real-time quantitative PCR (QRT-PCR) is rapid and yields good results though with the caveat that co-analyzing target protein levels via Western blotting or some other quantitative method yields the most convincing data. We prefer the primer/probe method of QRT-PCR though other varieties wherein fluorescent nucleotides are introduced into the amplification products are also effective and often less expensive.

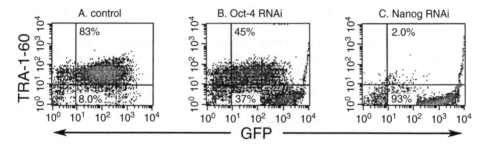

Figure 2   Human embryonic stem cells (WA09/H9) (Thomson *et al.*, 1998) were transduced with lentivirus vectors expressing shRNAs to target the self-renewal transcription factors hOct4 (NCBI sequence NM_002701) and hNanog (NCBI sequence NM_024865) (multiplicity of infection (MOI) = 50); for details see Zaehres *et al.* (2005). 7 days after transduction, cells were stained for the TRA-1-60 antigen as a marker associated with undifferentiated ES cells. The figure shows FACS histograms of the human embryonic stem cells infected with green fluorescent protein (GFP) expressing lentivirus vectors (>80% transduction efficiencies). The amount of GFP (*x*-axis) positive, TRA-1-60 (*y*-axis) negative cells are the result of differentiation and rise from 8% in the control vector transduced population (**A**) to 37% in the Oct4 knockdown population (**B**) and to 93% in the Nanog knockdown population (**C**). The RNA levels for Oct4 and Nanog were reduced accordingly below 10% as measured by quantitative real time PCR. Annexin-V as a marker of apoptotic cells was only slightly increased in the transduced cell populations.

Troubleshooting

My knockdown is not working as well as I had hoped. What can be done to create a more robust effect?

Typically, multiple siRNA constructs are used concurrently in order to obtain a robust knockdown. If two or three canonical motifs are present within the mRNA of interest, co-infection with all of them in an equimolar ratio will elicit a stronger response. In our experience, there has been significant variability (from 50% to nearly 100%) in the effectiveness of knockdown between targets and we have observed a trend that siRNAs targeting the 3' untranslated region show the greatest efficacy.

I am concerned that my construct might be generating off-target effects. How might this be determined?

While the concern that RNAi sometimes yields ambiguity due to the inadvertent knockdown of another, unintended RNAs (as sometimes occurs with other methods for manipulating the expression or products of a certain gene) we find that the inclusion of a few key controls is nearly always sufficient to evaluate such off-target effects. Typically, control RNAi experiments directed against a non-hESC gene (such as luciferase or GFP where appropriate) will provide a reasonable measure of specificity. As with any experiment, controls may be of an even greater variety and specificity including other genes or even genome-wide expression analysis. One caveat here of course is that such controls may be difficult to interpret as it may be impossible to distinguish off-target RNAi effects from *bona fide* genetic interactions due to the downstream effects of a specific knockdown.

The knockdown I wish to perform promotes cellular differentiation. What problems might this cause at later timepoints?

One caveat of differentiation-promoting changes in human ES cells is that members of the population maintaining pluripotency (by resisting the effects of RNAi for whatever reason) will generally grow more quickly than the differentiating cells. Even with a highly-efficient knockdown (in terms of the percentage of the population expressing the RNAi construct) the native human ES cells may overgrow the population in a relatively short period of time. One method for dealing with this is to use knockdown constructs that also contain a co-expressed GFP tag (and an upstream internal ribosomal entry sequence or IRES). Such systems permit fluorescent cell sorting in order to enrich for cells expressing the RNAi construct. We have employed such cell sorting-based systems for enrichment with good results. It is also possible to use a conditional RNAi based systems (Ventura *et al.*, 2004; Wiznerowicz and Trono, 2003), thereby allowing one to grow sufficient numbers of infected cells for analysis. Alternatively, an antibiotic selection marker (puro^R/hygro^R) co-expressing vector can be used to enrich for a pure knockdown population.

References, web resources, additional notes

Brummelkamp TR, R Bernards and R Agami. (2002). A system for stable expression of short interfering RNAs in mammalian cells. *Science* **296**: 550–553.

Cherry SR, D Biniszkiewicz, L van Parijs, D Baltimore and R Jaenisch. (2000). Retroviral expression in embryonic stem cells and hematopoietic stem cells. *Mol Cell Biol* **20**: 7419–7426.

Daheron L, SL Opitz, H Zaehres, WM Lensch, PW Andrews, J Itskovitz-Eldor and GQ Daley. (2004). LIF/STAT3 signaling fails to maintain self-renewal of human embryonic stem cells. *Stem Cells* **22**: 770–778.

Elbashir SM, J Harborth, W Lendeckel, A Yalcin, K Weber and T Tuschl. (2001). Duplexes of 21-nucleotide RNAs mediate RNA interference in cultured mammalian cells. *Nature* **411**: 494–498.

Fire A, S Xu, MK Montgomery, SA Kostas, SE Driver and CC Mello. (1998). Potent and specific genetic interference by double-stranded RNA in Caenorhabditis elegans. *Nature* **391**: 806–811.

Hyslop L, M Stojkovic, L Armstrong, T Walter, P Stojkovic, S Przyborski, M Herbert, A Murdoch, T Strachan and M Lako. (2005). Downregulation of NANOG induces differentiation of human embryonic stem cells to extraembryonic lineages. *Stem Cells* **23**: 1035–1043.

Lee NS, T Dohjima, G Bauer, H Li, MJ Li, A Ehsani, P Salvaterra and J Rossi. (2002). Expression of small interfering RNAs targeted against HIV-1 rev transcripts in human cells. *Nat Biotechnol* **20**: 500–505.

Lensch MW and GQ Daley. (2004). Origins of mammalian hematopoiesis: in vivo paradigms and in vitro models. *Curr Top Dev Biol* **60**: 127–196.

Lensch MW and GQ Daley. (2006). Scientific and clinical opportunities for modeling blood disorders with embryonic stem cells. *Blood* **107**: 2605–2612.

Li MJ and JJ Rossi. (2005). Lentiviral vector delivery of recombinant small interfering RNA expression cassettes. *Methods Enzymol* **392**: 218–226.

Lois C, EJ Hong, S Pease, EJ Brown and D Baltimore. (2002). Germline transmission and tissue-specific expression of transgenes delivered by lentiviral vectors. *Science* **295**: 868–872.

Miyagishi M and K Taira. (2002). U6 promoter-driven siRNAs with four uridine 3′ overhangs efficiently suppress targeted gene expression in mammalian cells. *Nat Biotechnol* **20**: 497–500.

Paddison PJ, AA Caudy, E Bernstein, GJ Hannon and DS Conklin. (2002). Short hairpin RNAs (shRNAs) induce sequence-specific silencing in mammalian cells. *Genes Dev* **16**: 948–958.

Paul CP, PD Good, I Winer and DR Engelke. (2002). Effective expression of small interfering RNA in human cells. *Nat Biotechnol* **20**: 505–508.

Pfeifer A, M Ikawa, Y Dayn and IM Verma. (2002). Transgenesis by lentiviral vectors: lack of gene silencing in mammalian embryonic stem cells and preimplantation embryos. *Proc Natl Acad Sci USA* **99**: 2140–2145.

Rubinson DA, CP Dillon, AV Kwiatkowski, C Sievers, L Yang, J Kopinja, DL Rooney, MM Ihrig, MT McManus, FB Gertler, ML Scott and L Van Parijs. (2003). A lentivirus-based system to functionally silence genes in primary mammalian cells, stem cells and transgenic mice by RNA interference. *Nat Genet* **33**: 401–406.

Sano M, Y Kato, H Akashi, M Miyagishi and K Taira. (2005). Novel methods for expressing RNA interference in human cells. *Methods Enzymol* **392**: 97–112.

Stewart SA, DM Dykxhoorn, D Palliser, H Mizuno, EY Yu, DS An, DM Sabatini, IS Chen, WC Hahn, PA Sharp, RA Weinberg and CD Novina. (2003). Lentivirus-delivered stable gene silencing by RNAi in primary cells. *Rna* **9**: 493–501.

Sui G, C Soohoo, B Affar el, F Gay, Y Shi, WC Forrester and Y Shi. (2002). A DNA vector-based RNAi technology to suppress gene expression in mammalian cells. *Proc Natl Acad Sci USA* **99**: 5515–5520.

Thomson JA, J Itskovitz-Eldor, SS Shapiro, MA Waknitz, JJ Swiergiel, VS Marshall and JM Jones. (1998). Embryonic stem cell lines derived from human blastocysts. *Science* **282**: 1145–1147.

Tiscornia G, O Singer and IM Verma. (2006). Production and purification of lentiviral vectors. *Nature Protocols* **1**: 241–245.

Vallier L, PJ Rugg-Gunn, IA Bouhon, FK Andersson, AJ Sadler and RA Pedersen. (2004). Enhancing and diminishing gene function in human embryonic stem cells. *Stem Cells* **22**: 2–11.

Ventura A, A Meissner, CP Dillon, M McManus, PA Sharp, L Van Parijs, R Jaenisch and T Jacks. (2004). Cre-lox-regulated conditional RNA interference from transgenes. *Proc Natl Acad Sci USA* **101**: 10380–10385.

Wiznerowicz M and D Trono. (2003). Conditional suppression of cellular genes: lentivirus vector-mediated drug-inducible RNA interference. *J Virol* **77**: 8957–8961.

Yu JY, SL DeRuiter and DL Turner. (2002). RNA interference by expression of short-interfering RNAs and hairpin RNAs in mammalian cells. *Proc Natl Acad Sci USA* **99**: 6047–6052.

Yuan B, R Latek, M Hossbach, T Tuschl and F Lewitter. (2004). siRNA Selection Server: an automated siRNA oligonucleotide prediction server. *Nucleic Acids Res* **32**: W130–W134.

Zaehres H, MW Lensch, L Daheron, SA Stewart, J Itskovitz-Eldor and GQ Daley. (2005). High-efficiency RNA interference in human embryonic stem cells. *Stem Cells* **23**: 299–305.

Zwaka TP and JA Thomson. (2003). Homologous recombination in human embryonic stem cells. *Nat Biotechnol* **21**: 319–321.

Web resources

siRNA Selection Program at the Whitehead Institute for Biomedical Research, Cambridge, MA:
http://jura.wi.mit.edu/bioc/siRNAext/
Lentivirus vectors described in Zaehres *et al.* (2005) are available through Addgene (Search for
   plasmids: daley):
http://www.addgene.org
Background material for the lentivirus siRNA expression vector system pLL3.7 at the Massachusetts Institute of Technology MIT Centre of Cancer Research:
http://web.mit.edu/jacks-lab/protocols/rnairesources.htm
http://web.mit.edu/jacks-lab/protocols/pSico.html
Broad Institute's (Cambridge, Massachusetts) RNAi Consortium shRNA Library:
http://www.broad.mit.edu/genome_bio/trc/

15, Part C

Generation of human gene reporters using bacterial artificial chromosome recombineering

ANDREW J. WASHKOWITZ

Columbia University, Sherman Fairchild Center, MC 2427, 1212 Amsterdam Ave, New York, NY 10027, USA

and

DAVID A. SHAYWITZ

Neuroendocrinology Division, Bullfinch 4, Massachusetts General Hospital, 55 Fruit Street, Boston, MA 02114, USA

Introduction

The study of gene function *in vivo* has been profoundly enhanced by the ability to insert a DNA sequence of interest into a specific chromosomal location (van der Weyden *et al.*, 2002; Capecchi, 2005). The inserted DNA often replaces all or part of an endogenous gene, in hope that the introduced DNA will be regulated in a parallel fashion. For example, if the inserted DNA encodes a fluorescent reporter such as GFP (or GFP-family proteins [Shaner *et al.* 2005]), then the introduction of GFP-encoding DNA in a specific developmentally-regulated locus can provide a useful readout of the expression of the endogenous gene (Yu *et al.*, 2003).

The ability to perform such gene targeting can be labor-intensive, and require extensive screening to detect what is often a rare homologous recombination event. A promising advance in this field has been the advent of bacterial artificial chromosome (BAC) recombineering, in which homologous recombination occurs in bacteria between the gene of interest (flanked by appropriate regions of homologous) and a BAC encompassing the target region (Copeland *et al.*, 2001; Yang and Seed, 2003; Valenzuela *et al.*, 2003; Testa *et al.*, 2004). In this scheme, homologous recombination occurs with relatively high frequency, and is comparatively simple to detect. The successfully "recombineered" BAC is subsequently purified and electroporated into mouse embryonic stem cells. The BAC will either incorporate randomly into the chromosomal DNA or will homologously recombine into the corresponding locus; since the large size of BACs allows for extensive regions of flanking homology (typically 50–100 kb), the frequency of BAC homologous recombination is expected to exceed that of conventional gene-replacement strategies. Of note, distinguishing between homologous BAC recombination and random BAC insertion is non-trivial, since the long flanking regions preclude use Southern-based approaches; that is, Southern analysis cannot distinguish these two categories of events. Instead, either quantitative PCR (qPCR) or fluorescent *in situ* hybridization (FISH) techniques have been employed (Yang and Seed, 2003; Valenzuela *et al.*, 2003).

While originally described for mouse embryonic stem cells, the BAC recombineering approach seems well-suited to address key questions in human stem cell biology; specifically, the ability to engineer fluorescent reporters that would signify the expression of developmental significant genes would facilitate the study of these genes, and might contribute to efforts designed to enhance the directed differentiation of human embryonic stem cells into specific cellular fates (Shaywitz and Melton, 2005). Moreover, the presence of long flanking regions both upstream and downstream of the gene insertion raised the possibility that even if a recombineered reporter BAC inserted randomly into the genome, the regulatory elements presumed to be present in the flanking regions might ensure "correct" reporter expression.

The protocol below describes the application of BAC recombineering to human embryonic stem cells.

Overview of protocol

Generation of recombineered BACs involves: (a) introduction of recombinase plasmid into the BAC-containing bacteria; (b) construction of suitable recombination template by introducing 5′ and 3′ homology arms into appropriate vector; (c) introduction of template into bacteria and selecting for recombinants; (d) confirmation of recombination by PCR; (e) recovery and preparation of recombineered BAC; (f) introduction of recombineered BAC into human ES cells; (g) antibiotic-based selection of BAC-containing colonies; (h) expansion and PCR validation of positive clones. The BAC-containing hES lines emerging from this procedure can then be subject to suitable functional testing to confirm accuracy of transgene expression. If desired, FISH or qPCR analysis can also be performed to distinguish random BAC integration from homologous recombination.

Materials, reagents, and equipment

BAC containing the gene of interest (BACPAC Resource Center, Children's Hospital of Oakland Research Institute: http://bacpac.chori.org)
Arabinose-inducible recombinase plasmid (pBAD; Yang and Seed, 2003)
Chemically competent bacteria (Strategene #200236)

Elecrocompetent bacteria (Invitrogen #18312-017)

L (+) arabinose (Sigma #10839)

500 mL disposable flasks (Nalgene #4113-0500)

BioRad Genepulser II System (Biorad)

0.2m gap electrocuvettes (Biorad #165-2086)

QIAprep Spin Miniprep Kit (Qiagen #27104)

Nucleobond BAC Maxi Kit (Clonetech #K3008-1)

Homing Endonuclease PI-Sce1 (NEB#RO696L)

HUES human embryonic stem cells (this technique has not been tested on any lines other than the Harvard University lines)

10 cm tissue culture dishes (Nunc #150318)

6-well tissue culture dishes (Nunc #140675)

24-well tissue culture dishes (Nunc #142475)

Antibiotic resistant mouse embryonic fibroblasts (Open Biosystems #MES3948, Specialty Media PMEF-H [hygromycin resistant] or Specialty Media PMEF-N [neomycin resistant])

Antibiotic stocks (G-418: Sigma #A1720, Hygromycin B: Sigma #H3274, Puromycin: Sigma #P8833)

Phosphate buffered saline (Invitrogen/Gibco #70013-032)

0.05 % trypsin (Invitrogen/Gibco #25300-054)

Protocol

A. Preparation of bacteria

1. Identify BAC containing the genomic region of interest (we use the UCSC genome browser at http://genome.ucsc.edu/, formatted for BAC display); typically, we have selected BACs that contain large flanking regions both upstream and downstream of region of interest.

2. Obtain BAC-containing *E. Coli* from either the BACPAC Resource Center at the Children's Hospital of Oakland Research Institute http://bacpac.chori.org/) or a commercial supplier such as Invitrogen. Confirm that the proper BAC has been shipped using PCR.

3. Introduce arabinose-inducable recombinase plasmid (containing the bacteriophage lambda *red* α and β genes under the control of the araBAD promoter; Yang and Seed, 2003) to BAC-containing bacteria:

   a. Innoculate a single colony of BAC containing bacteria in 2.5 mL LB + chloramphenicol, 20 mg/mL ("Chlor20"), grow overnight at 37°C shaking at 250 rpm.

   b. Subculture bacteria 1:200 into 50 mL LB + Chlor20, grow until OD_{600} is between 0.5 and 0.6 (usually 4–5 hr).

   c. Transfer culture into 50 mL polypropylene tube, incubate on ice for 10 min.

d. Centrifuge culture at 2,600 g for 10 min at 4°C.

e. Aspirate supernatant, resuspend pellet in 10 mL chilled 0.1 M CaCl$_2$.

f. Centrifuge culture at 2,600 g for 10 min at 4°C.

g. Aspirate supernatant, resuspend pellet in 0.5 mL chilled 0.1 M CaCl$_2$.

h. Mix 100 µL resuspension with 2 µL plasmid DNA (100–500 ng), incubate on ice for 10 min.

i. Heat shock mixture at 42°C for 90 s, recover briefly on ice, add 1 mL Luria Broth (LB), shake at 37°C for 1 hr.

j. Plate in dilutions on plates containing LB + Chlor 20 + ampicillin, 100 mg/mL ("Amp100"), assuming plasmid carries ampicillin-resistance gene; grow overnight. Surviving colonies represent bacteria containing both BAC and recombinase-encoding plasmid, and are ready to receive recombination template.

B. Preparation of recombination template vector

The template vectors that we developed (such as pTosca, **Figure 1**) consisted of the following basic features, in sequence: cloning site 1 (for 5′ homology arm), reporter gene (without promoter), poly A, FRT, genes providing eukaryotic antibiotic resistance (to either neomycin, hygromycin, or puromycin) and prokaryotic antibiotic resistance (typically kanamycin), preceded by appropriate promoters, second FRT, cloning site 2 (for 3′ homology arm).

The homologous arms were used were typically obtained by high-fidelity PCR amplification of BAC DNA, and were generally 500–1,000 bp in length, using primers designed according to the site of desired recombination. We typically engineered "ATG knock-ins," and usually removed only the first 18 nucleotides of endogenous coding sequence. Thus, the 5′ arm is designed to include the DNA immediately preceding the initial ATG, while the 3′ arm is designed to include DNA starting at approximately

Design of Template Vector

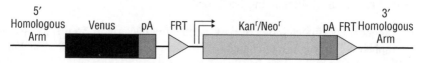

Figure 1 Design of template vector. The template vector consists of the following components: a 5′ homologous arm, a reporter gene followed by a poly-adenylated region, an FRT-flanked antibiotic resistance cassette including both bacterial and eukaryotic promoters, and a 3′ homologous arm.

nucleotide 19; the PCR primers must also incorporate the restriction enzyme sites needed to facilitate subsequent cloning to the template vector. A reporter construct engineered in this fashion is expected to be expressed in a fashion closely paralleling that of the endogenous gene.

C. Recombination of the BAC

Day 1:

1. Start 3 mL culture of BAC/recombinase containing bacteria in LB + chlor20 + amp100.

2. Digest 2–3 μg of template vector to liberate recombination template of the reporter construct out of the recombination vector using two unique restriction sites on the outsides of the 5' and 3' arm. Cut to completion, usually 3–4 hr at 37°C.

3. Load digest sample on 1 % agarose gel and run at 20–30 V overnight.

Day 2:

1. Sterile filter 150 mL LB + chlor20 + amp100 into a 500 mL disposable flask.

2. Innoculate 1.5 mL bacteria into the 150 mL LB; shake at 37°C.

3. Extract DNA of interest from preparative gel. Ultimately, purified DNA should be in volume of 40 μL sterile-filtered EB.

4. When the OD_{600} of your culture reaches 0.2 (about 2 hr depending on concentration of 3 mL culture), induce recombinase expression by addition to 750 μL sterile filtered 20 % L(+)-arabinose (arabinose should be added at 1:200).

5. Let the culture continue to grow until it reaches $OD_{600} \approx$ 0.6–0.8. It is important to harvest the bacteria when they reach this concentration, as in our experience a significantly lower frequency of recombination is obtained if the OD_{600} is appreciably above or below this range.

6. When the culture reaches the necessary OD_{600}, put the flasks on ice for 15–30 min.

7. Next, aliquot the culture into four 50 mL conical tubes: two should have 50 mL culture, one should have 25 mL culture and one should have 10 mL culture. This will allow you to electroporate three samples of the bacteria representing 100 mL of the original culture (the 250 mL tubes will be combined together), 25 mL of the original culture, and 10 mL of the original culture.

8. Centrifuge at 2,600 g for 15 min at 4°C, then carefully aspirate supernatant and resuspend each pellet in 1 mL sterile water. Resuspension should be done as

gently as possible, preferably by gently shaking the tubes while submerged in an ice–water slurry.

9. When pellets are resuspended, add 9 mL sterile water to each tube to bring the total volume to 10 mL.

10. Centrifuge at 2,600 g for 15 min at 4°C, then carefully aspirate the supernatant as soon as the spin is finished.

11. Resuspend each pellet in 1 mL sterile water as above. When pellet is resuspended, add 9 mL sterile water to bring the total volume to 10 mL.

12. Centrifuge at 2,600 g for 15 min at 4°C, then carefully aspirate supernatant. Resuspend the pellets in whatever residual supernatant is left in the tube. You will need 100 µL culture per electroporation.

13. Aliquot 100 µL culture into 1.7 mL polypropylene tubes. Each polypropylene tube will represent one electroporation.

14. Add 8 µL purified template DNA to each tube.

15. Electroporate samples in 0.2 cm cuvettes at the following setting: 2.3 kV, 200 Ω, 25 µF. Time constants are typically between 4.0–4.5 ms. Add 1 mL SOC immediately after elecroporation.

16. Let cultures recover by shaking at 37°C for 1–2 hr.

17. Plate each sample on LB-Chlor20-Kanamycin 25 µg/mL ("Kan25") agar plates. Let grow ON at 37°C.

Day 3: Harvest colonies for subsequent verification.

D. Verification of recombined BAC

- Potential recombinants are screened by PCR using both 5′ and 3′ primer sets; each primer sets includes an internal primer, corresponding to the template DNA, and an external primer, corresponding to DNA present slightly outside of the expected region of recombination. Thus, only successful recombinants should yield a band, which should be a predictable size. We generally validate both 5′ and 3′ regions, although we have not encountered an example of recombination in only one of these regions.

- To purify the recombineered BAC away from the recombinase-encoding plasmid, the BAC is first isolated by alkaline lysis of the bacteria followed by phenol-chloroform extraction:

    1. 3 mL of recombineered BAC-containing bacteria are grown in LB + Chlor20 at 37°C shaking overnight.

2. 1.7 mL of bacteria is transferred to a polypropylene tube and centrifuged at 14,000 g.

3. Medium is aspirated and pellet is resuspended in 250 µL P1 (Qiagen miniprep kit).

4. Add 250 µL P2 (Qiagen miniprep kit), mix gently and incubate at room temperature for 5 min.

5. Add 350 µL N3 (Qiagen miniprep kit), mix gently and centrifuge at 14,000 g for 10 min.

6. Transfer supernatant to a fresh polypropylene tube containing 510 µL isopropanol, mix gently and incubate for 10 min at room temperature.

7. Centrifuge at 14,000 g for 20 min at 4°C.

8. Aspirate supernatant, resuspend pellet in 300 µL EB (Qiagen miniprep kit).

9. Add 300 µL pre-warmed phenol, mix gently, centrifuge 14,000 g for 10 min at room temperature.

10. Transfer 250 µL of the top layer of the aqueous phase into a fresh polypropylene tube, add 20 µL 3M sodium acetate (pH = 5.2) and 600 µL chilled 100% ethanol. Mix well and incubate at −80°C for 30 min.

11. Centrifuge 14,000 g for 15 min at 4°C.

12. Aspirate supernatant, wash with 70% ethanol, centrifuge 14,000 g for 15 min at 4°C.

13. Aspirate the supernatant, briefly air dry the pellet and resuspend in 20 µL distilled water.

- Retransform purified BAC into electrocompetent DH10β (Invitrogen) by electroporation:

    1. Mix 10 µL BAC miniprep DNA with 100 µL competent bacteria in a 0.2 cm gap cuvette.

    2. Electroporate bacteria at 2.3 kV, 25 µF, 200 Ω using a Biorad Gene Pulser (Biorad). Time constants are usually in the 4.0–4.5 ms range.

    3. Add 1,000 µL SOC immediately after electroporation, transfer to 14 mL polypropylene test tube, and shake at 37°C for 1 hr.

    4. Plate bacteria in varying concentrations on LB + Chlor20 plates.

- Transformants are subsequently tested for growth on LB + Amp100 to verify ampicillin sensitivity, thus confirming absence of recombinase-encoding plasmid.

E. Preparation of BAC for electroporation

- After purification, BAC is prepared using the Nucleobond BAC Maxi Kit (Clonetech K3008-1). Typically, 5,000 mL of culture are used and approximately 300–600 μg of DNA are isolated. The BAC is subsequently linearized by overnight digestion with the endonuclease PI-Sce1 (NEB #RO696L).

- Linearized BAC is precipitated using ethanol and resuspended in PBS at a concentration of 200 μg/mL. Due to the size of the BAC as well as its concentration, resuspension can be difficult; it is helpful to incubate the DNA at 37°C. It is essential to be careful when handling the BAC as manual DNA shearing is a problem.

F. Electroporation of ES cells

- Note: the following protocol has been developed specifically for HUES cell lines, which were adapted to trypsin during their derivation (Cowan *et al.*, 2004); other hES cell lines might require protocol modification.

- Human ES cells are cultured on mouse embryonic fibroblasts (MEFs) as described (http://www.mcb.harvard.edu/melton/hues/HUES_manual.pdf). When cells are confluent ($\sim 1 \times 10^7 - 1.5 \times 10^7$), they are vigorously trypsinized using 0.05 % trypsin (Gibco) for approximately 5–10 min.

- Trypsinized cells are quenched with hES media and centrifuged at 500–600 g for 5 min. Cells are washed with 10 mL phosphate buffered saline (PBS) (Gibco) and centrifuged at 500–600 g for 5 min.

- Cells are resuspended in at a concentration of approximately 3×10^7 cells/mL hES media.

- Immediately prior to electroporation, 0.3 mL PBS (Gibco) containing 60 μg linearized BAC is added to each 0.5 mL aliquot of resuspended cells.

- 800 μL aliquots of cells and linearized BAC are transferred to a 0.4 cm gap cuvette (BioRad 165–2088) and exposed to a single 320 V 200 μF pulse at room temperature using the BioRad Gene Pulser II system. Time Constants are typically in the 1.8–2.2 ms range.

- Following a brief recovery, the cells are resuspended in 20 mL of human ES media and distributed to two culture dishes containing MEFs resistant to the appropriate antibiotic (Specialty Media/Chemicon).

G. Selection

- Human ES cells are allowed to grow for 2–3 days or until distinct colonies start to form. Before selection, there should be hundreds or even thousands of colonies per plate. The vast majority of these colonies will not contain the BAC and will be killed by antibiotic.

- Cells are selected by culturing in the presence of appropriate antibiotic (Neomycin 50μg/mL, Hygromycin 25μg/mL, or Puromycin 1μg/mL) for 2 days. During this time, there should be massive cell death and cells should be washed once with PBS prior to performing daily media changes.

- Cells are then cultured in antibiotic-free human ES cell medium for one day, followed by one more day of antibiotic-containing media. After the last day of selection, there should be very few visible colonies. However, there will be several small cell clumps that will grow into colonies.

- Cells are then cultured in antibiotic-free human ES cell medium until they are large enough to be picked. It is typical to get 10–40 colonies per 10 cm plate.

- Cells should be picked with a 200 μL pipette tip and a 200 μL pipette and immediately incubated in a 100 μL droplet of 0.05 % trypsin (Gibco) for 5–10 min. Cells and trypsin should be transferred to one well of a 24-well MEF-containing cell culture dish.

H. Expansion and verification

- Cells should be cultured with human ES cell medium until they are sub-confluent (5–7 days) and then split 1:2 to two replica 24-well MEF-containing cell culture dishes; one of the plates should be used to freeze and expand the cell line while the other should be used for genotyping.

- To quickly recover DNA from the clones, cells should be washed with PBS and then lysed with 0.5 mL lysis buffer (100 mM Tris-HCl pH = 8.5, 5 mM EDTA, 0.2 % SDS, 200 mM NaCl, 100 μg Proteinase K per milliliter) overnight at 55°C. One volume of isopropanol is added to precipitate DNA. DNA is then transferred to a polypropylene tube and washed in 70 % ethanol. DNA is then resuspended in 500 μL 10 mM Tris-HCl pH = 7.5.

Troubleshooting

I can't seem to get the recombination in bacteria to happen. There are no colonies after plating on LB + chlor20 + kan25 overnight and I think something is wrong. What do I do?

> This recombination step is difficult because it is so reliant on the exact growth phase that the bacteria are in when they are electroporated. It is absolutely essential to begin cooling and centrifuging the bacteria as soon as they reach $OD_{600} = 0.6$–0.8 as measured by spectrometer. Also, it is essential that the proper L(+) arabinose

is added at the proper time. Another potential problem involves the linearization of the recombination vector. You must use unique restriction enzymes to cut the homology and reporter portions out of the recombination vector. If problems persist, try using more bacteria in the electroporation (collect 150 mL or 200 mL instead of 100 mL) or more cut recombination vector (4.0 µg instead of 3.0 µg). If both of these conditions are met and there are still problems, try to use longer arms of homology in the recombination vector.

The colonies that come up after the recombination in bacteria show no band after PCR with primers internal and external to the reporter construct. Is there really nothing there?

While it is possible that the recombination has not happened, it is also possible that your PCR reaction is simply not working correctly. If you use primers specific for just the reporter or just the antibiotic resistance cassette, you can discern whether the problem is the recombination or simply the PCR verification. If you get bands using primers specific for the reporter or the antibiotic cassette, try selecting new primers external to the reporter construct.

I can't get enough recombined BAC from the BAC maxi kit you suggested, how do I improve yield?

While other large construct DNA kits or BAC maxi kits can be used, the Nucleobond kit has shown higher yield at a greater purity than any others. Because the BAC is so long, it puts a strain on the bacteria's reproduction. Try growing your bacteria for 1.5–2 times as long as you usually do. Also, after lysing the bacteria, make sure that you handle the lysate very gently as the BAC is large and can easily be sheared by excess mechanical stress. Another way to improve yield is to heat the elution buffer to 60°C before elution. Finally, try using more columns with fewer bacteria in each. Our experience has shown that about 625 mL of bacteria grown for 36 hr from a 1:1000 subculture yield about 60 µg.

I am getting very few resistant ES colonies after electroporation and selection: what am I doing wrong?

The ES selection is difficult because different cell lines respond differently to the selection. Further complicating matters is the fact that some antibiotics are harsher than others. While the "2 days on, 1 day off, 1 day on" selection scheme is recommended, it is by no means the only scheme that works. Above all, it is essential that you culture the cells in antibiotic until you begin to see significant cell death, and then culture them in regular ES media to allow them to recover sufficiently. Then, culture them in antibiotic again until the colonies left are only a few cells large, and finally allow them to recover and grow in regular ES medium. Be especially careful when using hygromycin and puromycin as the ES cells seem to be especially sensitive to them. If you are still getting few colonies, it is recommended that you electroporate more than one plate of ES cells. Our experience has shown that electroporating six 10 cm ES culture dishes with 360 µg of linearlized BAC will typically yield more than 80–100 colonies.

My resistant colonies are not indicating the presence of the reporter when I genotype them. Is this normal?

It is completely normal to observe that some antibiotic resistant colonies do not carry the reporter. Expect about 50 % of the resistant colonies to carry the reporter.

The clones that have come up positive for presence of the reporter do not seem to exhibit correct expression of the reporter. What should I do?

Unfortunately, we have commonly observed this in our experience. It is essential to have multiple clones to screen, as some will not express the reporters in the anticipated fashion, perhaps reflecting a suboptimal site of integration. It is also possible that the reporter is functioning properly, but is expressed at such a low level that it is hard to see with the naked eye. Try alternative approaches such as RT-PCR, *in situ* hybridization, or western blot to confirm that the problem is level of expression, rather than absence of expression. It may also be desirable to design your reporter so that it is coupled to a nuclear localization signal, so that the reporter will be concentrated in the nucleus of the cell.

References

Capecchi MR (2005). Gene targeting in mice: functional analysis of the mammalian genome for the twenty-first century. *Nat Rev Genet* **6**: 507–512.

Copeland NG, NA Jenkins and DL Court. (2001). Recombineering: A powerful new tool for mouse functional genomics. *Nat Rev Genet* **2**: 769–779

Cowan CA, I Klimanskaya, J McMahon, J Atienza, J Witmyer, JP Zucker, S Wang, CC Morton, AP McMahon, D Powers and DA Melton. (2004). Derivation of embryonic stem-cell lines from human blastocysts. *New Engl J Med* **350**: 1353–1356.

Shaner NC, PA Steinbach and RY Tsien. (2005). A guide to choosing fluorescent proteins. *Nat Methods* **2**: 905–909.

Shaywitz DA and DA Melton. (2005). The molecular biography of the cell. *Cell* **120**: 729–731.

Testa G, K Vintersten, Y Zhang, V Benes, JP Muyrers and AF Stewart. (2004). BAC engineering for the generation of ES cell-targeting constructs and mouse transgenics. *Methods Mol Biol* **256**: 123–139.

Valenzuela DM, AJ Murphy, D Frendewey, NW Gale, AN Economides, W Auerbach, WT Poueymirou, NC Adams, J Rojas, J Yasenchak, R Chernomorsky, M Boucher, AL Elsasser, L Esau, J Zheng, JA Griffiths, X Wang, H Su, Y Xue, MG Dominguez, I Noguera, R Torres, LE Macdonald, AF Stewart, TM DeChiara and GD Yancopoulos. (2003). High-throughput engineering of the mouse genome coupled with high-resolution expression analysis. *Nat Biotechnol* **21**: 652–659.

Van der Weyden L, DJ Adams and A Bradley. (2002). Tools for targeted manipulation of the mouse genome. *Physiol Genomics* **11**: 133–164.

Yang Y and B Seed. (2003). Site-specific gene targeting in mouse embryonic stem cells with intact bacterial artificial chromosomes. *Nat Biotechnol* **21**: 447–451.

Yu YA, AA Szalay, G Wang and K Obert. (2003). Visualization of molecular and cellular events with green fluorescent proteins in developing embryos: a review. *Luminescence* **18**: 1–18.

Afterword

It is 25 years since the derivation of the first murine embryonic stem (ES) cells was reported, and it is just 8 years since the first human ES cells were generated. As with all new discoveries, especially those with the perceived importance for advances in human medicine, there tends to be considerable general interest, which has both positive and negative consequences. In the case of human ES cell research, there are also the important ethical issues that have to be taken into consideration, which involve voices from other quarters, such as religious and political. The intense public interest can provide the positive energy to drive the field forward but it also creates potential pitfalls if all that appears in print is accepted uncritically.

There has evidently been considerable investment of resources already in research on human ES cells, both by various states and private donors. This is partly because there is a belief that this area of research provides opportunities to develop cures for many diseases, but there is a need for calm reflection. Sufficient time has now elapsed, and the mystique surrounding human ES cells is somewhat diminished. I present here my personal expectations of how this field will progress. One thing I can safely state is that the precise outcome of research in this field is unpredictable. Whatever turns out to be the case, there is a general agreement that research on human ES cells is exciting, and this is attracting many talented young scientists into the field. It is absolutely necessary to pursue fundamental research alongside target-orientated work towards generating possible cures and prevention of diseases.

Understanding pluripotent stem cells

To begin with, it is evident that there are different types of pluripotent stem cells, which while they are similar are by no means identical. Even though certain key properties are shared by all pluripotent stem cells, such as the expression of pluripotency-specific genes, there are critical differences. It is, for example, clear that the murine ES cells differ in some respects fundamentally from human ES cells, which makes uncritical extrapolations from studies on murine ES to human ES cells somewhat unreliable. For a start, we lack knowledge of how the transient population of pluripotent primitive ectoderm cells in the inner cell mass or epiblast are captured *in vitro* and converted into self-renewing pluripotent stem cells. Indeed, we also do not fully understand the mechanism that is responsible for the transition from totipotency to pluripotency, which occurs during preimplantation development. A better understanding of these aspects is necessary as it may pave the way to generating ES cells from many other mammalian species, which would provide useful comparative information.

Potential of human embryonic stem cell research

The availability of human ES cells provides a unique opportunity to study aspects of human development that would otherwise be impossible. An example where the use of human ES cells can contribute significantly to fundamental knowledge is on the germ cell lineage, which would otherwise remain largely unexplored. If we can generate primordial germ cells and even gametes reliably from human ES cells, this would advance knowledge on one of the most fascinating lineages that generates totipotency and transmits genetic information to all subsequent generations. Furthermore, if it becomes possible to generate human oocytes, they could be used for nuclear transplantation for reprogramming somatic nuclei to pluripotency, even if these oocytes are imperfect in other respects.

An area of research on human ES cells that occupies significant time and resources is the quest to generate diverse cell types. Such work has captured the public imagination, particularly in the context of cell replacement therapy. This area of research may indirectly generate other outcomes such as increased knowledge of the properties of different cells, which may prove useful towards developing better drugs. Many of the current approaches in this area are designed to emulate the way cell fate decisions are made during development, which is accomplished by the use of a variety of signaling molecules. To emulate nature in this way is difficult, and artifacts will likely be encountered. Of course, this approach is not unworthy of effort. It is possible that we may not need to try to mimic nature and follow the step-wise events that occur in the developing embryo. It is possible that as we gain detailed knowledge of the mechanisms of cell-fate determination, we may be able to trigger desirable ends by manipulating intrinsic switches, for example, through a combined use of transcriptional factors and epigenetic modifiers.

Is pluripotency essential?

This raises another important issue, which is whether it is essential and necessary to start with the "ground state" as represented by pluripotent cells for the directed differentiation of specific cell types. From advances in the field of genomic reprogramming of somatic cells, there are some encouraging signs that progress can be made to confer pluripotency on somatic cells, possibly even without the use of oocytes. The use of oocytes is in any case difficult both because of ethical considerations and because it is difficult to obtain sufficient numbers of them for such work, unless of course we can generate them directly from human ES cells as discussed above. To convert somatic cells directly into pluripotent stem cells efficiently and reliably, will require manipulation of their epigenetic state before inducing a change in the transcriptional network through the introduction of key pluripotency-specific genes. The modification of the epigenetic status of somatic cells would be needed to make these cells more responsive to the exogenous transcription factors. If such an approach becomes feasible, it may also be possible to re-direct one type of somatic cell to re-differentiate into another cell type without first making them pluripotent. Concerning this approach, much can be learnt from the phenomenon of transdetermination from other organisms. In this, as in other respects, work on basic developmental biology of non-mammalian organisms is important as some of them exhibit profound capacity for regeneration, which can contribute significantly to studies on human ES cells.

Directly converting adult somatic cells into pluripotent stem cells, could open the possibility of generating human ES cells from many individuals with predisposition to different diseases. As we now know, there are significant and unexpected variations in the individual human genomes. Whereas previously it was assumed that the DNA sequence of individual humans is 99.9% similar, surprisingly, we now find that the genomes of individuals vary considerably due to gains and losses of significant segments of DNA resulting in large variations in gene copy numbers. This is an unexpected development that could have a big impact on our understanding of human diseases. For this reason, derivation of human ES cells from large number of individuals with different genetic and ethnic backgrounds may provide an important tool to investigate how these differences affect the phenotypic properties of different cell types.

Benefits of human ES cell research

There are aspects of studies on human ES cells, such as their capacity for self-renewal and the underlying mechanisms that are responsible for it, as well as aspects of epigenetic reprogramming of somatic cells, which may provide insights into the nature of cancer stem cells. This could turn out to be a significant addition to our knowledge that can contribute to the causes and cures for cancer.

It is likely that it will be possible to develop clear-cut cell-based models by manipulating human ES cells to study various disease processes. This approach could have a big impact on assays which either mimic normal or aberrant cellular processed underlying diseases. These models would be attractive to the pharmaceutical industry, as they could use them to search for chemicals that can prevent aberrant cellular processes. Such an approach could add significantly to the development of human medicines to prevent or cure human diseases.

There is little doubt that research on human ES cells is exciting even if we may not yet know where it will take us. As is amply clear from the history of scientific enterprise and enquiry, some of the knowledge generated from these investigations will surprise us.

M. Azim SURANI
Wellcome Trust Cancer Research UK Gurdon Institute,
Tennis Court Road,
Cambridge CB2 1QN,
UK

Index